钱敏生学术论文集

马尽文　于剑／主编

北京大学出版社
PEKING UNIVERSITY PRESS

图书在版编目(CIP)数据

程乾生学术论文集/马尽文，于剑主编. —北京：北京大学出版社，2015.9
ISBN 978-7-301-23561-4

Ⅰ.①程… Ⅱ.①马… ②于… Ⅲ.①数学—文集 Ⅳ.O1-53

中国版本图书馆 CIP 数据核字(2015)第 223213 号

书　　　　名：程乾生学术论文集
著作责任者：马尽文　于剑　主编
责 任 编 辑：潘丽娜
标 准 书 号：ISBN 978-7-301-23561-4
出 版 发 行：北京大学出版社
地　　　　址：北京市海淀区成府路 205 号　100871
网　　　　址：http://www.pup.cn　新浪微博：@北京大学出版社
电 子 信 箱：zpup@pup.cn
电　　　　话：邮购部 62752015　发行部 62750672　编辑部 62752021
印　　刷　者：北京大学印刷厂
经　　销　者：新华书店
　　　　　　　787 毫米×1092 毫米　16 开本　17.75 印张　309 千字　插页 4 页
　　　　　　　2015 年 9 月第 1 版　2015 年 9 月第 1 次印刷
定　　　　价：50.00 元

未经许可，不得以任何方式复制或抄袭本书之部分或全部内容。
版权所有，侵权必究
举报电话：010 - 62752024　电子信箱：fd@pup.pku.edu.cn
图书如有印装质量问题，请与出版部联系，电话：010 - 62756370

1986年,程乾生(左二)在美国访问期间,与丁石孙校长(中间)及其他学者

1997年7月,程乾生与博士毕业生

序

为《程乾生学术论文集》作序,我感到欣慰,在此我向邀请人马尽文教授表示感谢.程乾生教授作为中国电子学会、中国仪器仪表学会信号处理分会的创始人之一,在我们的学会中一直是最活跃和有见解的学会领导成员,每次学术活动中他都承担着信号处理理论学组的工作,在把握学术水平方面起到重要的作用.作为来自数学学界的信号处理学者,我们都非常尊重他.他那积极发表意见的形象,还是经常出现在我们的记忆中.

在信号处理的研究方面,程乾生教授是从事基础理论的工作.他在时间序列、谱估计、模式识别、属性理论及应用等方面都做了大量的工作.这次论文集的出版可以为后人了解和学习程乾生教授的工作起到引导作用,鼓励我们继续他的方向努力下去.

程乾生教授很重视人才培养.他精心培养研究生,为他们的发展创造优良的条件.我有幸收受了他推荐的博士后李宏伟博士,在我实验室工作了两年,现在他已经是教授了.程老师的另一名博士,在我们北京交通大学已经是学科的带头人,拥有了很好的学术梯队.他自己精心挑选的学术骨干已经成功地发展起来,北京大学信号处理学科后继有人,一直保持了学术上的高水平.

我衷心地预祝,通过本书的出版和北京大学师生的努力,由程乾生教授开创的学术研究梯队能为中国信号处理的学术水平提高和应用成果的推广取得更多出色的成果,为信号处理学会做出更多的贡献.

袁保宗
2015 年 6 月

目 录

信号处理

纯相位序列的能量传递性质 ·· 1

滤波因子的均匀收敛和均方收敛性质 ··· 11

Multidimensional All-Pass Filters and Minimum-Phase Filters ·········· 22

Simultaneous Wavelet Estimation and Deconvolution of Reflection Seismic Signals ·· 27

Nonlinear Fusion Filters Based on Prediction and Smoothing ·········· 45

An Alternative Switching Criterion for Independent Component Analysis (ICA) ··· 50

Independent Particle Filters ·· 57

Particle Filters for Maneuvering Target Tracking Problem ················ 81

Nonlinear Discriminant Analysis on Embedded Manifold ··················· 95

谱估计与分析

谱估计中的最佳高分辨时窗函数 ··· 119

关于多维平稳序列奇异性和 WOLD 分解的谱表示 ················· 135

Z-Transform Models and Data Extrapolation Formulas in the Maximum Entropy Methods of Power Spectral Analysis ··············· 147

Rank of a Class of Autocorrelation Matrixes in Spectral Estimation ·········· 153

多谱估计的参数方法 ·· 156

时间序列分析

The Convolution-Type Matrix and the Property of the Complex Space l_2 ·········· 165

Parameter Estimation in Exponential Models ··············· 180

On Time-Reversibility of Linear Processes ··············· 195

Almost Sure Convergence Analysis of Mixed Time Averages and kth-Order Cyclic Statistics ··············· 200

模式识别与属性数学

有序样品聚类的相关序列法 ··· 210

一种新的样品聚类方法——差异序列法 …………………………………… 216

属性识别理论模型及其应用 ……………………………………………… 221

属性层次模型 AHM——一种新的无结构决策方法 …………………… 231

Fusion Prediction Based on the Attribute Clustering Network and the Radial Basis Function ……………………………………………………………… 236

Analysis of the Weighting Exponent in the FCM …………………… 245

附　录 …………………………………………………………………… 259

　　程乾生教授生平简介 ……………………………………………… 259

　　程乾生论著目录 …………………………………………………… 261

编后记 …………………………………………………………………… 277

信号处理

纯相位序列的能量传递性质

舒立华[①]

北京大学数学力学系

在处理地球物理勘探数据及其他数据中，经常遇到时间序列的能量延迟问题[1]。通常对这个问题的讨论是直接建立在解析函数的边界性质之上[2]。对能量延迟问题的讨论，纯相位序列起着重要的作用。我们认为，直接从能量传递角度分析纯相位序列的性质是必要的。本文第一部分介绍有关概念；第二部分讨论纯相位序列的能量传递性质。

1. 能量有限的物理可实现序列、纯相位序列

定义 时间序列 h_n 称为物理可实现序列，如果

$$h_n = 0, \quad \text{当} \; n < 0 \; \text{时}. \tag{1.1}$$

时间序列 h_n 称为能量有限序列，如果

$$\sum_{n=-\infty}^{\infty} |h_n|^2 < +\infty.$$

当 h_n 为物理可实现序列时，上式变为

$$\sum_{n=0}^{\infty} |h_n|^2 < +\infty. \tag{1.2}$$

h_n 的 Z 变换为

$$H(Z) = \sum_{n=0}^{\infty} h_n Z^n. \tag{1.3}$$

h_n 的频谱为

$$H(\omega) = H(Z)|_{Z=e^{-i\omega}} = \sum_{n=0}^{\infty} h_n e^{-in\omega} = A(\omega)e^{i\Phi(\omega)}, \tag{1.4}$$

[①] 作者程乾生教授以笔名舒立华发表。

其中 $A(\omega)=|H(\omega)|$ 为 h_n 的振幅谱,或称 h_n 的增益;$\Phi(\omega)$ 为 h_n 的相位谱.

序列、Z 变换、频谱之间皆有一一对应关系.

序列 h_n 和频谱 $H(\omega)$ 有如下关系:

$$h_n = \frac{1}{2\pi}\int_{-\pi}^{\pi} H(\omega)e^{in\omega}d\omega. \tag{1.5}$$

频谱 $H(\omega)$ 和能量 $\sum_{n=0}^{\infty}|h_n|^2$ 有如下关系:

$$\sum_{n=0}^{\infty}|h_n|^2 = \frac{1}{2\pi}\int_{-\pi}^{\pi}|H(\omega)|^2 d\omega. \tag{1.6}$$

以上关系由傅氏级数理论即可知.

定义 时间序列 g_n 称为纯相位序列,如果

$$\left.\begin{array}{l} g_n = 0, \quad \text{当 } n<0 \text{ 时}; \\ |G(\omega)| = \left|\sum_{n=0}^{\infty} g_n e^{-in\omega}\right| = 1, (a,e). \end{array}\right\} \tag{1.7}$$

由(1.6)知,纯相位序列 g_n 的能量为 1,即

$$\sum_{n=0}^{\infty}|g_n|^2 = 1. \tag{1.8}$$

定理 1.1

① 若 $g(n)$ 为纯相位序列,则 $|g(0)|\leqslant 1$. 特别地,当 $|g(0)|=1$ 时,有 $g(n)=0$,当 $n\geqslant 1$ 时.

② 设 $G_j(Z), 1\leqslant j\leqslant N$,为纯相位序列的 Z 变换,若 $\prod_{j=1}^{N} G_j(Z) = e^{i\beta}$,则 $G_j(Z) = e^{i\beta_j}$,其中 β 和 β_j 皆为实数,且 $\sum_{j=1}^{N}\beta_j = \beta + 2k\pi$,$k$ 为整数.

证明

① 由(1.8)知,我们的结论是明显的.

② Z 变换 $G_j(Z)$ 在 $Z=0$ 的值为 $g_j(0)$,因此 $\prod_{j=1}^{N}|g_j(0)| = 1$. 根据①,$|g_j(0)|\leqslant 1$,所以 $|g_j(0)|=1$,当 $n\geqslant 1$ 时 $g_j(n)=0$. 于是 $G_j(Z)=g_j(0)=e^{i\beta_j}$,$\beta_j$ 为实数,且有 $e^{i(\beta_1+\beta_2+\cdots+\beta_N)}=e^{i\beta}$,因此 $\sum_{j=1}^{N}\beta_j = \beta + 2k\pi$. 证毕.

由解析函数的边界性质知(参看文献[3]),任何一个能量有限的物理可实现序列 h_n 的 Z 变换 $H(Z)$(即 H_2 类函数)可表示为

$$H(Z) = G(Z)H_0(Z), \tag{1.9}$$

$$G(Z)=e^{i\lambda}\times Z^m\times\prod_{k=1}^{\infty}\frac{\alpha_k-Z}{1-\bar{\alpha}_k Z}\cdot\frac{|\alpha_k|}{\alpha_k}\times\exp\frac{1}{2\pi}\int_{-\pi}^{\pi}\frac{e^{i\varphi}+Z}{e^{i\varphi}-Z}d\psi(\varphi),\quad(1.10)$$

$$H_0(Z)=\exp\frac{1}{2\pi}\int_{-\pi}^{\pi}\ln|H(\varphi)|\frac{e^{-i\varphi}+Z}{e^{-i\varphi}-Z}d\varphi,\quad(1.11)$$

其中 λ 为一实数,

m 为非负整数,

α_k 为复数序列, $|\alpha_k|<1$, $\sum_{k=1}^{\infty}(1-|\alpha_k|)<+\infty$,

$\psi(\varphi)$ 为一个不上升的函数,具有几乎处处等于 0 的导数,

$|H(\omega)|$ 为序列 h_n 的振幅谱.

不难知, $G(Z)$ 为纯相位序列的 Z 变换(即只需验证 $G(Z)|_{Z=e^{-i\omega}}$ 能表成 $e^{i\Phi(\omega)}$ 的形式就行了), $e^{i\lambda}H_0(Z)$ 为最小相位滞后(简称最小相位) Z 变换.最小相位滞后概念不是本文讨论重点,这里不做介绍了,详细可参阅文献[4].

2. 纯相位序列的能量传递性质

如果一个系统或一个滤波器的脉冲响应函数为纯相位序列,则这个系统或这个滤波器称为纯相位系统或纯相位滤波器.用纯相位滤波器进行滤波,我们称之为纯相位滤波.

纯相位滤波的能量传递性质,分总能量传递性质和部分能量传递性质两个方面.总能量传递性质是容易得到的(见定理 2.1),部分能量性质则不是那么直接能得到的.但是,总能量与部分能量既是矛盾的又是统一的,在一定条件下,部分能量就转化为总能量(参看定理 2.2 及其证明),同时,根据对物理可实现序列进行滤波的具体特点(参见(2.6)),我们就可以由总能量传递性质导出部分能量传递性质.下面进行具体分析.

纯相位滤波有个实质性的特点,即总能量不变性:任何信号经过纯相位滤波后总能量不变.这是因为纯相位滤波不改变输入信号的振幅谱,而振幅谱唯一地确定了总能量的缘故.见下面定理.

定理 2.1 物理可实现时间序列 g_n 为纯相位序列的充分必要条件是:任何信号经过 g_n 滤波后总能量不变,即对任何输入序列 x_n,有输出序列 $y_n=\sum_{k=0}^{n}g_k x_{n-k}$,必有

$$\sum_{n=0}^{\infty}|x_n|^2=\sum_{n=0}^{\infty}|y_n|^2.\quad(2.1)$$

证明 设 x_n, g_n, y_n 的频谱分别为 $X(\omega), G(\omega), Y(\omega)$. 按照褶积滤波的性质有 $Y(\omega) = G(\omega) X(\omega)$. 根据能量和频谱的关系(1.6)式,(2.1)式可写为

$$\frac{1}{2\pi}\int_{-\pi}^{\pi} |X(\omega)|^2 d\omega = \frac{1}{2\pi}\int_{-\pi}^{\pi} |G(\omega)|^2 \cdot |X(\omega)|^2 d\omega, \tag{2.2}$$

把它改写为

$$\int_{-\pi}^{\pi} (1 - |G(\omega)|^2) |X(\omega)|^2 d\omega = 0.$$

由实变函数的理论,要求对任何 $|X(\omega)|^2$ 上式都成立,充分必要条件是

$$1 - |G(\omega)|^2 = 0,$$

即

$$|G(\omega)| = 1, \quad (a.e).$$

再由纯相位序列定义便知,本定理是成立的.

信号经过纯相位滤波后,除保持总能量不变(见(2.1))以外,还具有关于部分能量延迟的特殊性质.

定义 对任何一个物理可实现序列 x_n,我们称

$$\sum_{n=0}^{N} |x_n|^2$$

为序列 x_n 的部分能量.

定理 2.2 对任何 $N+1$ 个数 $(x_0, x_1, \cdots, x_N)$ 和纯相位序列 g_n,有

$$\sum_{n=0}^{N} |x_n|^2 = \sum_{n=0}^{\infty} |\tilde{y}_n|^2, \tag{2.3}$$

其中

$$\tilde{y}_n = \sum_{k=0}^{N} x_k g_{n-k}. \tag{2.4}$$

证明 把序列 $(x_0, x_1, \cdots, x_N, 0, 0, \cdots)$ 作为输入信号,g_n 为滤波因子,由定理 2.1 的(2.1)即得(2.3). 证毕.

由(2.3)直接得到不等式

$$\sum_{n=0}^{N} |x_n|^2 \geqslant \sum_{n=0}^{N} |\tilde{y}_n|^2. \tag{2.5}$$

当输入信号为 x_n,滤波器为纯相位序列 g_n,输出信号为 y_n,则

$$y_n = \sum_{k=0}^{\infty} g_k x_{n-k} = \sum_{k=0}^{n} g_k x_{n-k} = \sum_{k=0}^{n} x_k g_{n-k}.$$

当 $n \leqslant N$ 时,

$$\tilde{y}_n = \sum_{k=0}^{N} x_k g_{n-k} = \sum_{k=0}^{n} x_k g_{n-k},$$

因此有
$$\tilde{y}_n = y_n, \quad \text{当 } n \leqslant N \text{ 时}. \tag{2.6}$$

由(2.5)和(2.6)可得下面定理 2.3.

定理 2.3 设输入为物理可实现序列 x_n,滤波器为纯相位序列 g_n,输出为 y_n,则对一切 N 有

$$\sum_{n=0}^{N} |x_n|^2 \geqslant \sum_{n=0}^{N} |y_n|^2. \tag{2.7}$$

定理 2.3 说明了纯相位滤波具有部分能量延迟性质,由(2.7)知,纯相位滤波后的部分能量 $\sum_{n=0}^{N} |y_n|^2$ 比滤波前的相应部分能量 $\sum_{n=0}^{N} |x_n|^2$ 要小,所差的能量被延迟到输出序列的后面部分去了,这是因为信号在滤波前和滤波后总能量相等的缘故(见定理 2.1).

当在(2.7)中取等号时,作为滤波因子的纯相位序列必须有特殊的结构,定理 2.4 回答这个问题.

定理 2.4 在定理 2.3 的条件下,对某个 N 等式

$$\sum_{n=0}^{N} |x_n|^2 = \sum_{n=0}^{N} |y_n|^2 > 0 \tag{2.8}$$

成立的充分必要条件是:纯相位序列 g_n 的 Z 变换为

$$G(Z) = \frac{y_0 + y_1 Z + \cdots + y_N Z^N}{x_0 + x_1 Z + \cdots + x_N Z^N}, \tag{2.9}$$

其中 $G(Z)$ 在单位圆内及其上是解析的.

证明 必要性 当 $n \leqslant N$ 时,$y_n = \tilde{y}_n$. 由(2.8)知,$\sum_{n=0}^{N} |x_n|^2 = \sum_{n=0}^{N} |\tilde{y}_n|^2$. 将此等式与(2.3)相比较,就可得到 $\tilde{y} = 0$,当 $n \geqslant N+1$. 因此 $\tilde{y}_n$(见(2.4))的 Z 变换 $\tilde{Y}(Z) = \tilde{y}_0 + \tilde{y}_1 Z + \cdots + \tilde{y}_N Z^N = y_0 + y_1 Z + \cdots + y_N Z^N$. 又由于输入 $(x_0, x_1, \cdots, x_N)$ 的 Z 变换为 $\tilde{X}(Z) = x_0 + x_1 Z + \cdots + x_N Z^N$,还有 $\tilde{Y}(Z) = G(Z) \tilde{X}(Z)$,便得到(2.9).因为 g_n 是物理可实现的,故 $G(Z)$ 必须在单位圆内解析,又因为 g_n 的能量是有限的,故要求 $G(Z)$ 在单位圆周上无极点,也即解析.

充分性 设输入序列 Z 的变换为 $\tilde{X}(Z) = \sum_{n=0}^{N} x_n Z^n$,于是得输出的 Z 变换为 $\tilde{Y}(Z) = G(Z) \tilde{X}(Z) = \sum_{n=0}^{N} y_n Z^n$,由(2.3)知(2.8)成立.证毕.

推论 2.1 在定理 2.3 的条件下,若 $x_N \neq 0$,则

$$\sum_{n=0}^{N-1}|x_n|^2 = \sum_{n=0}^{N-1}|y_n|^2 > 0 \quad \text{与} \quad \sum_{n=0}^{N}|x_n|^2 = \sum_{n=0}^{N}|y_n|^2 \quad (2.10)$$

同时成立的充分必要条件是

$$G(Z) = e^{i\beta}, \quad \text{其中 } \beta \text{ 为实数}.$$

证明 设(2.10)成立,按定理 2.4,

$$G(Z) = \frac{y_0 + \cdots + y_{N-1}Z^{N-1}}{x_0 + \cdots + x_{N-1}Z^{N-1}} = \frac{y_0 + \cdots + y_{N-1}Z^{N-1} + y_N Z^N}{x_0 + \cdots + x_{N-1}Z^{N-1} + x_N Z^N}.$$

因为当 $\dfrac{A}{B} = \dfrac{A+B}{B+D}$ ($D \neq 0$)时,有 $\dfrac{A}{B} = \dfrac{C}{D}$,所以

$$G(Z) = \frac{y_N Z^N}{x_N Z^N} = \frac{y_N}{x_N}.$$

由于 $|G(\omega)| = 1$,故有 $\left|\dfrac{y_N}{x_N}\right| = 1$,$G(Z) = e^{i\beta}$,$\beta$ 为实数. 必要性证毕.

设 $G(Z) = e^{i\beta}$. 根据定理 1.1,$g_0 = e^{i\beta}$,$g_n = 0$(当 $n \neq 0$ 时). 这时

$$y_n = \sum_k g_k x_{n-k} = g_0 x_n = e^{i\beta} x_n.$$

对任何 N,(2.10)都成立. 充分性证毕.

推论 2.2 在定理 2.3 的条件下,若 $x_0 \neq 0$,则 $|x_0|^2 = |y_0|^2$ 成立的充分必要条件是

$$G(Z) = e^{i\beta}, \quad \beta \text{ 为实数}.$$

证明 在定理 2.4 中,令 $N=0$,便得本推论.

为了说明定理 2.4 和推论 2.1,我们举一例.

例 滤波器为 $G(Z) = \dfrac{b+aZ}{a+bZ}$,其中 $|a| > |b| > 0$,输入为 $X(Z) = \sum_{n=0}^{\infty} x_n Z^n = (a+bZ) + Z^4(a+bZ)$,则输出为 $Y(Z) = \sum_{n=0}^{\infty} y_n Z^n = G(Z)X(Z) = (b+aZ) + Z^4(b+aZ)$. 输入和输出的部分能量情况为

$$|x_0|^2 = |a|^2 > |y_0|^2 = |b|^2,$$

$$\sum_{n=0}^{N}|x_n|^2 = \sum_{n=0}^{N}|y_n|^2 = |a|^2 + |b|^2, \quad N = 1,2,3,$$

$$\sum_{n=0}^{4}|x_n|^2 = |a|^2 + |b|^2 + |a|^2 > \sum_{n=0}^{4}|y_n|^2 = |a|^2 + |b|^2 + |b|^2,$$

$$\sum_{n=0}^{N}|x_n|^2 = \sum_{n=0}^{4}|y_n|^2 = 2(|a|^2 + |b|^2), \quad N \geq 5.$$

(2.11)

这个例子说明推论 2.1 的条件 $x_N \neq 0$ 是必要的, 否则推论是不成立的. 在这个例子中, 当 $N=3$ 时, (2.10) 是成立的, 但 $G(Z) \neq e^{i\beta}$, 原因就是 $x_3 = 0$.

定理 2.3 说明了纯相位滤波具有部分能量延迟性质, 下面我们证明, 部分能量延迟性质也是纯相位滤波器的实质性质.

定理 2.5 设 g_n 为物理可实现序列, 具有单位能量, 即 $\sum_{n=0}^{\infty} |g_n|^2 = 1$. 则 g_n 为纯相位序列, 即 g_n 的频谱 $G(\omega)$ 满足 $|G(\omega)| = 1 (a,e)$ 的充分必要条件是: 对任何物理可实现输入序列 x_n, 经 g_n 滤波后输出为 $y_n = \sum_{k=0}^{n} g_k x_{n-k}$, 部分能量被延迟了, 即

$$\sum_{n=0}^{N} |x_n|^2 \geqslant \sum_{n=0}^{N} |y_n|^2, \quad N \geqslant 0. \tag{2.12}$$

证明 必要性 由定理 2.3 即知.

充分性 用反证法. 假定 $|G(\omega)| \neq 1$.

设输入满足

$$|X(\omega)|^2 = \begin{cases} 2, & \text{当 } |G(\omega)|^2 - 1 > 0, \\ 1, & \text{当 } |G(\omega)|^2 - 1 \leqslant 0. \end{cases}$$

因为 g_n 具有单位能量, 即

$$\frac{1}{2\pi} \int_{-\pi}^{\pi} |G(\omega)|^2 d\omega = 1,$$

而 $\frac{1}{2\pi} \int_{-\pi}^{\pi} d\omega = 1$, 所以有

$$\frac{1}{2\pi} \int_{-\pi}^{\pi} (|G(\omega)|^2 - 1) d\omega = 0.$$

函数 $|G(\omega)|^2 - 1$ 的值有正有负, 我们把正值放大两倍, 负值不变, 那么积分值必大于 0, 即

$$\frac{1}{2\pi} \int_{-\pi}^{\pi} (|G(\omega)|^2 - 1) |X(\omega)|^2 d\omega > 0,$$

也即

$$\frac{1}{2\pi} \int_{-\pi}^{\pi} |G(\omega)|^2 |X(\omega)|^2 d\omega > \frac{1}{2\pi} \int_{-\pi}^{\pi} |X(\omega)|^2 d\omega,$$

写成能量形式就是

$$\sum_{n=0}^{\infty} |y_n|^2 > \sum_{n=0}^{\infty} |x_n|^2.$$

我们总可选取某个 N_0，使

$$\sum_{n=0}^{N_0}|y_n|^2 > \sum_{n=0}^{N_0}|x_n|^2,$$

而这与(2.12)矛盾，因此，必有 $|G(\omega)|=1(a,e)$。证毕。

在分析了纯相位滤波的部分能量延迟性质以后，我们自然会了解最小能量延迟概念。

定义 一个能量有限的物理可实现序列 $h_1(n)$ 称为最小能量延迟(简称最小延迟)序列，如果对任何物理可实现序列 $h(n)$，只要 $h(n)$ 和 $h_1(n)$ 的振幅谱相同，皆有

$$\sum_{n=0}^{N}|h_1(n)|^2 \geqslant \sum_{n=0}^{N}|h(n)|^2, \quad N \geqslant 0. \tag{2.13}$$

定理 2.6 设 $h_1(n)$ 为最小能量延迟序列，g_n 为纯相位序列，$h(n)$ 为物理可实现序列，且有关系 $h_1(n)=\sum_{k=0}^{n}g_k h(n-k)$，用 Z 变换形式写就是 $H_1(Z)=G(Z)H(Z)$，则 $G(Z)=e^{i\beta}$（β 为实数），$h(n)$ 为最小能量延迟序列。

证明 根据定理 2.3 有 $\sum_{n=0}^{N}|h(n)|^2 \geqslant \sum_{n=0}^{N}|h_1(n)|^2$，又由于 $h_1(n)$ 是最小能量延迟序列，所以关系式(2.13)成立。比较上面两个关系式，我们得 $\sum_{n=0}^{N}|h_1(n)|^2 = \sum_{n=0}^{N}|h(n)|^2 (N \geqslant 0)$。这说明 $h(n)$ 也是最小能量延迟序列。按照推论 2.1，则有 $G(Z)=e^{i\beta}$（β 为实数）。证毕。

我们给出一个关于最小能量延迟序列的定理。

定理 2.7 设 $h_0(n)$ 是与 Z 变换 $H_0(Z)$（见(1.11)）相应的时间序列，则振幅谱为 $|H(\omega)|$ 的物理可实现序列 $h_1(n)$ 是最小能量延迟序列，充分必要条件是

$$h_1(n) = e^{i\beta} h_0(n), \quad \beta \text{ 为实数}. \tag{2.14}$$

证明 必要性 按照(1.9)，$h_1(n)$ 的 Z 变换为 $H_1(Z)=G(Z)H_0(Z)$，其中 $G(Z)$ 为纯相位序列的 Z 变换。根据定理 2.6，知 $G(Z)=e^{i\beta}$，也即(2.14)成立。

充分性 任何一个振幅谱为 $|H(\omega)|$ 的物理可实现序列 $h(n)$，它的 Z 变换皆可表成(1.9)式。根据定理 2.3，有

$$\sum_{n=0}^{N}|h_0(n)|^2 \geqslant \sum_{n=0}^{N}|h(n)|^2, \quad N \geqslant 0.$$

这说明 $h_0(n)$ 是最小能量延迟序列。又因为 $h_1(n)=e^{i\beta}h_0(n)$ 的部分能量完全一样，因此，$h_1(n)$ 为最小能量延迟序列。证毕。

最后，我们指出一个重要事实：经纯相位滤波前后的信号具有能量延迟性质，但是，具有能量延迟性质的两个信号并不都是经纯相位滤波前后的信号. 具体意思如下. 设 x_n, y_n 为能量有限的物理可实现序列. 如果 y_n 是 x_n 经纯相位滤波 (g_n) 后之输出，即

$$y_n = \sum_{k=0}^{n} g_k x_{n-k}, \qquad (2.15)$$

其中 g_n 为纯相位序列，则必有

$$\sum_{n=0}^{N} |x_n|^2 \geqslant \sum_{n=0}^{N} |y_n|^2, \quad N \geqslant 0. \qquad (2.16)$$

这由定理 2.3 即可知. 但是，反之则不一定成立，意即若 (2.16) 成立，则 (2.15) 不一定成立. 我们只需举一例就行了.

例 设

$$X(Z) = \frac{\frac{1}{4} - Z}{1 - \frac{1}{4}Z} = \sum_{n=0}^{\infty} x_n Z^n,$$

$$F(Z) = \frac{\frac{1}{2} - Z}{1 - \frac{1}{2}Z} = \sum_{n=0}^{\infty} f_n Z^n,$$

$$Y(Z) = F(Z)F(Z) = \sum_{n=0}^{\infty} y_n Z^n.$$

显然 $X(Z), F(Z), Y(Z)$ 都是纯相位序列的 Z 变换，它们的振幅谱是相同的，且为 1. 为了讨论它们的部分能量，我们先给出一般的公式：对 $-1 < \alpha < 1$ 有

$$\frac{\alpha - Z}{1 - \alpha Z} = \sum_{n=0}^{\infty} a_n Z^n, \quad a_n = \begin{cases} \alpha, & n = 0, \\ \alpha^{n+1} - \alpha^{n-1}, & n > 0. \end{cases}$$

$$\sum_{n=0}^{N} |a_n|^2 = \begin{cases} \alpha^2, & N = 0, \\ 1 - \alpha^{2N}(1 - \alpha^2), & N > 0. \end{cases}$$

按照这个公式，x_n 的部分能量为

$$\sum_{n=0}^{N} |x_n|^2 = \begin{cases} \dfrac{1}{2^4}, & N = 0, \\ 1 - \dfrac{1}{2^{4N}} \cdot \dfrac{15}{2^4}, & N > 0. \end{cases}$$

f_n 的部分能量为

$$\sum_{n=0}^{N} |f_n|^2 = \begin{cases} \dfrac{1}{2^2}, & N = 0, \\ 1 - \dfrac{1}{2^{2N}} \cdot \dfrac{3}{2^2}, & N > 0. \end{cases}$$

y_n 的部分能量有如下关系:

$$|y_0|^2 = |f_0^2|^2 = \frac{1}{2^4},$$

$$\sum_{n=0}^{N}|y_n|^2 \leqslant \sum_{n=0}^{N}|f_n|^2$$

(这是因为 $Y(Z)$ 是 $F(Z)$ 经纯相位滤波器 $F(Z)$ 滤波的结果,由定理 2.3 即得此不等式).

当 $N > 1$ 时,$\frac{1}{2^{4N}} \cdot \frac{15}{4^4} < \frac{1}{2^{2N}} \cdot \frac{3}{2^2}$,所以 $1 - \frac{1}{2^{4N}} \cdot \frac{15}{4^4} > 1 - \frac{1}{2^{2N}} \cdot \frac{3}{2^2}$,也即

$$\sum_{n=0}^{N}|x_n|^2 > \sum_{n=0}^{N}|f_n|^2 \geqslant \sum_{n=0}^{N}|y_n|^2;$$

当 $N = 0$ 时,$|x_0|^2 = \frac{1}{2^4} = |y_0|^2$.

这说明 y_n 的部分能量比 x_n 相应的部分能量要小,即 (2.16) 成立. 但是,不存在纯相位序列的 Z 变换 $G(Z)$,使 $Y(Z) = G(Z)X(Z)$,这是因为作为能量有限的物理可实现序列 y_n 的 Z 变换表示式是唯一的. 在这里,直接可看出 $G(Z)X(Z)$ 在 $Z = \frac{1}{4}$ 时为零,而 $Y(Z)$ 不为零,因此,对一切 $G(Z)$,

$$Y(Z) \neq G(Z)X(Z).$$

以上讨论的是序列的能量传递性质. 本文所用的方法,也可以用到连续度量的情形,得到相似的能量传递性质.

参考文献

[1] Robinson E. A., Multichannel Z-transforms and minimum-delay, *Geophysics*, 31(1966), 482—500.

[2] Robinson E. A., Random Wavelets and Cybernetic systems, *MRC Technical Summary Report*, 195, October, 1960.

[3] И. И. 普里瓦洛夫, 解析函数的边界性质, 科学出版社, 1956.

[4] 燃料化学工业部石油地球物理勘探局计算中心站, 北京大学数学力学系等, 地震勘探数字技术(第二册), 科学出版社, 1974.

原文载于《数学学报》,第 17 卷,第 1 期,1974 年 3 月,20—27.

滤波因子的均匀收敛和均方收敛性质

程 乾 生

北京大学数学系

Abstract: It is well known that filtering plays an important part in signal processing. The filter factors can be determined in differend ways. For example, the lengths of filter factors may be different, and this gives rise to the stability problem of filter factors. In this paper, the concepts of uniform convergence and mean square convergence of filter factors are introduced, and the necessary and sufficient conditions for uniform convergence and mean square convergence are presented. Besides, the limit properties of filter factors and errors in least square filtering are discussed.

本文讨论滤波因子的稳定性问题，引入了滤波因子的均匀收敛和均方收敛概念，给出了均匀收敛和均方收敛的充分必要条件．最后，我们还讨论了最小平方滤波因子和误差的极限性质．

1. 能量有限信号及其滤波问题

数列 $x(t)(t=0,\pm 1,\cdots)$ 称为能量有限信号，假定 $x(t)$ 满足

$$\sum_{t=-\infty}^{+\infty}|x(t)|^2<\infty. \tag{1.1}$$

所有的能量有限信号所组成的集合记为

$$l_2(-\infty,+\infty)=\{x(t)|x(t)满足(1.1)\}. \tag{1.2}$$

在 $l_2(-\infty,+\infty)$ 中定义内积

$$(x(t),y(t))=\sum_{t=-\infty}^{\infty}x(t)\overline{y(t)}, \tag{1.3}$$

则 $l_2(-\infty,+\infty)$ 为希尔伯特空间．$l_2(-\infty,+\infty)$ 中元素 $x(t)$ 的范数 $\|x(t)\|$ 定义为

$$\|x(t)\|^2=(x(t),x(t))=\sum_{t=-\infty}^{+\infty}|x(t)|^2. \tag{1.4}$$

数列 $h(t)$ 称为有限长信号，如果存在整数 $l,m,l\leqslant m$，使

$$h(t) = \begin{cases} h(t), & l \leqslant t \leqslant m, \\ 0, & \text{其他}. \end{cases} \quad (1.5)$$

对能量有限信号 $x(t)$，记 $\mathscr{L}_x$ 为 $x(t+n)$ 产生的线性闭包，即

$$\mathscr{L}_x = \mathscr{L}(x(t+n), n = 0, \pm 1, \cdots). \quad (1.6)$$

由线性闭包的定义可知，若 $y(t)$ 满足

$$y(t) \in \mathscr{L}_x, \quad (1.7)$$

则总存在有限长信号 $h_n(t)$ 使

$$\|x(t) * h_n(t) - y(t)\| \to 0 \quad (n \to +\infty), \quad (1.8)$$

其中

$$x(t) * h_n(t) = \sum_{s=-\infty}^{+\infty} h_n(s) x(t-s). \quad (1.9)$$

在信号数字处理中[2]，我们常把 (1.8) 中的 $x(t)$ 称为输入信号，$y(t)$ 称为希望输出信号，$h_n(t)$ 称为有限长滤波因子.

如果我们用不同的方法求得两列有限长滤波因子 $h_n^{(j)}(t)$

$$h_n^{(j)}(t) = \begin{cases} h_n^{(j)}(t), & l_n^{(j)} \leqslant t \leqslant m_n^{(j)}, \\ 0, & \text{其他}, \end{cases} \quad j = 1, 2, \quad (1.10)$$

使得

$$\|x(t) * h_n^{(j)}(t) - y(t)\| \to 0 \quad (n \to +\infty), \quad j = 1, 2, \quad (1.11)$$

那么，$h_n^{(1)}(t)$ 与 $h_n^{(2)}(t)$ 是否很接近呢？在一般情况下，很难保证这一点。下面的例子说明，即使对最小延迟信号 $x(t)$，也可以使 $h_n^{(1)}(t)$ 与 $h_n^{(2)}(t)$ 相差很大.

例 令

$$x(t) = \begin{cases} 1, & t = 0, 1, \\ 0, & \text{其他}, \end{cases}$$

$$y(t) = \begin{cases} 1, & t = 0, \\ 0, & \text{其他}, \end{cases}$$

$$h_n^{(1)}(t) = \begin{cases} h_n^{(1)}(t), & 0 \leqslant t \leqslant n, \\ 0, & \text{其他}, \end{cases}$$

$$h_n^{(2)}(t) = \begin{cases} h_n^{(2)}(t), & -(n+1) \leqslant t \leqslant n, \\ 0, & \text{其他}, \end{cases}$$

其中 $n = 1, 2, \cdots$. 取 $h_n^{(j)}(t)$ 为使 $\|x(t) * h_n^{(j)}(t) - y(t)\|^2$ 达最小的最小平方滤波因子. 由于 $x(t)$ 是最小延迟信号[2]，(1.11) 式成立，且

$$\lim_{n\to+\infty} h_n^{(1)}(t) = \begin{cases} (-1)^t, & t\geq 0, \\ 0, & t<0. \end{cases} \quad (1.12)$$

$h_n^{(2)}(t)$ 满足最小平方滤波方程

$$\begin{bmatrix} r_{xx}(0) & r_{xx}(1) & \cdots & r_{xx}(2n+1) \\ r_{xx}(1) & r_{xx}(0) & \cdots & r_{xx}(2n) \\ \vdots & \vdots & & \vdots \\ r_{xx}(2n+1) & r_{xx}(2n) & \cdots & r_{xx}(0) \end{bmatrix} \begin{bmatrix} h_n^{(2)}(-(n+1)) \\ \vdots \\ h_n^{(2)}(-1) \\ h_n^{(2)}(0) \\ \vdots \\ h_n^{(2)}(n) \end{bmatrix} = \begin{bmatrix} r_{yx}(-(n+1)) \\ \vdots \\ r_{yx}(-1) \\ r_{yx}(0) \\ \vdots \\ r_{yx}(n) \end{bmatrix} = \begin{bmatrix} 0 \\ \vdots \\ 1 \\ 1 \\ \vdots \\ 0 \end{bmatrix},$$

其中 $r_{xx}(t)$ 为 $x(t)$ 的自相关函数，$r_{yx}(t)$ 为 $y(t)$ 与 $x(t)$ 的互相关函数，根据陶布利兹方程的性质[2]，有

$$h_n^{(2)}(t) = h_n^{(2)}(-(t+1)), \quad 0\leq t\leq n. \quad (1.13)$$

由 (1.12) 和 (1.13) 知，对任何 $t\geq 0$，当 $n\to+\infty$ 时，$h_n^{(1)}(t)-h_n^{(2)}(t)$ 和 $h_n^{(1)}(-(t+1))-h_n^{(2)}(-(t+1))$ 都不能同时趋于零.

上面例子说明，即使对两个最小平方滤波因子 $h_n^{(1)}(t)$ 和 $h_n^{(2)}(t)$，它们之间的关系也不是稳定的，意即它们之差不是对任何 t 都趋于零的.

为了刻画滤波因子的稳定性，我们提出两个概念，即下面的两个定义.

定义 1 设 $x(t)$ 为能量有限信号，$y(t)$ 满足 (1.7). 如果对任何满足 (1.10)，(1.11) 的滤波因子 $h_n^{(1)}(t)$ 和 $h_n^{(2)}(t)$，皆有

$$\frac{1}{2\pi}\int_{-\pi}^{\pi}\left|\sum_{s=l_n^{(1)}}^{m_n^{(1)}} h_n^{(1)}(s)e^{-is\omega} - \sum_{s=l_n^{(2)}}^{m_n^{(2)}} h_n^{(2)}(s)e^{-is\omega}\right|d\omega \to 0 \quad (n\to+\infty), \quad (1.14)$$

则称 $x(t)$ 的滤波因子具有均匀收敛性.

定义 2 设 $x(t)$ 为能量有限信号，$y(t)$ 满足 (1.7). 如果对任何满足 (1.10)，(1.11) 的滤波因子 $h_n^{(1)}(t)$ 和 $h_n^{(2)}(t)$，皆有

$$\|h_n^{(1)}(t)-h_n^{(2)}(t)\| \to 0 \quad (n\to+\infty), \quad (1.15)$$

则称 $x(t)$ 的滤波因子具有均方收敛性.

现在对滤波因子的均匀收敛性和均方收敛性做一说明.
由于

$$|h_n^{(1)}(t)-h_n^{(2)}(t)|=\left|\frac{1}{2\pi}\int_{-\pi}^{\pi}e^{it\omega}\Big(\sum_{s=l_n^{(1)}}^{m_n^{(1)}}h_n^{(1)}(s)e^{-is\omega}-\sum_{s=l_n^{(2)}}^{m_n^{(2)}}h_n^{(2)}(s)e^{-is\omega}\Big)d\omega\right|$$

$$\leqslant \frac{1}{2\pi}\int_{(-\pi)}^{\pi}\left|\sum_{s=l_n^{(1)}}^{m_n^{(1)}}h_n^{(1)}(s)e^{-is\omega}-\sum_{s=l_n^{(2)}}^{m_n^{(2)}}h_n^{(2)}(s)e^{-is\omega}\right|d\omega, \quad (1.16)$$

如果 $x(t)$ 的滤波因子具有均匀收敛性,则对任何 t,$h_n^{(1)}(t)-h_n^{(2)}(t)$ 当 $n\to+\infty$ 时一致收敛到零.

根据巴什瓦尔等式有

$$\|h_n^{(1)}(t)-h_n^{(2)}(t)\|^2=\frac{1}{2\pi}\int_{-\pi}^{\pi}\left|\sum_{s=l_n^{(1)}}^{m_n^{(1)}}h_n^{(1)}(s)e^{-is\omega}-\sum_{s=l_n^{(2)}}^{m_n^{(2)}}h_n^{(2)}(s)e^{-is\omega}\right|^2 d\omega. \quad (1.17)$$

由(1.17)和许瓦兹不等式可知,若(1.15)成立,则(1.14)成立.这说明,若 $x(t)$ 的滤波因子具有均方收敛性,则必具有均匀收敛性.

现在我们给出滤波因子具有均匀收敛性和均方收敛性的等价条件.

定理 1 1) $x(t)$ 的滤波因子具有均匀收敛性的充分必要条件是:对任何满足(1.8)的有限长滤波因子 $h_n(t)$

$$h_n(t)=\begin{cases}h_n(t), & l_n\leqslant t\leqslant m_n,\\ 0, & \text{其他},\end{cases} \quad (1.18)$$

皆有

$$\lim_{j,k\to+\infty}\frac{1}{2\pi}\int_{-\pi}^{\pi}\left|\sum_{s=l_j}^{m_j}h_j(s)e^{-is\omega}-\sum_{s=t_k}^{m_k}h_k(s)e^{-is\omega}\right|d\omega=0. \quad (1.19)$$

2) $x(t)$ 的滤波因子具有均方收敛性的充分必要条件是:对任何满足(1.8)的有限长滤波因子 $h_n(t)$(见(1.18))皆有

$$\lim_{j,k\to+\infty}\|h_j(t)-h_k(t)\|=0. \quad (1.20)$$

证明 1) 必要性 用反证法.设 $h_n(t)$ 满足(1.18)和(1.8),但(1.19)不成立.于是存在 $\varepsilon>0$ 和整数列 $j_1<k_1<j_2<k_2<\cdots<j_n<k_n<\cdots$,使得

$$\frac{1}{2\pi}\int_{-\pi}^{\pi}\left|\sum_{s=l_{j_n}}^{m_{j_n}}h_{j_n}(s)e^{-is\omega}-\sum_{s=l_{k_n}}^{m_{k_n}}h_{k_n}(s)e^{-is\omega}\right|d\omega\geqslant\varepsilon, \quad n\geqslant 1. \quad (1.21)$$

令

$$h_n^{(1)}(t)=h_{j_n}(t), \quad h_n^{(2)}(t)=h_{k_n}(t).$$

由 $h_n(t)$ 的性质知，$h_n^{(1)}(t)$ 和 $h_n^{(2)}(t)$ 满足(1.10),(1.11). 但由(1.21)知,(1.14)是不成立的. 这与定义 1 矛盾. 由此可知, 必要性成立.

充分性 设 $h_n^{(1)}(t)$ 和 $h_n^{(2)}(t)$ 满足(1.10),(1.11).

令

$$h_n(t) = \begin{cases} h_k^{(1)}(t), & n = 2k-1, \\ h_k^{(2)}(t), & n = 2k, \end{cases} \quad k \geq 1, \quad (1.22)$$

易知 $h_n(t)$ 满足(1.18)和(1.8), 因此(1.19)成立. 由(1.22)和(1.19)知,(1.14)成立. 按定义 1, $x(t)$ 的滤波因子具有均匀收敛性, 充分性成立.

2) 的证明方法与 1) 类似, 这里不再细述. 证毕.

下面我们讨论滤波因子具有均匀收敛性和均方收敛性的充分必要条件.

2. 滤波因子具有均匀收敛性和均方收敛性的充要条件

我们知道, 能量有限信号 $x(t)$ 和它的频谱 $X(e^{-i\omega})$ 有密切的关系. $X(e^{-i\omega})$ 满足

$$\begin{cases} X(e^{-i\omega}) \in L_2(-\pi, \pi), \\ x(t) = \dfrac{1}{2\pi}\displaystyle\int_{-\pi}^{\pi} X(e^{-i\omega}) e^{-it\omega} d\omega. \end{cases} \quad (2.1)$$

我们把 $X(e^{-i\omega})$ 记为

$$X(e^{-i\omega}) = \sum_{t=-\infty}^{+\infty} x(t) e^{-it\omega}. \quad (2.2)$$

设能量有限信号 $x(t)$ 和 $y(t)$ 的频谱分别为 $X(e^{-i\omega})$ 和 $Y(e^{-i\omega})$, $x(t)$ 和 $y(t)$ 的褶积 $x(t) * y(t)$ 为

$$x(t) * y(t) = \sum_{s=-\infty}^{+\infty} y(s) x(t-s). \quad (2.3)$$

则有

$$x(t) * y(t) = \frac{1}{2\pi}\int_{-\pi}^{\pi} X(e^{-i\omega}) Y(e^{-i\omega}) e^{it\omega} d\omega. \quad (2.4)$$

根据巴什瓦尔等式有

$$\sum_{t=-\infty}^{+\infty} |x(t)|^2 = \frac{1}{2\pi}\int_{-\pi}^{\pi} |X(e^{-i\omega})|^2 d\omega. \quad (2.5)$$

以上各关系式由富氏级数理论即可知.

定理 2 设 $x(t)$ 为能量有限信号, 它的频谱为 $X(e^{-i\omega})$. 则 $x(t)$ 的滤波因子具有均匀收敛性的充分必要条件是

$$\frac{1}{|X(e^{-i\omega})|^2} \in L_1(-\pi,\pi). \tag{2.6}$$

证明 充分性 设 $y(t), h_n^{(1)}(t), h_n^{(2)}(t)$ 满足 $(1.7), (1.10), (1.11)$，它们的频谱分别为 $Y(e^{-i\omega}), H_n^{(1)}(e^{-i\omega}), H_n^{(2)}(e^{-i\omega})$. 由 (2.6) 知

$$\frac{Y(e^{-i\omega})}{X(e^{-i\omega})} \in L_1(-\pi,\pi).$$

根据许瓦兹不等式可得

$$\int_{-\pi}^{\pi} \left| H_n^{(j)}(e^{-i\omega}) - \frac{Y(e^{-i\omega})}{X(e^{-i\omega})} \right| d\omega = \int_{-\pi}^{\pi} |X(e^{-i\omega})H_n^{(j)}(e^{-i\omega}) - Y(e^{-i\omega})| \frac{1}{|X(e^{-i\omega})|} d\omega$$

$$\leq \left[\int_{-\pi}^{\pi} |X(e^{-i\omega})H_n^{(j)}(e^{-i\omega}) - Y(e^{-i\omega})|^2 d\omega \int_{-\pi}^{\pi} \frac{1}{|X(e^{-i\omega})|^2} d\omega\right]^{\frac{1}{2}}$$

$$= \|x(t) * h_n^{(j)}(t) - y(t)\| \left[\int_{-\pi}^{\pi} \frac{1}{|X(e^{-i\omega})|^2} d\omega\right]^{\frac{1}{2}} \to 0 \quad (n \to +\infty), \quad j=1,2.$$

再由

$$\int_{-\pi}^{\pi} |H_n^{(1)}(e^{-i\omega}) - H_n^{(2)}(e^{-i\omega})| d\omega \leq \sum_{j=1}^{2} \int_{-\pi}^{\pi} \left| H_n^{(j)}(e^{-i\omega}) - \frac{Y(e^{-i\omega})}{X(e^{-i\omega})} \right| d\omega$$

知 (1.14) 成立. 因此，$x(t)$ 的滤波因子具有均匀收敛性.

必要性 记

$$\mathcal{M} = \left\{ \sum_{s=1}^{m} b_s e^{-is\omega} \right\}, \tag{2.7}$$

即 $\mathcal{M}$ 为有限三角多项式全体. 由 $x(t)$ 的滤波因子具有均匀收敛性可知，必存在正数 λ 使下式成立

$$\int_{-\pi}^{\pi} |H(e^{-i\omega})| d\omega \leq \lambda \left[\int_{-\pi}^{\pi} |X(e^{-i\omega})|^2 |H(e^{-i\omega})|^2 d\omega\right]^{\frac{1}{2}}, \quad 其中 H(e^{-i\omega}) \in \mathcal{M}. \tag{2.8}$$

（用反证法可证 (2.8) 是成立的）

现在我们要证明，对任何有界可测函数 $G(\omega)$，(2.8) 式也成立，即

$$\int_{-\pi}^{\pi} |G(\omega)| d\omega \leq \lambda \left[\int_{-\pi}^{\pi} |X(e^{-i\omega})|^2 |G(\omega)|^2 d\omega\right]^{\frac{1}{2}} \tag{2.9}$$

成立.

设有界可测函数 $G(\omega)$ 满足

$$|G(\omega)| \leq \beta. \tag{2.10}$$

根据可测函数性质（可参看 [3]，124 页定理 3）易知，存在 $[-\pi,\pi]$ 上的连续函数 $B_n(\omega)$ 满足

$$\begin{cases} |B_n(\omega)| \leqslant \beta, \\ \lim_{n\to+\infty} B_n(\omega) = G(\omega), \quad a.e. \end{cases} \tag{2.11}$$

令

$$C_n(\omega) = \begin{cases} B_n(\omega), & \omega \in \left[-\pi, \pi - \frac{1}{n}\right], \\ \left[B_n\left(\pi - \frac{1}{n}\right) - B_n(-\pi)\right]n(\pi - \omega) + B_n(-\pi), & \omega \in \left(\pi - \frac{1}{n}, \pi\right]. \end{cases} \tag{2.12}$$

由(2.11)和(2.12)知,$C_n(\omega)$满足

$$\begin{cases} C_n(\omega) 在 [-\pi, \pi] 上连续, 且 C_n(-\pi) = C_n(\pi), \\ \lim_{n\to+\infty} C_n(\omega) = G(\omega), \quad a.e., \\ |C_n(\omega)| \leqslant \beta. \end{cases} \tag{2.13}$$

根据维尔斯脱拉斯定理(见[3],130 页),对 $\frac{1}{n}$,存在三角多项式 $H_n(e^{-i\omega})$ $\in \mathcal{M}$ 使

$$|C_n(\omega) - H_n(e^{-i\omega})| < \frac{1}{n}. \tag{2.14}$$

由(2.13)和(2.14)知,$H_n(e^{-i\omega})$满足

$$\begin{cases} \lim_{n\to+\infty} H_n(e^{-i\omega}) = G(\omega), \quad a.e., \\ |H_n(e^{-i\omega})| \leqslant \beta + 1. \end{cases} \tag{2.15}$$

我们知道,$H_n(e^{-i\omega})$使(2.8)成立.根据勒贝格控制收敛定理(见[3]180 页定理 1),由(2.15)知(2.9)是成立的.

取

$$G(\omega) = \chi\{X(e^{-i\omega}) = 0\},$$

则由(2.9)可知

$$\mu(X(e^{-i\omega}) = 0) = 0, \tag{2.16}$$

其中 μ 表示勒贝格测度,χ_e 表示集 e 的特征函数

$$\chi_e = \begin{cases} 1, & \omega \in e, \\ 0, & \omega \notin e. \end{cases}$$

取

$$G(\omega) = \frac{1}{|X(e^{-i\omega})|^2} \chi_{\{|X(e^{-i\omega})| > \varepsilon\}},$$

其中 $\varepsilon > 0$. 则由(2.9)得

$$\int_{\{|X|>\varepsilon\}} \frac{1}{|X|^2} d\omega \leqslant \lambda \Big[\int_{\{|X|>\varepsilon\}} \frac{1}{|X|^2} d\omega\Big]^{\frac{1}{2}},$$

即

$$\int_{\{|X|>\varepsilon\}} \frac{1}{|X|^2} d\omega \leqslant \lambda^2. \tag{2.17}$$

由于 ε 是任意正数,由(2.16)和(2.17)便知(2.6)成立. 证毕.

定理 3 设 $x(t)$ 为能量有限信号,它的频谱为 $X(e^{-i\omega})$. 则 $x(t)$ 的滤波因子具有均方收敛性的充分必要条件是:存在常数 c 使

$$|X(e^{-i\omega})| \geqslant c > 0, \quad a.e. \tag{2.18}$$

证明 充分性 设 $y(t), h_n^{(1)}(t), h_n^{(2)}(t)$ 满足(1.7), (1.10)和(1.11), 它的频谱分别为 $Y(e^{-i\omega}), H_n^{(1)}(e^{-i\omega}), H_n^{(2)}(e^{-i\omega})$.

令

$$H(e^{-i\omega}) = \frac{Y(e^{-i\omega})}{X(e^{-i\omega})}. \tag{2.19}$$

由(2.18)知

$$|H(e^{-i\omega})| \leqslant \frac{1}{c} |Y(e^{-i\omega})|.$$

所以 $H(e^{-i\omega}) \in L_2(-\pi, \pi)$.

设 $H(e^{-i\omega})$ 所对应的信号为 $h(t)$.

$$\begin{aligned}
\|h_n^{(j)}(t) - h(t)\|^2 &= \frac{1}{2\pi}\int_{-\pi}^{\pi} \Big| H_n^{(j)}(e^{-i\omega}) - \frac{Y(e^{-i\omega})}{X(e^{-i\omega})} \Big|^2 d\omega \\
&\leqslant \frac{1}{c^2} \frac{1}{2\pi} \int_{-\pi}^{\pi} |X(e^{-i\omega})H_n^{(j)}(e^{-i\omega}) - Y(e^{-i\omega})|^2 d\omega \\
&= \frac{1}{c^2} \|x(t) * h_n^{(j)}(t) - y(t)\|^2 \to 0.
\end{aligned}$$

因此

$$\|h_n^{(1)}(t) - h_n^{(2)}(t)\| \to 0.$$

这说明(1.15)成立, $x(t)$ 的滤波因子具有均方收敛性.

必要性 由 $x(t)$ 的滤波因子具有均方收敛性可知(用反证法),必存在正数 λ 使下式成立

$$\Big[\int_{-\pi}^{\pi} |H(e^{-i\omega})|^2 d\omega\Big]^{\frac{1}{2}} \leqslant \lambda \Big[\int_{-\pi}^{\pi} |X(e^{-i\omega})|^2 |H(e^{-i\omega})|^2 d\omega\Big]^{\frac{1}{2}}, \text{ 其中 } H(e^{-i\omega}) \in \mathcal{M}. \tag{2.20}$$

与定理 2 的证明方法完全相同,由(2.20)可推出,对任何有界可测函数

$G(\omega)$ 皆有

$$\int_{-\pi}^{\pi}|G(\omega)|^2 d\omega \leqslant \lambda^2 \int_{-\pi}^{\pi}|X(e^{-i\omega})|^2|G(\omega)|^2 d\omega,$$

即

$$\int_{-\pi}^{\pi}|G(\omega)|^2[\lambda^2|X(e^{-i\omega})|^2-1]d\omega \geqslant 0. \quad (2.21)$$

取

$$G(\omega) = \chi_{\{\lambda^2|X|^2-1<0\}},$$

由(2.21)可知

$$\mu(\lambda^2|X(e^{-i\omega})|^2-1<0) = 0,$$

此即表明

$$|X(e^{-i\omega})|^2 \geqslant \frac{1}{\lambda^2}(a.e.),$$

也即(2.18)成立. 证毕.

定理 2 和定理 3 从理论上告诉我们,只有当(2.18)或(2.6)成立时,滤波因子才具有稳定性. 这有重要的实际意义. 在求最小平方滤波因子时,为使因子稳定采用白噪化方法,把输入信号的频谱加以改造使之满足(2.18). 关于这方面的问题,可参看文献[2].

3. 最小平方滤波的极限性质

在这一节,我们讨论最小平方滤波因子和误差的极限性质. 我们在一般的条件下给出结论,而省略证明.

设 $x_j(t), y_j(t), j=1,2,\cdots,J$,为能量有限信号,相应的频谱分别为 $X_j(e^{-i\omega}), Y_j(e^{-i\omega})$.

设

$$h_{l,m}(t) = \begin{cases} h(t), & l \leqslant t \leqslant m, \\ 0, & \text{其他} \end{cases} \quad (3.1)$$

是使

$$Q(l,m) = \sum_{j=1}^{J}\sum_{t=-\infty}^{\infty}[x_j(t)*h_{l,m}(t)-y_j(t)]^2$$

达最小值 $Q_{\min}(l,m)$ 的最小平方滤波因子.

设

$$\ln \sum_{j=1}^{J} | X_j(e^{-i\omega}) |^2 \in L_1(-\pi,\pi), \tag{3.2}$$

$$U(Z) = \exp \frac{1}{2\pi} \int_{-\pi}^{\pi} \ln \sqrt{\sum_{j=1}^{J} | X_j(e^{-i\varphi}) |^2} \frac{e^{-i\varphi}+Z}{e^{-i\varphi}-Z} d\varphi,$$

$$W(e^{-i\omega}) = \sum_{j=1}^{J} Y_j(e^{-i\omega}) \overline{X_j(e^{-i\omega})} \frac{1}{U(e^{-i\omega})}.$$

设 $u(t)$ 和 $w(t)$ 是分别由 $U(e^{-i\omega})$ 和 $W(e^{-i\omega})$ 确定的信号, $h_l(t)(t \geqslant 0)$ 由关系式

$$\sum_{s=0}^{t} h_l(s) u(t-s) = w(t+l), \qquad t \geqslant 0$$

唯一确定.

定理 4 设 (3.2) 成立, 则

$$\lim_{m \to +\infty} Q_{\min}(l,m) = \sum_{j=1}^{J} \sum_{t=-\infty}^{\infty} y_j^2(t) - \sum_{t=l}^{\infty} w^2(t),$$

$$\lim_{m \to +\infty} h_{l,m}(t) = h_l(t-l), \quad t \geqslant l,$$

$$\lim_{\substack{l \to -\infty \\ m \to +\infty}} Q_{\min}(l,m) = \frac{1}{2\pi}\int_{-\pi}^{\pi} \sum_{j=1}^{J} | Y_j(e^{-i\omega}) |^2 d\omega - \frac{1}{2\pi}\int_{-\pi}^{\pi} \frac{\left| \sum_{j=1}^{J} Y_j(e^{-i\omega}) \overline{X_j(e^{-i\omega})} \right|^2}{\sum_{j=1}^{J} | X_j(e^{-i\omega}) |^2} d\omega.$$

定理 5 设 $\ln \sum_{j=1}^{J} | X_j(e^{-i\omega}) |^2 \notin L_1(-\pi,\pi)$ 且 $\sum_{j=1}^{J} | X_j(e^{-i\omega}) |^2 \not\equiv 0$, 则

$$\lim_{m \to +\infty} Q_{\min}(l,m) = \lim_{\substack{l \to -\infty \\ m \to +\infty}} Q_{\min}(l,m)$$

$$= \frac{1}{2\pi}\int_{-\pi}^{\pi} \sum_{j=1}^{J} | Y_j(e^{-i\omega}) |^2 d\omega - \frac{1}{2\pi}\int_{-\pi}^{\pi} | V(e^{-i\omega}) |^2 d\omega,$$

其中

$$V(e^{-i\omega}) = \begin{cases} \dfrac{\sum_{j=1}^{J} Y_j(e^{-i\omega}) \overline{X_j(e^{-i\omega})}}{\sqrt{\sum_{j=1}^{J} | X_j(e^{-i\omega}) |^2}}, & \sum_{j=1}^{J} | X_j |^2 > 0, \\ 0, & \sum_{j=1}^{J} | X_j |^2 = 0. \end{cases}$$

参考文献

[1] P. L. Duren, Theory of H^p Space, Acadmic Press, 1970.
[2] 程乾生,信号数字处理的数学原理,石油工业出版社,1979.
[3] И. П. 那汤松著,徐瑞云译,实变函数论(上册),高等教育出版社,1957.

原文载于《应用数学学报》,第 6 卷,第 3 期,1983 年 7 月,267—275.

Multidimensional All-Pass Filters and Minimum-Phase Filters

Cheng Qiansheng(程乾生)

Department of Mathematics, Peking University

The 1-dimensional all-pass filter energy delay problem was studied by Robison[1]. His method is based on the canonical factorization theorem. However, multidimensional analytic function is so complicated that is does not admit any canonical factorization representation like 1-dimensional one. So the multidimensional energy delay problem has been little studied. In this paper, by the method introduced in [2] or [5], multidimensional all-pass filters are characterized, and a multidimensional energy delay theorem is presented. By this theorem, the problem of turning a finite-order filter into a minimum-phase filter is discussed.

The problems in multidimensional case and 2-dimensional case are essentially the same. For ease of representation, we consider only the 2-dimensional case.

We denote by l_2^+ the space which consists of $h(m,n)$, where $h(m,n)$ is a 2-dimensional sequence and satisfies

$$\begin{cases} h(m,n) = 0, \quad m < 0 \text{ or } n < 0, \\ \sum_{m=0}^{\infty}\sum_{n=0}^{\infty} |h(m,n)|^2 < \infty. \end{cases} \quad (1)$$

The z-transformation of $h(m,n)$ is defined as

$$H(z_1, z_2) = \sum_{m=0}^{\infty}\sum_{n=0}^{\infty} h(m,n) z_1^m z_2^n, \quad (2)$$

the frequence spectrum of $h(m,n)$ as $H(e^{-i\omega_1}, e^{-i\omega_2})$, and the amplitude spectrum as $|H(e^{-i\omega_1}, e^{-i\omega_2})|$. When $x(m,n)$ and $y(m,n)$ belong to l_2^+, the convolution of $x(m,n)$ and $y(m,n)$ is defined as

$$x(m,n) * y(m,n) = \sum_{s=0}^{\infty} \sum_{t=0}^{\infty} x(s,t) y(m-s, n-t). \tag{3}$$

Definition 1. Let $g(m,n) \in l_2^+$. $g(m,n)$ is called an all-pass sequence or an impulse response of an all-pass filter if

$$|G(e^{-i\omega_1}, e^{-i\omega_2})| = 1 \quad (\text{a. e.}). \tag{4}$$

From the Parseval equation it follows that the total energy of an all-pass sequence $g(m,n)$ equals one, that is,

$$\sum_{m=0}^{\infty} \sum_{n=0}^{\infty} |g(m,n)|^2 = \frac{1}{4\pi^2} \int_{-\pi}^{\pi} \int_{-\pi}^{\pi} |G(e^{-i\omega_1}, e^{-i\omega_2})| \, d\omega_1 d\omega_2 = 1. \tag{5}$$

Applying Eq. (5), we can easily get the following theorem.

Theorem 1. (i) *Let $g(m,n)$ be an all-pass sequence, then $|g(0,0)| \leqslant 1$. In particular, if $|g(0,0)| = 1$, then $G(z_1, z_2) = g(0,0)$.*

(ii) *Let $G_j(z_1, z_2), 1 \leqslant j \leqslant J$, be z-transformations of all-pass sequences, and $\prod_{j=1}^{J} G_j(z_1, z_2) = e^{i\beta}$ (β is a real number), then $G_j(z_1, z_2) = e^{i\beta_j}$, where β_j are real numbers, $\sum_{j=1}^{J} \beta_j = \beta + 2k\pi$, k is an integer.*

We are now ready to characterize all-pass filters.

Theorem 2 (All-pass filter energy delay theorem). *Let $g(m,n)$ be an all-pass filter, $x(m,n) \in l_2^+$ be an input sequence and $y(m,n) = g(m,n) * x(m,n)$ be an output sequence. Then for any $M \geqslant 0$, $N \geqslant 0$,*

$$\sum_{m=0}^{M} \sum_{n=0}^{N} |y(m,n)|^2 \leqslant \sum_{m=0}^{M} \sum_{n=0}^{N} |x(m,n)|^2. \tag{6}$$

Proof. Set

$$x_{MN}(m,n) = \begin{cases} x(m,n), & 0 \leqslant m \leqslant M, 0 \leqslant n \leqslant N, \\ 0, & \text{others}, \end{cases} \tag{7}$$

$$y_{MN}(m,n) = g(m,n) * x_{MN}(m,n). \tag{8}$$

Applying the Parseval equation, from (8) and (4) it follows that

$$\sum_{m=0}^{\infty} \sum_{n=0}^{\infty} |y_{MN}(m,n)|^2 = \frac{1}{4\pi^2} \int_{-\pi}^{\pi} \int_{-\pi}^{\pi} |Y_{MN}(e^{-i\omega_1}, e^{-i\omega_2})|^2 d\omega_1 d\omega_2$$

$$= \frac{1}{4\pi^2} \int_{-\pi}^{\pi} \int_{-\pi}^{\pi} |X_{MN}(e^{-i\omega_1}, e^{-i\omega_2})|^2 d\omega_1 d\omega_2$$

$$= \sum_{m=0}^{M} \sum_{n=0}^{N} |x(m,n)|^2. \tag{9}$$

According to the definition of convolution, from $y(m,n)=g(m,n)*x(m,n)$ and (8) we find

$$y_{MN}(m,n)=y(m,n), \quad 0\leqslant m\leqslant M, 0\leqslant n\leqslant N. \tag{10}$$

(9) and (10) yield (6). Theorem 2 is proved.

Theorem 3. *Let* $g(m,n)\in l_2^+$ *and* $\sum_{m=0}^{\infty}\sum_{n=0}^{\infty}|g(m,n)|^2=1$. *Then in order to make* $g(m,n)$ *an impulse response of an all-pass filter, it is necessary and sufficient that for any* $x(m,n)\in l_2^+$, $M\geqslant 0, N\geqslant 0$, *we have*

$$\sum_{m=0}^{M}\sum_{n=0}^{N}|y(m,n)|^2 \leqslant \sum_{m=0}^{M}\sum_{n=0}^{N}|x(m,n)|^2, \tag{11}$$

where $y(m,n)=g(m,n)*x(m,n)$.

Proof. Necessity. From Theorem 2 it follows that (11) always holds.

Sufficiency. We assume that $|G(e^{-i\omega_1},e^{-i\omega_2})|\not\equiv 1$(a. e.).

Set

$$|\hat{X}(e^{-i\omega_1},e^{-i\omega_2})|^2=\begin{cases}2, & |G(e^{-i\omega_1},e^{-i\omega_2})|^2-1>0,\\ 1, & |G(e^{-i\omega_1},e^{-i\omega_2})|^2-1\leqslant 0.\end{cases}$$

Since

$$\frac{1}{4\pi^2}\int_{-\pi}^{\pi}\int_{-\pi}^{\pi}|G(e^{-i\omega_1},e^{-i\omega_2})|^2 d\omega_1 d\omega_2 = \sum_{m=0}^{\infty}\sum_{n=0}^{\infty}|g(m,n)|^2 = 1,$$

$$\frac{1}{4\pi^2}\int_{-\pi}^{\pi}\int_{-\pi}^{\pi} d\omega_1 d\omega_2 = 1,$$

we know that

$$\frac{1}{4\pi^2}\int_{-\pi}^{\pi}\int_{-\pi}^{\pi}(|G(e^{-i\omega_1},e^{-i\omega_2})|^2-1)d\omega_1 d\omega_2 = 0.$$

Hence

$$\frac{1}{4\pi^2}\int_{-\pi}^{\pi}\int_{-\pi}^{\pi}(|G(e^{-i\omega_1},e^{-i\omega_2})|^2-1)|\hat{X}(e^{-i\omega_1},e^{-i\omega_2})|^2 d\omega_1 d\omega_2 > 0.$$

Since $|\hat{X}(e^{-i\omega_1},e^{-i\omega_2})|^2$ is a bounded function, we always find a trigonometric polynomial $X(e^{-i\omega_1},e^{-i\omega_2})$ such that

$$\frac{1}{4\pi^2}\int_{-\pi}^{\pi}\int_{-\pi}^{\pi}(|G(e^{-i\omega_1},e^{-i\omega_2})|^2-1)|X(e^{-i\omega_1},e^{-i\omega_2})|^2 d\omega_1 d\omega_2 > 0,$$

that is

$$\frac{1}{4\pi^2}\int_{-\pi}^{\pi}\int_{-\pi}^{\pi}|G(e^{-i\omega_1},e^{-i\omega_2})|^2 |X(e^{-i\omega_1},e^{-i\omega_2})|^2 d\omega_1 d\omega_2$$

$$> \frac{1}{4\pi^2}\int_{-\pi}^{\pi}\int_{-\pi}^{\pi}|X(e^{-i\omega_1},e^{-i\omega_2})|^2 d\omega_1 d\omega_2.$$

It can be rewritten as

$$\sum_{m=0}^{\infty}\sum_{n=0}^{\infty}|y(m,n)|^2 > \sum_{m=0}^{\infty}\sum_{n=0}^{\infty}|x(m,n)|^2.$$

We can always mark off M_0 and N_0, so that

$$\sum_{m=0}^{M_0}\sum_{n=0}^{N_0}|y(m,n)|^2 > \sum_{m=0}^{M_0}\sum_{n=0}^{N_0}|x(m,n)|^2.$$

But this is contrary to (11). Hence, the sufficiency holds. And the theorem is proved.

We shall now discuss the problem of minimum-phase filters.

Definition 2. A filter is said to be a minimum-phase one if its z-transformation $H(z_1, z_2)$ is continuous and has no zero on $D_1 = \{|z_1|\leqslant 1, |z_2|\leqslant 1\}$, and it is analytic in $\{|z_1|<1, |z_2|<1\}$.

Definition 3. If an impulse response of a filter satisfies

$$x(m,n) = \begin{cases} x(m,n), & 0 \leqslant m \leqslant M, 0 \leqslant n \leqslant N, \\ 0, & others, \end{cases}$$

then the filter is called an (M,N)-order filter, $x(m,n)$ as an (M,N)-order sequence.

Theorem 4. Let $y(m,n)$ be an (M,N)-order sequence, and $x(m,n)$ be an impulse response of a minimum filter. If $|Y(e^{-i\omega_1},e^{-i\omega_2})| = |X(e^{-i\omega_1},e^{-i\omega_2})|$, then $x(m,n)$ is an (M,N)-order sequence too.

Proof. From definition 2, $X(z_1, z_2)$ is analytic and has no zero in $\{|z_1|<1, |z_2|<1\}$, so

$$A(z_1, z_2) \triangleq \frac{1}{X(z_1, z_2)} = \sum_{m=0}^{\infty}\sum_{n=0}^{\infty} a(m,n) z_1^m z_2^n$$

is analytic in $\{|z_1|<1, |z_2|<1\}$. Furthermore, since $X(z_1, z_2)$ is continuous and has no zero on D_1, $a(m,n) \in l_2^+$.

Set

$$g(m,n) = y(m,n) * a(m,n).$$

Because

$$|G(e^{-i\omega_1},e^{-i\omega_2})| = |Y(e^{-i\omega_1},e^{-i\omega_2})|\frac{1}{|X(e^{-i\omega_1},e^{-i\omega_2})|} = 1,$$

$g(m,n)$ is an all-pass sequene.

We note
$$Y(z_1, z_2) = G(z_1, z_2) X(z_1, z_2),$$
thus
$$y(m,n) = g(m,n) * x(m,n).$$
From Theorem 2, it follows that
$$\sum_{m=0}^{M} \sum_{n=0}^{N} |y(m,n)|^2 \leqslant \sum_{m=0}^{M} \sum_{n=0}^{N} |x(m,n)|^2. \tag{12}$$
Since $y(m,n)$ is an (M, N)-order sequence and the amplitude spectrum of $y(m,n)$ is the same as $x(m,n)$, according to the Parseval equation, we have
$$\sum_{m=0}^{M} \sum_{n=0}^{N} |y(m,n)|^2 = \sum_{m=0}^{\infty} \sum_{n=0}^{\infty} |x(m,n)|^2. \tag{13}$$
By comparison between (12) and (13), it follows that $x(m,n) = 0$ when $(m, n) \notin \{(l,k) | 0 \leqslant l \leqslant M, 0 \leqslant k \leqslant N\}$. This shows that $x(m,n)$ is an (M,N)-order sequence. Theorem 4 is proved.

Woods gave in [3] an example of a $(1,1)$-order filter whose z-transformation is
$$Y(z_1, z_2) = (1 \quad z_1) \begin{pmatrix} \frac{1}{4} & 0 \\ 1 & \frac{1}{4} \end{pmatrix} \begin{pmatrix} 1 \\ z_2 \end{pmatrix}.$$

He proved by direct calculation that there does not exist any $(1,1)$-order minimum-phase filter such that its amplitude spectrum is the same as $|Y(e^{-i\omega_1}, e^{-i\omega_2})|$. Theorem 4 indicates more deeply that there does not exist any finite- or infinite- order minimum phase filter such that its amplitude spectrum coincides with $|Y(e^{-i\omega_1}, e^{-i\omega_2})|$.

References

[1] Robison, E. A., *Random Wavelets and Cybernetic Systems*, Stechert-Hafner, New York, 1962.
[2] 程乾生, 信号数字处理的数学原理, 石油工业出版社, 北京, 1979.
[3] Woods, J. W., *IEEE.*, GE-12(1974), 3:104—105.
[4] Mitra, S. K. & Ekstrom, M. P., *Two-Dimensional Digital Signal Processing*, Hutchinson & Ross. Inc., Dowden, 1978.
[5] 舒立华, 数学学报, 17(1974), 20—27.

原文载于《科学通报》, Vol. 28, No. 5(1988.5), 588—591.

Simultaneous Wavelet Estimation and Deconvolution of Reflection Seismic Signals

Cheng Qiansheng[1], Chen Rong[2], and Li Ta-Hsin[2], *Member*, *IEEE*
1. Department of Mathematics, Peking University;
2. Department of Statistics, Texas A&M University

Abstract: In this paper, the problem of simultaneous wavelet estimation and deconvolution is investigated with a Bayesian approach under the assumption that the reflectivity obeys a Bernoulli-Gaussian distribution. Unknown quantities, including the seismic wavelet, the reflection sequence, and the statistical parameters of reflection sequence and noise are all treated as realizations of random variables endowed with suitable prior distributions. Instead of deterministic procedures that can be quite computationally burdensome, a simple Monte Carlo method, called Gibbs sampler, is employed to produce random samples iteratively from the joint posterior distribution of the unknowns. Modifications are made in the Gibbs sampler to overcome the ambiguity problems inherent in seismic deconvolution. Simple averages of the random samples are used to approximate the minimum mean-squared error (MMSE) estimates of the unknowns. Numerical examples are given to demonstrate the performance of the method.

I. Introduction

In reflection seismology experiments (e.g., [16]), a test signal (or seismic source) is sent to probe the layered earth and the reflected signal (seismic trace) is recorded by a geophone or hydrophone at the surface. Geophysical structure of the earth is then investigated through an analysis of the reflectivity from deep layers of the earth. The true reflectivity, however, is not directly accessible; instead, the recorded signal is a smeared version of the reflectivity, caused by reverberations due to the surface layers. One of the important objectives in seismic data processing is to undo the effects of the degradation in order to recover the true reflectivity. This usually requires a certain *deconvolution* technique, since the received signal can be regarded as a convolution of the reflection sequence with an unknown seismic wavelet. As a common practice. the seismic wavelet is often modeled as a deterministic finite moving-average filter, while the reflectivity is assumed to be a sequence of random variables with scattered nonzero values (e. g. , [16]). In most of the available deconvolufion techniques, the reflection sequence is further assumed to be *white* so that the random variables in the sequence are mutually independent (e. g. , [14],[15]).

To successfully recover the reflection sequence, it is crucial for the deconvolution algorithms to incorporate as much prior knowledge about the reflectivity as possible. Wiggins [18], for example, proposed a criterion that describes the "simplicity of appearance" of the reflection sequence, resulting in the well-known minimum entropy method (see also [2] and [4]). Probabilistic models for the reflection sequence are also helpful, among which the *Bernoulli-Gaussian white sequence model* has been proved very successful (e. g. , [10], [14], [15], [17]).

In the Bernoulli-Gaussian white sequenee model, the occurrence of nonzero values in the reflection sequence is governed by a Bernoulli law while the magnitude of nonzero values by an independent Gaussian distribution (e. g. , [14], [15]). With suitable choice of parameters, the Bernoulli-Gaussian model can imitate the appearance of reflection sequences characterized by the sparseness of nonzero values. The problem, then, becomes that of simultaneous estimation of the seismic wavelet, the statistical parameters, and the reflection sequence. Since ambient noise is always present in the recorded signal, the statistical parameters to be estimated also include those that describe the probability distribution of the noise.

To estimate these unknown quantities, maximum likelihood (ML) approach has been shown to yield satisfactory results (e. g. , [14], [15]); yet, to obtain the ML estimates numerically involves highly complicated iterative algorithms and detection schemes that can be quite computationally expensive (e. g. , [9], [11], [14], [15]). Besides, the convergence of these algorithms to desired solutions is not always guaranteed, depending on the choice of initial guess.

Under the Bernoulli-Gaussian white sequence model, the present paper deals with the deconvolution problem from a *Bayesian* point of view. The unknown quantities are treated as random variables with suitable prior distribution and inferences about the unknowns obtained on the basis of their posterior distribution. This approach is found to be able to incorporate nonminimum-phase wavelet and even colored reflection sequences into a unified framework and make effective use of prior information about the unknown quantities.

Instead of directly computing the Bayesian estimates, we employ a Monte Carlo procedure—the *Gibbs sampler*—to iteratively generate random samples from the joint posterior distribution of the unknowns. Bayesian estimators, such as the minimum mean-squared error (MMSE) estimator and the maximum *a posteriori* (MAP) estimator, of the unknown quantities can be easily approximated using the random samples. For instance, to find the MMSE estimate of the reflection sequence, one simply calculates the average of the corresponding components in the sample, and thus avoids the burdensome computation of the conditional expectation of the reflection sequence given the recorded signal.

The Gibbs sampler has been applied with success to image restoration problems when the degradation filter and the statistical parameters of the signal and noise are known *a priori* (e. g., [7]). Extensions to the cases of *unknown* degradation filters and statistical parameters have also been developed recently by Chen and Li [1] and Liu and Chen [13] for the deconvolution of discrete-valued input signals (see also [12]). Our experiment in this paper on simulated data further proves the suitability of the Gibbs sampler in seismic deconvolution.

The rest of the paper is organized as follows. Section II provides the mathematical formulation of the problem. Section III introduces the Bayesian approach and the prior distributions to be used. Then the joint posterior and conditional posterior distributions of the unknowns to be used in the Gibbs sampler are derived. Section IV briefly reviews the Gibbs sampler in general and details its implementation in our particular problem. Some simulation results are presented in Section V.

II. Formulation of the Problem

Given a seismic trace, the observed data, $\{x_1, \cdots, x_n\}$, can be modeled as a convolution of the unknown seismic wavelet $\{\phi_i\}$ with the unknown reflection sequence $\{u_t\}$, namely

$$x_t = \sum_{i=0}^{q} \phi_i u_{t-i} + \varepsilon_t, \qquad (1)$$

where the ε_t stand for the additive measurement error and noise. Assume that the ε_t are mutually independent and have a common Gaussian distribution $N(0, \sigma^2)$ with mean zero and a certain unknown variance σ^2.

It is clear that some information about the reflection sequence must be provided in order to distinguish it from the seismic wavelet. The Bernoulli-Gaussian white sequence model ([11], [14], [15]) is one of the well-established models in this regard. The model assumes that $\{u_t\}$ is a white sequence whose common probability distribution is a mixture of Bernoulli and Gaussian distributions. In other words, the distribution of u_t takes the form

$$\eta \delta(u_t = 0) + (1-\eta) N(0, r^2), \qquad (2)$$

where $\eta \in (0, 1)$ is the probability that $u_t = 0$. According to this model, the nonzero values of u_t follow a Gaussian distribution with mean zero and a certain variance r^2. Both η and r^2 are unknown in practice, and hence, to be estimated along with the reflection sequence from the observed data.

The seismic wavelet $\{\phi_i\}$ is a broad-band sequence often modeled as being deterministic with finite length (e.g., [8], [14], [15]). The order q of the seismic wavelet is assumed to be available in the sequel. The causality assumption of $\{\phi_i\}$, i.e., the assumption that $\phi_i = 0$ for any $i < 0$, is not crucial since a noncausal wavelet of finite length [17] can be easily transformed into a causal one by index substitution (both i and t). The scale ambiguity in $\{\phi_i\}$, however, is crucial to the existence of unique solution. In fact, since multiplying the ϕ_i by any nonzero constant does not change the problem if the reflectivity is re-scaled accordingly, the deconvolution problem cannot be uniquely solved without imposing extra conditions that eliminate the ambiguity. To this end, we assume that the wavelet is normalized with $\phi_0 = 1$, although other conditions, such as the total power of the seismic wavelet, also help to achieve the same objective.

The ultimate goal of simultaneous wavelet estimation and deconvolution is to estimate $\{\phi_i\}$, η, r^2, σ^2, as well as $\{u_1, \cdots, u_n\}$, solely on the basis of $\{x_1, \cdots, x_n\}$. Clearly, there are more unknowns to be estimated than the observed data. This partially explains the difficulties that arise in the maximum likelihood method in seeking to maximize a function of such enormous dimension. The Bayesian approach, on the other hand, helps to ease the problem, especially when combined with Gibbs sampling.

III. The Bayesian Approach

To deal with the aforementioned problem of simultaneous wavelet estimation and deconvolution, the Bayesian approach assumes that the unknown quantities are realizations of random variables governed by certain *prior* probability distributions. This assumption is particularly reasonable for the seismic wavelet since the reverberations in the surface layers of the earth are indeed random in nature [14]. For other parameters, the imposed prior distributions simply reflect the degree of uncertainty of belief about the values of the parameters in a mathematical way that can be easily coped with by the Bayes theorem.

In the Bayesian paradigm, inferences about the unknowns are made by investigating their joint *posterior* probability distribution that reflects the gain of knowledge from the observed data toward the unknowns. Widely-used Bayesian point estimates include the conditional expectation, known as the minimum mean-squared error (MMSE) estimator, and the mode of the posterior distribution, known as the maximum *a posteriori* (MAP) estimator; the former minimizes the mean-squared error criterion while the latter intends to provide the maximum certainty about the parameter in a way similar to the maximum likelihood principle.

A. Prior Distributions

A unique feature of Bayesian approach rests on its ability to incorporate prior information about unknown parameters in the form of *prior probability distribution*. A suitable choice of priors is essential not only to the accommodation of prior information but to the insurance of simplicity of the resulting posterior distributions as well. Roughly speaking, the priors should contain predetermined parameters that reflect the degree of uncertainty toward the unknowns. If no prior information is available, flat or "nearly" flat priors should be used. The prior distributions should also be simple enough so as to minimize the computational burden of the posterior distributions. Conjugate priors are commonly used for this purpose.

A *conjugate* family of prior distributions is closed under sampling, i. e. , if the prior distribution is in the family, the posterior distribution given a sample will also be in the family, regardless of the observed values in the sample and the sample size. Hence, if a conjugate prior is used, the posterior distribution will have the same form with updated information reflected by a change in the parameters of the distribution. For example, the beta distribution is a conjugate family for samples from a Bernoulli distribution—when (binary) data are observed from a Bernoulli (θ) distribution and θ follows a Beta (a,b) prior, the posterior distribution of θ given the data will always be a beta distribution. In fact, the posterior distribution is Beta ($a+X$, $b+Y$), where X and Y are the numbers of 0's and 1's, respectively, in the data set.

For most commonly-used distribution families, a conjugate prior also ensures that the resulting MMSE estimator is consistent (e. g. , [3], p. 335). The MMSE estimator may be biased if the mean of the prior does not coincide with the parameter. However, as the sample size grows, the effect of the prior distribution fades away and the bias eventually vanishes.

Based on these principles, the following priors are employed in our problem of deconvolution and all of them are conjugate priors. Other choices are possible but may not lead to computationally tractable posteriors.

1) Assume that η has a prior beta distribution $\pi(\eta) \sim$ Beta (a,b), where $a>0$ and $b>0$ are predetermined parameters that control the shape of the distribution.

2) Assume that r^2 is a known constant. Another possibility is to regard r^2 as a random variable with an inverted chi-square distribution so that $\mu\gamma/r^2 \sim \chi^2(\mu)$ for some $\gamma>0$ and $\mu>0$. For simplicity, we shall not pursue this option in the present paper. It should be noted that in practice r^2 may vary with time so that a time-dependent model for r^2 is more realistic. For instance, r^2 may obey a gain law such as exponential decay with some unknown parameters. If r^2 is allowed to change with time, the problem becomes more complicated, though still solvable. The difficulty comes from

the fact that the conditional posterior distribution of the parameters in r^2 may not have a simple form and hence more sophisticated sampling procedures such as the rejection method have to be employed to generate random samples for these parameters. At this stage of development, stationarity of the data record is assumed.

3) Given η and r^2, the reflection sequence $\{u_t\}$ is assumed to be i. i. d. (independent and identically distributed) with a common distribution of the form (2). For convenience, it is also assumed that $u_t = 0$ with probability one for $t = -q+1, \cdots, -1, 0$.

4) With the restriction $\phi_0 = 1$, assume that the rest parameters $\Phi := [\phi_1, \cdots, \phi_q]^T$ in the seismic wavelet obey the multivariate Gaussian law of the form $\pi(\Phi) \sim N_q(\Phi_0, \Sigma_0)$ for some predetermined mean vector Φ_0 and covariance matrix Σ_0.

5) The noise variance σ^2 is assumed to have an inverted chi-square prior distribution $\pi(\sigma^2) \sim \chi^{-2}(\nu, \lambda)$, i. e. $\nu\lambda/\sigma^2 \sim \chi^2(\nu)$, for some known parameters $\lambda > 0$ and $\nu > 0$.

Typical graphs of the beta and inverted chi-square distributions with various parameters can be found in Figs. 1 and 2, respectively. These graphs help to understand the impact of each parameter on the shape and center of the priors. Since the degree of uncertainty about the unknowns is controlled by these parameters, one is always advised to experiment with a few combinations of the parameter values until satisfactory results can be obtained.

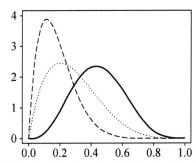

Fig. 1　Prior probability density of η, the Beta(a,b) distribution: solid line for $a=4$ and $b=5$, dotted line for $a=2$ and $b=5$, and dashed line for $a=2$ and $b=9$

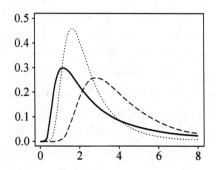

Fig. 2 Prior probability density of σ^2, the $\chi^{-2}(\nu,\lambda)$ distribution: solid line for $\nu=3$ and $\lambda=2$, dotted line for $\nu=9$ and $\lambda=2$, and dashed line for $\nu=9$ and $\lambda=3.5$

B. Posterior Distributions

Equipped with these priors, one can easily derive the *conditional posterior distributions* for a later use in the Gibbs sampler. To this end, it is crucial to note that under the assumption that $\{\varepsilon_t\}$ is Gaussian white noise the joint distribution of $X := \{x_1, \cdots, x_n\}$, Φ, $U := \{u_{1-q}, \cdots, u_n\}$, σ^2, and η takes the form of

$$\pi(X, \Phi, U, \sigma^2, \eta)$$
$$= \pi(X|\Phi, U, \sigma^2, \eta)\pi(\Phi)\pi(U|\eta)\pi(\sigma^2)\pi(\eta)$$
$$= \frac{1}{(2\pi\sigma^2)^{n/2}} \exp\left\{-\frac{1}{2\sigma^2}\sum_{t=1}^{n}(x_t - \sum_{i=1}^{q}\phi_i u_{t-1})^2\right\} \times$$
$$\pi(\Phi)\pi(U|\eta)\pi(\sigma^2)\pi(\eta). \tag{3}$$

This expression leads to the following conditional posterior distributions.

1) Let $U_t := [u_{t-1}, \cdots, u_{t-q}]^T$; then
$$\pi(\Phi|\text{rest}) \sim N_q(\Phi_*, \Sigma_*), \tag{4}$$
where "rest" $:= (X, U, \sigma^2, \eta)$,
$$\Sigma_*^{-1} := \frac{1}{\sigma^2}\sum_{t=1}^{n} U_t U_t^T + \Sigma_0^{-1}$$
and
$$\Phi_* := \Sigma_*\left[\frac{1}{\sigma^2}\sum_{t=1}^{n}(x_t - u_t)U_t + \Sigma_0\Phi_0\right].$$

2) Let $s^2 := \sum_{t=1}^{n}(x_t - \sum_{i=0}^{q}\phi_i u_{t-i})^2$; then
$$\pi(\sigma^2|\text{rest}) \sim \chi^{-2}(\nu_*, \lambda_*),$$
where "rest" $:= (X, \Phi, U, \eta)$, $\nu_* := \nu + n$, and $\lambda_* := \nu\lambda + s^2$.

3) For any fixed $j \in \{1, \cdots, n\}$, it is not difficult to show from (3) that

$$\pi(u_j \mid \text{rest}) \propto$$

$$\left[\eta \delta(u_j=0) + (1-\eta)\frac{1}{\sqrt{2\pi r^2}}\exp\left\{\frac{-\left(\frac{u_j}{r}\right)^2}{2}\right\}\right] \times \exp\left\{-\frac{1}{2\sigma^2}\sum_{t=1}^{j+q}\left(x_t - \sum_{i=0}^{q}\phi_i u_{t-i}\right)^2\right\}.$$

Let $\varepsilon_t' := x_t - \sum'\phi_i u_{t-i}$ where the sum $\sum'$ is over $i=0,1,\cdots,q$ such that $i \neq t-j$; then, after some straightforward manipulations, one obtains

$$\pi(u_j \mid \text{rest}) \sim \eta_j \delta(u_j=0) + (1-\eta_j) N(u_j^*, r_j^2),$$

where

$$\eta_j := \frac{\eta}{\eta + (1-\eta)\frac{r_j}{r}\exp\left\{\frac{(u_j^*/r_j)^2}{2}\right\}},$$

$$u_j^* := \frac{r^2 \sum_{i=0}^{q}\varepsilon_{j+i}'\phi_i}{r^2 \sum_{i=0}^{q}\phi_i^2 + \sigma^2},$$

and

$$r_j^2 := \frac{r^2 \sigma^2}{r^2 \sum_{i=0}^{q}\phi_i^2 + \sigma^2}.$$

It is worthwhile to note that using standard routines one can easily generate random samples from these conditional posterior distributions. This turns out to be crucial to successful application of the Gibbs sampler to the deconvolution problem. We also remark that the proposed approach can be easily extended to handle colored Gaussian noise (e.g., a MA(q_e) process). In this case, the probability density function $\pi(X \mid \Phi, U, \sigma^2, \eta)$ in (3) needs slight modification (though still an exponential function) in order to incorporate the correlation structure of the noise. The other calculations are similar. For simplicity, however, we use the Gaussian white noise model in developing our algorithms.

IV. The Gibbs Sampler

Because of its high complexity, direct inference from the joint posterior distribution of the unknow quantities in our problem is extremely difficult, if not impossible. The Gibbs sampler serves as a means to circumvent the difficulties.

Rather than pinpoint the optimal solutions, as many numerical optimization algorithms do, the Gibbs sampler generates a Markov chain of random samples whose equilibrium distribution coincides with the desired joint posterior distribution of the unknowns. It is mostly helpful when the joint posterior distribution is complicated while the conditional posterior distributions are simple. This is indeed the case in our problem.

A. A Generic Example

To be more specific, let us consider a generic example of making inference about three unknown variables/vectors z_1, z_2, and z_3 from the data vector x. Suppose the joint posterior distribution of (z_1, z_2, z_3) given x is known to be $p(z_1, z_2, z_3 | x)$. The objective of Gibbs sampling is to generate random samples of (z_1, z_2, z_3) according to the distribution $p(z_1, z_2, z_3 | x)$. A commonly-used procedure of Gibbs sampling proceeds iteratively as follows: Starting with a trivial set of initial values $(z_{1,0}, z_{2,0}, z_{3,0})$ and $m=0$:

1) Step 1. Generate $z_{1,m+1}$ from $p(z_1 | z_2 = z_{2,m}, z_3 = z_{3,m}, x)$;
2) Step 2. Generate $z_{2,m+1}$ from $p(z_2 | z_1 = z_{1,m+1}, z_3 = z_{3,m}, x)$;
3) Step 3. Generate $z_{3,m+1}$ from $p(z_3 | z_1 = z_{1,m+1}, z_2 = z_{2,m+1}, x)$;
4) Step 4. Set $m = m+1$ and go to Step 1.

In so doing, the Gibbs sampler produces a sample Markov chain $\{(z_{1,m}, z_{2,m}, z_{3,m}), m=0,1,\cdots\}$ whose limiting (or equilibrium) distribution as $m \to \infty$ can be shown to agree with the desired posterior distribution $p(z_1, z_2, z_3 | x)$ under some regularity conditions. For details, see [5] and [7].

To ensure the convergence in practice, the first N samples in sequence $\{(z_{1,m}, z_{2,m}, z_{3,m}), m=0,1,\cdots\}$ are usually discarded and the remaining ones, $(z_{1,N+m}, z_{2,N+m}, z_{3,N+m})$ for $m=1,\cdots,M$, can be regarded as independent samples governed by the distribution $p(z_1, z_2, z_3 | x)$. To eliminate possible correlations between the samples, one can also save one sample from every k samples to obtain $(z_{1,N+mk}, z_{2,N+mk}, z_{3,N+mk})$ for $m=1,\cdots,M$. From the saved samples, inference can be made about the unknowns. For instance, the MMSE estimate of z_1, namely $E(z_1 | x)$, can be approximated by averaging $z_{1,N+m}$ (or, $z_{1,N+mk}$) for $m=1,\cdots,M$.

B. Remarks on Implementation

In our problem of deconvolution, the Gibbs sampler is implemented by using the conditional posterior distribution in Section III. To accelerate the convergence, the reflection sequence is "visited" five times more often than the other parameters due to the high correlation between successive samples of the reflection sequence.

It is also worth pointing out that even with the constraint $\phi_0=1$ there are still possible ambiguities in the deconvolution problem, of which the shift ambiguity is especially prominent in some cases. For example, if the leading coefficients $\phi_0,\cdots,\phi_k$, (for some $k<q$ with $\phi_k\approx 1$), are relatively small in the seismic wavelet, the Gibbs sampler may not be able to effectively distinguish the time-shifted model

$$y_t = \sum_{j=0}^{q} \phi'_j x'_{t-j} + \varepsilon_t$$

from the true model in (1), where $\phi'_j := \phi_{j+k}$ for $j=0,1,\cdots,q$ (assuming $\phi_j=0$ for $j>q$) and $x'_t := x_{t-k}$ for $t=0,1,\cdots,q$. In such cases, the posterior distribution of the unknowns becomes approximately a *mixture* of several distributions—each corresponding to a possible time delay—that play the same role as local equilibrium in optimization problems. As a result, the convergence speed of the Gibbs sampler may be considerably decreased.

To overcome this problem, we employ the following *constrained* Gibbs sampler as reported in [1]. Since in many cases the seismic wavelet has a unique maximum in absolute value, one can impose a constraint so that the largest $|\phi_j|$ appears at a known location. For instance, assuming that $|\phi_k|>|\phi_j|$ for some $1\leq k\leq q$ and all $j\neq k$ forces the largest wavelet coefficient to appear at lag k. With this restriction, the prior of Φ becomes $\pi(\Phi) \propto N_q(\Phi_0, \Sigma_0) I(|\phi_k|>|\phi_j|, j=1,\cdots,q$ and $j\neq k)$. It is easy to show that the conditional posterior distribution of Φ given the rest unknowns remains the same as in (4) except for an extra factor $I(|\phi_k|>|\phi_j|, j=1,\cdots,q$ and $j\neq k)$. An easy way to sample from this constrained Gaussian distribution is to use the *rejection method*, i.e., when a sample is drawn from the unconstrained distribution (4), it is checked to see if the constraint is satisfied; if not, the sample is rejected and a new sample is generated. This procedure continues until a sample that statisfies the constraint is obtained. One can also restart the Gibbs sampler when it takes too long (e.g., 1000 runs) for the rejection method to generate a desired sample. In this case, it seems plausible to start with the last rejected Φ and shift it forward or backward (with vacancies filled by zeros) so that the resulting wavelet satisfies the constraint.

V. Simulation Results

Some simulations are carried out to test the proposed method. The true seismic wavelet $\{\phi_i, i=1,\cdots,40\}$ is assumed to be of the form $\phi_i:=\phi_i^*/\phi_0^*$, where

$$\phi_i^*:=\sin\frac{\pi(i+1)}{6.4}\exp\{-0.12|i-9|\}. \quad (5)$$

The reflection sequence $\{u_t\}$ is generated from a Bernoulli-Gaussian distribution (2) with $\eta=0.9$. Three different values of r^2 are considered, namely $r^2=25, 4$, and 0.25, corresponding a singal-to-noise ratio (SNR) of 26, 18.6, and 4 dB, respectively. The wavelet estimation and deconvolution results are shown in Figs. 3-5 for the three cases. The observed signal is obtained from (1) for $t=1,\cdots,2000$, where $\{\varepsilon_t\}$ is Gaussian white noise with zero mean and unit variance $\sigma^2=1$. Fig. 3(a) shows the observed signals for the case of SNR=26 dB; Fig. 3(b) is the reflection sequence recovered on the basis of the MMSE criterion with the dots representing the true reflectivity; and Fig. 3(c) shows the MMSE estimate for the seismic wavelet along with the true wavelet. Figures 4 and 5 are organized in a similar way. In each of the cases, the Gibbs sampler is carried out with 4000 iterations and the last 1000 samples are averaged to obtain the MMSE estimates.

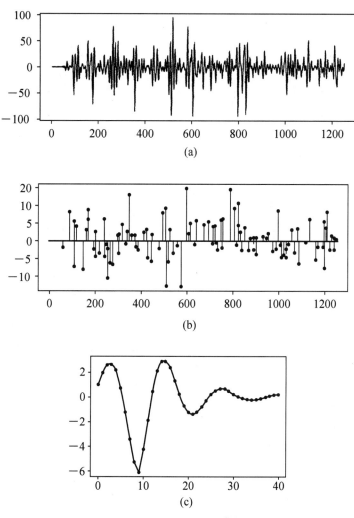

Fig. 3 (a) Observed time series with SNR=26 dB;
(b) MMSE estimates of reflectivity from the data in (a), where the dots stand for true values and the solid lines for estimates;
(c) MMSE estimates of the seismic wavelet in (5) from the data in (a), where true values of the wavelet are indicated by dots and their estimates by solid curve

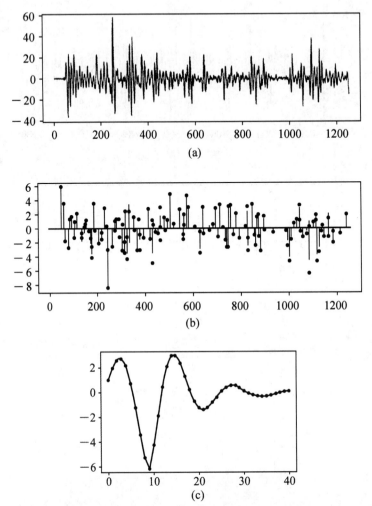

Fig. 4 (a) Observed time series with SNR=18.6 dB;
(b) MMSE estimates of reflectivity from the data in (a), where the dots stand for true values and the solid lines for estimates;
(c) MMSE estimates of the seismic wavelet in (5) from the data in (a), where true values of the wavelet are indicated by dots and their estimates by solid curve

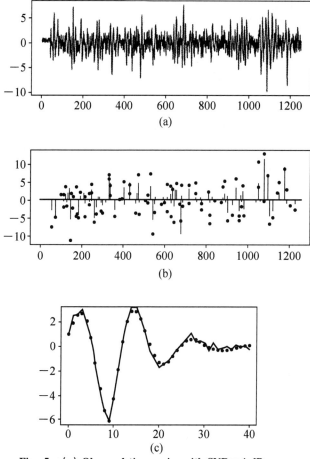

Fig. 5 (a) Observed time series with SNR=4 dB;
(b) MMSE estimates of reflectivity from the data in (a), where the dots stand for true values(×10) and the solid lines for estimates;
(c) MMSE estimates of the seismic wavelet in (5) from the data in (a), where true values of the wavelet are indicated by dots and their estimates by solid curve

Note that checking for convergence in the Gibbs sampler is in general an open problem to be tackled with rigorous methods. Most of the time, the convergence depends on the specific problem to which the Gibbs sampler is applied, although several methods have been proposed recently for checking the convergence in general situations (e. g. , [6]). One of the methods proposes to generate multiple sequences and to compare these sequences using some summary statistics in order to determine whether the Gibbs sampler has converged. This, however, requires additional computing time and memory. In our example, experiments show that 4000 iterations are enough for convergence.

In the above simulation, we assume that the location of the largest value of the seismic wavelet is known *a priori* in order to remove the shift ambiguities in the deconvolution problem. The prior distribution for Φ is assumed to be multivariate normal with mean zero and a diagonal covariance matrix with 100 as diagonal elements, namely $\Phi_0=0$ and $\Sigma_0=100\mathrm{I}$. The prior distribution for η is assumed to be Beta(50,10), i. e. , $a=50$ and $b=10$. The order q of the seismic wavelet and the variance r^2 of the reflectivity are assumed to be known and the corresponding true values are used in the Gibbs sampler. The prior distribution for σ^2 is assumed to be $\chi^{-2}(2,0.3)$, i. e. , we take $\nu=2$ and $\lambda=0.3$. The initial values in the Gibbs sampler are taken to be $\sigma^2=1$, $u_t=0$, for $t=1,\cdots,2000$, and $\eta=0.9$.

As seen from these results, the proposed method is able to recover the reflectivity almost perfectly when the signal-to-noise ratio is high or moderate. For SNR as low as 4 dB, the method still provides satisfactory estimates for most of the reflection coefficients, especially for those with large values, although the performance generally deteriorates with the decrease of SNR.

The computing time of the proposed procedure is minimum due to the fact that all the conditional posterior distributions from which we generate random samples are very simple so that the random number generation can be done rather rapidly. The 5000 iterations in the above example took less than one minute to complete on a Sun SPARC10 workstation. More careful implementation of the method may help to further increase the speed.

Ⅵ. Concluding Remarks

In this paper, we presented a new method for the simultaneous wavelet estimation and deconvolution of reflection seismic signals. The method employs the celebrated Bernoulli-Gaussian white sequence model for the reflectivity. In addition to the reflectivity sequence and the wavelet, statistical parameters in the reflectivity and the noise models are also estimated from the data under a Bayesian framework. The minimum mean-squared error (MMSE) estimators of these unknowns are calculated using a Monte Carlo technique called the Gibbs sampler. Numerical experiments on simulated data have shown that the method is able to provide rather precise estimates even under low signal-to-noise ratios. Further research should extend this method to other reflectivity models (e. g. , [4], [15]) and to multichannel seismic signals (e. g. , [10]).

Acknowledgement

The authors would like to thank three anonymous referees for their insightful comments and suggestions.

References

[1] R. Chen and T. H. Li, "Blind restoration of linearly degraded discrete signals by Gibbs sampler," *IEEE Trans. Signal Processing*, vol. 43. no. 10, pp. 2410—2413, 1995.

[2] Q. Cheng, "Maximum standardized cumulant deconvolution of non-Gaussian linear processes," *Ann. Statist.*, vol. 18, pp. 1745—1783, 1990.

[3] M. H. DeGroot, *Probability and Statistics*, 2nd ed. Amsterdam: Addison-Wesley, 1986.

[4] D. Donoho, "On minimum entropy deconvolution," in *Applied Time Series Analysis*, II, D. F. Findley, Ed. New York: Academic, 1981.

[5] A. E. Gelfand and A. F. M. Smith, "Sampling-based approaches to calculating marginal densities," *J. Amer. Statist. Assoc.*, vol. 85, pp. 398—409, 1990.

[6] A. Gelman and D. B. Rubin, "Inference from interactive simulation using multiple sequences," *Statist. Sci.*, vol. 7, pp. 457—472, 1992.

[7] S. Geman and D. Geman, "Stochastic relaxation, Gibbs distribution, and the Bayesian restoration of images," *IEEE Trans. Pattern Anal. Machine Intell.*, vol. PAMI-6, pp. 721—741, 1984.

[8] Y. Goussard, G. Demoment, and F. Monfront, "Maximum *a posteriori* detection-estimation of Bernoulli-Gaussian processes," *Mathematics in Signal Processing* II, G. McWhirter, Ed. Oxford: Clarendon, 1990, pp. 121—138.

[9] J. Goutsias and J. M. Mendel, "Maximum likelihood deconvolution: An optimization theory perspective," *Geophysics*, vol. 51, pp. 1206—1220, 1986.

[10] J. Idier and Y. Goussard, "Multichannel seismic deconvolution," *IEEE Trans. Geosci. Remote Sensing*, vol. 31, pp. 961—979, 1993.

[11] J. Kormylo and J. M. Mendel, "Maximum likelihood detection and estimation of Bernoullii-Gaussian processes," *IEEE Trans. Inform. Theory*, vol. IT-28, pp. 482—488, 1982.

[12] T. H. Li, "Blind deconvolution of discrete-valued signals," in *Proc. 27th Asilomar*

Conf. Signals, Systems, Computers, Pacific Grove, CA, Nov. 1993, vol. 2, pp. 1240—1244.

[13] J. S. Liu and R. Chen, "Blind deconvolution via sequential imputation," *J. Amer. Statist. Assoc.*, vol. 90. pp. 567—576, 1995.

[14] J. M. Mendel, *Optimal Seismic Deconvolution: An Estimation-Based Approach*. New York: Academic, 1983.

[15] ——, *Maximum-Likelihood Deconvolution: A Journey into Model-Based Signal Processing*. New York: Springer-Verlag, 1990.

[16] E. A. Robinson, "Dynamic predictive deconvolution," *Geophys. Prospecting*, vol. 23, pp. 780—798, 1975.

[17] T. J. Yu, P. C. Müller, G. Z. Dai, and C. F. Liu, "Minimum-variance deconvolution for noncausal wavelets," *IEEE Trans. Geosci. Remote Sensing*, vol. 32, pp. 513—524, 1994.

[18] R. A. Wiggins, "Minimum entropy deconvolution," *Geoexploration*, vol. 16, pp. 21—35, 1978.

原文载于 IEEE Transactions on Geoscience and Remote Sensing, VoL. 34, No. 2, March 1996, 377—384.

Nonlinear Fusion Filters Based on Prediction and Smoothing

Cheng Qiansheng, Zhou Xiaobo, Sun Xichen

Department of Mathematics, Peking University

Abstract: To attenuate white noise, nonstationary noise and impulse noise are important for signal processing. In this letter, we present nonlinear fusion filters (NFF) based on prediction and smoothing. By means of least square fitting of a polynomial, we define and give the operators of left prediction and right prediction, left smoothing and right smoothing, central smoothing and cross-validation smoothing. In simulated experiments, it is shown that the present method is an effective one.

Keywords: prediction; smoothing; nonlinear fusion filters

There are many signals which are smooth with the change of time, but occasionally there are fast changes from one signal level to another. For example, detection of climate jump points is one of the kernel questions in studying global climate change[1]. A video signal. if its all edges are blurred, is not acceptable. It is necessary to preserve jump points and edges. and to attenuate noise. Some nonlinear filters have been developed in these applications[2]. In the papers, Heinonen et al.[3] and Flaig et al.[4] give predictive FIR-Median Hybrid filter (FMH), and Affine Order-Statistic Filter (AOSF), respectively.

A new principle for spatially adaptive estimation can be based on recently developed wavelet analysis. It can be considered as a breakthrough of Fourier analysis. Orthonormal bases of compactly supported wavelets provide a powerful complement to traditional Fourier methods: they permit an analysis of a signal or an image into localized oscillating methods. Donoho et al.[5] give wavelet shrinkage algorithm (WSA). From then on, the nonlinear wavelet analysis has started.

However, FMH does not sufficiently use continuity of signals, WSA cannot attenuate impulse noise. In this letter, we give nonlinear fusion filters (NFF) based on prediction and smoothing. In simulated experiments, it is shown that the present method is effective.

1. NFF

Assume that we observe the values of a response variable X at m predetermined values of an independent variable t. The resulting bivariate observations $(t_1, x_1), \cdots, (t_m, x_m)$ follow the model

$$x_i = f(t_i) + \varepsilon_i + \eta_i, \qquad (1)$$

where $i=1, 2, \cdots, m$, ε_i is a white noise, η_i a sparse impulse series, and f an unknown regression function that we wish to estimate.

(i) Operators of prediction and smoothing. By means of least square fitting of a polynomial, we can define and give the operators of left prediction and right prediction, left smoothing and right smoothing, central smoothing and cross-validation[6] smoothing ①. For example, we define the cross-validation[6] smoothing of the $x(n)$ as follows:

$$x_{\text{CV}}(n) = \sum_{i=-K, i \neq 0}^{K} g_i \times y(n-i), \qquad (2)$$

where K is smoothing bandwidth, g_i is CV operator which depends on CV smoothing polynomial degrees.

(ii) NFF. We use $\bar{x}_-(n)$, $\bar{x}_+(n)$ to denote left smoothing value and the right smoothing value of $x(n)$, $\tilde{x}_-(n)$, $\tilde{x}_+(n)$ the left predictive value and the right predictive value of $x(n)$, $x_0(n)$ the central smoothing value and $x_{\text{CV}}(n)$ denotes cross-validtion smoothing value. We use $(N_1, K_1; N_2, K_2; N_3, K_3; N_4, K_4)$ to denote predictive bandwidth, predictive polynomial degrees, smoothing bandwidth, CV smoothing bandwidth, smoothing polynomial degrees, central smoothing bandwidth, central smoothing polynomial degrees, CV smoothing bandwidth, and CV smoothing degrees.

---

① Cheng, Q. S., Zhou, X. B., Nonlinear fusion filters based on prediction and smoothing, Research Report No. 57, Institute of Mathematics and School of Mathematical Sciences, Peking University, Beijing, China, 1998.

NFF is defined as follows. In order to denote conveniently, we remark:

$$\mu = \min(|\bar{x}_-(n) - \tilde{x}_-(n)|, |\tilde{x}_+(n) - \tilde{x}_+(n)|,$$
$$|\tilde{x}_-(n) - \tilde{x}_+(n)|, |\bar{x}_-(n) - \bar{x}_+(n)|).$$

If
$$|\tilde{x}_-(n) - \tilde{x}_+(n)| = \mu,$$
then NFF:
$$y(n) = \mathrm{MED}(\tilde{x}_-(n), x_{\mathrm{CV}}(n), \tilde{x}_+(n));$$

If
$$|\bar{x}_-(n) - \tilde{x}_-(n)| = \mu,$$
then NFF:
$$y(n) = \mathrm{MED}(\bar{x}_-(n), x_0(n), \tilde{x}_-(n));$$

If
$$|\bar{x}_+(n) - \tilde{x}_+(n)| = \mu,$$
then NFF:
$$y(n) = \mathrm{MED}(\bar{x}_+(n), x_0(n), \tilde{x}_+(n));$$

If
$$|\bar{x}_-(n) - \bar{x}_+(n)| = \mu,$$
then NFF:
$$y(n) = \mathrm{MED}(\bar{x}_-(n), x_{\mathrm{CV}}(n), \bar{x}_+(n)),$$

where MED represents the median of the samples.

2. Examples Comparison

In this letter, we reproduce the pictures corresponding to the example, following recipe of Donoho et al.[5]. The function is HeaviSine (Fig. 1(a)).

We first add the white noise with unit variance by scale and some impulse noise to the function (Fig. 1(b)). Then we consider the methods NFF and WSA. Fig. 1(c) and (d) are the corresponding curves. The corresponding mean square errors (MSE) are listed in table 1. The graphs show that NFF can attenuate the nonstationary noise. Donoho et al.'s WSA is an effective method for white noise, but it cannot remove the impulse noise from Fig. 1(d). From the MSE in table 1, we also find this result.

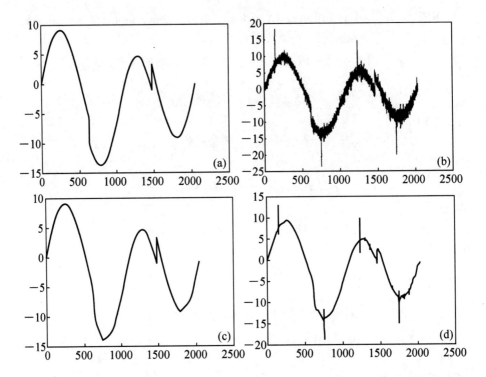

Fig. 1 (a) Typical simulated data set with sample size $n=2048$ (HeaviSine); (b) noise HeaviSine; (c) typical output of the NFF filter; (d) typical output of the WSA filter

Table 1 Mean square errors $\|\hat{f}-f\|_{2,n}^{2}/n$ in Fig. 1

Method	NFF	WSA	AOSF
HeaviSine	0.133250	0.490327	0.139700

3. Conclusion

In this letter, we present an adapted NFF which can attenuate nonstationary noise and impulse noise, and maintain the jump information. By using the method, we can give statistical properties of estimation of signals.

Note. Although its performance is close to that of NFF, AOSF needs to cost more time in looking for the best γ (see ref. [4]). In Table 1, the best γ of WOSAF is 10.

References

[1] Cheng, Q. S., Zhou, X. B., Climate change by cluster analysis, Acta Geophysica Sinica, 1998, 41(3): 308.

[2] Pitas, I., Venetsanoponlous, A., Order statistics in digital image processing, Pro. IEEE, 1992, 80: 1893.

[3] Heinonen, P., Neuvo, Y., FIR-median hybrid filters with predictive FIR substructures, IEEE Trans. ASSP, 1988, 36:892.

[4] Flaig, A., Arce, G. R., Barner, K. E., Affine order-statistic filters: "Medianization" of linear FIR filters, IEEE Trans. Signal Processing, 1998, 46(8): 2101.

[5] Donoho, D. L., Johnstone, I. M., Ideal spatial adaption via wavelet shrinkage, Biometrika, 1994, 81:425.

[6] Eubank, R., Spline Smoothing and Nonparametric Regression, New York: Dekker, 1988.

原文载于 Chinese Science Bulletin, VoL. 45, No. 18, September 2000, 1726—1728.

An Alternative Switching Criterion for Independent Component Analysis (ICA)

Gao Dengpan, Ma Jinwen, Cheng Qiansheng

Department of Information Science, School of Mathematical Sciences and LMAM, Peking University

Abstract: In solving the problem of noiseless independent component analysis (ICA) in which sources of super- and sub-Gaussian coexist in an unknown manner, one can be lead to a feasible solution using the natural gradient learning algorithm with a kind of switching criterion for the model probability distribution densities to be selected as super- or sub-Gaussians appropriately during the iterations. In this letter, an alternative switching criterion is proposed for the natural gradient learning algorithm to solve the noiseless ICA problem with both super- and sub-Gaussian sources. It is demonstrated by the experiments that this alternative switching criterion works well on the noiseless ICA problem with both super- and sub-Gaussian sources.

Keywords: independent component analysis; switching criterion; mixed signals; sub- and super-Gaussian; kurtosis

1. Introduction

The noisy or general independent component analysis (ICA)[2] aims at blindly separating the independent sources $\mathbf{s}$ form the observations $\mathbf{x} = \mathbf{A}\mathbf{s} + \mathbf{v}$ via

$$\mathbf{y} = \mathbf{W}\mathbf{x}, \quad \mathbf{x}, \mathbf{v} \in \mathbb{R}^m, \quad \mathbf{y} \in \mathbb{R}^n, \quad \mathbf{W} \in \mathbb{R}^{n \times m}, \tag{1}$$

where $\mathbf{A}$ is a mixing matrix, $\mathbf{v}$ is a vector of uncorrelated noise terms, and $\mathbf{W}$ is the so-called de-mixing matrix to be estimated. Because the estimation of a de-mixing matrix $\mathbf{W}$ in the presence of noise is rather difficult, the majority of past research has been devoted to the noiseless or classical ICA problem where $\mathbf{v} = \mathbf{0}$[6]. In fact, in some cases, the noise of the ICA problem can be neglected. In other cases, we can measure or model this environmental noise so that it can be filtered by some finite-duration impulse-response filter [5]. Moreover, when $m > n$, we can use the PCA technique to construct the signal subspace corresponding to the principal components associatde with the n largest eigenvalues of the automated autocorrelation matrix of $\mathbf{x}$ and filter out the noise by projecting the observations $\mathbf{x}$ on it. Therefore, the noiseless ICA problem is fundamental to the general one and has been already applied to biomedical signal and image processing[8].

Methodically, the noiseless ICA problem is generally solved via a layer of n processing neurons under the basic requirement that the recovered $\mathbf{y}$ should be componentwise independent, i. e., $p(y)=\Pi_{i=1}^{n} p_i(y_i)$. Actually, it has been extensively studied from different directions (e. g., [1,3,4]). However, all the approaches are equivalent to minimizing the following objective function:

$$D = -H(\mathbf{y}) - \sum_{i=1}^{n} \int p(y_i;\mathbf{W}) \log p_i(y_i) dy_i, \qquad (2)$$

where $H(\mathbf{y}) = -\int p(y) \log p(y) dy$ is the entropy of $\mathbf{y}$, $p_i(y_i)$ is the pre-determined model probability density function (pdf) (i. e., the activation function of neuron i) and $p(y_i;\mathbf{W})$ is the pdf based on $\mathbf{y}=\mathbf{Wx}$.

In the literature, how to choose the model pdf's $p_i(y_i)$ remains a key issue for Eq. (2)-based ICA algorithms. In general, any descent learning algorithm, such as the natural or relative gradient one [1,4], can work only in the cases that the components of $\mathbf{s}$ are either all super-Gaussians or all sub-Gaussians. In order to solve the general ICA problem, XU et al. [12] summarized the one-bit-matching conjecture which states that "all the sources can be separated as long as there is a one-to-one same-sign-correspondence between the kurtosis signs of all source pdf's and the kurtosis signs of all model pdf's". Recently, Liu et al. [10] proved this conjecture by globally minimizing the objective function under certain assumptions on the model pdf's. Ma et al. [11] further proved the conjecture by locally minimizing the same objective function on the two-source ICA problems. Moreover, Lee et al. [9] already proposed a switching criterion that can be embedded in the natural gradient learning algorithm to try to make the model pdf's match the source pdf's in the light of the one-bit-matching condition. Actually, the switching criterion can allocate the sign of the kurtosis of each model pdf at each iteration of the natural gradient learnign algorithm via a sufficient condition guaranteeing the asymptotic stability of the feasible solution of the ICA problem. It was demonstrated in simulation experiments that the natural gradient learning algorithm with this switching criterion can solve the ICA problem successfully in the cases of both super- and sub-Gaussian sources.

In this letter, we propose an alternative switching criterion directly from the point of view of the one-bit-matching principle, which is demonstrated for separating well the sources of both super- and sub-Gaussians from their linearly mixed signals.

2. The Alternative Switching Criterion

The natural gradient learning algorithm with a switching criterion can be expressed as follows (refer to [9] for details):

$$\Delta \mathbf{W} = \eta [\mathbf{I} - \mathbf{K}\tanh(y)y^T - yy^T]\mathbf{W}, \qquad (3)$$

where $\eta > 0$ is the learning rate, $\mathbf{K} = \text{diag}[k_1, \cdots, k_n]$ is considered as a switching criterion represented by an n-order diagonal matrix with $k_i = 1$ and -1 for the super- and sub-Gaussian of model pdf $p_i(y_i)$, respectively, and $\tanh(y) = [\tanh(y_1), \tanh(y_2), \cdots, \tanh(y_n)]^T$ with $\tanh(y_i)$ being the hyperbolic tangent function of y_i, and $m = n$. The model pdfs of sub- and super-Gaussians are actually selected as

$$p_{\text{sub}}(u) = \frac{1}{2}[p_{N(1,1)}(u) + p_{N(-1,1)}(u)], \quad p_{\text{super}}(u) \propto p_{N(0,1)}(u)\text{sech}^2(u), \qquad (4)$$

respectively, where $p_{N(\mu,\sigma^2)}(u)$ is the Gaussian or normal density with mean μ and variance σ^2, and $\text{sech}(u)$ is the hyperbolic secant function. The switching criterion proposed by Lee et al. [9] is given by

$$k_i = \text{sign}(E\{\text{sech}^2(y_i)\}E\{y_i^2\} - E\{[\tanh(y_i)]y_i\}), \quad i = 1, 2, \cdots, n, \qquad (5)$$

which is just a sufficient condition for the asymptotic stability of the solution of the ICA problem obtained by Cardoso and Laheld [4].

We now believe that the one-bit-matching conjecture on the natural gradient learning algorithm is true and construct a switching criterion from the point of view of the one-bit-matching condition alternatively. When the one-bit-matching condition is satisfied, the natural gradient learnign algorithm will be finally stable at a feasible solution. i.e., with a certain permutation, the sign of kurtosis of model pdf $p_i(y_i)$ will be equal to the sign of kurtosis of the source pdf of s_i or equivalently the $p(y_i, \mathbf{W})$ of the resulting output y_i since $y_i = \lambda_i s_i$ with $\lambda_i \neq 0$. Thus, we have a sufficient condition on the stable and feasible solution of the ICA as follows:

$$\text{sign}(\kappa(p_i(y_i))) = \text{sign}(\kappa(p(y_i, \mathbf{W})))$$

$$= \text{sign}(E\{y_i^4\} - 3E^2\{y_i^2\}), \quad i = 1, 2, \cdots, n, \tag{6}$$

where $\kappa(p_i(y_i))$ is the kurtosis of model pdf $p_i(y_i)$, while $\kappa(p(y_i, \mathbf{W})) = E\{y_i^4\} - 3E^2\{y_i^2\}$ is the kurtosis of $p(y_i, \mathbf{W})$ or the output signal y_i. So, we can design the sign of kurtosis of model pdf $p(y_i)$ as the sign of kurtosis of output y_i online during the iterations of the natural gradient learning algorithm so that the algorithm will tend to a feasible solution. Hence, we have the following switching criterion of k_i in Eq. (3) sa follows:

$$k_i = \text{sign}(E\{y_i^4\} - 3E^2\{y_i^2\}), \quad i = 1, 2, \cdots, n. \tag{7}$$

Since $E(y_i^2), E(y_i^4)$ can be easily estimated from a sample data set of $\mathbf{x}$ via the relation $\mathbf{y} = \mathbf{W}\mathbf{x}$ in the noiseless case, this switching criterion or rule can be easily implemented by the estimation of the kurtosis of y_i during the iteration of the natural gradient learning algorithm. In fact, this kind of switching criterion was already suggested in the learning of the deflationary exploratory projection pursuit network [7].

3. Simulation Results

We conducted the experiments on the noiseless ICA problem of five independent sources in which there are three super-Gaussian sources generated from the exponential distribution $E(0.5)$, the Gamma distribution $\gamma(1,4)$, and the Chi-square distribution $\chi^2(6)$, respectively, and two sub-Gaussian sources generated from the β distribution $\beta(2,2)$ and the Uniform distribution $U([0,1])$, respectively. From each distribution, 100 000 i.i.d. samples were generated to form a source. The linearly mixed signals were then generated from the five source signals in parallel via the following mixing matrix:

$$\mathbf{A} = \begin{bmatrix} 0.3987 & 0.1283 & 0.0814 & 0.7530 & 0.1213 \\ 0.1991 & 0.2616 & 0.1248 & 0.5767 & 0.7344 \\ 0.7607 & 0.4484 & 0.3950 & 0.7119 & 0.2145 \\ 0.4609 & 0.0275 & 0.0745 & 0.5171 & 0.0674 \\ 0.8369 & 0.5794 & 0.7752 & 0.4330 & 0.2373 \end{bmatrix}. \tag{8}$$

The learning rate was selected as $\eta = 0.001$ and the natural gradient learning algorithm operated in the adaptive mode and was stopped when all the 100 000 data points of the mixed signals had been passed only once through the updating or learning rule Eq. (3) with the switching criterion.

The result of the natural gradient learning algorithm with the alternative switching criterion Eq. (7) is given in Eq. (9), while the result of the natural gradient learning algorithm with the original switching criterion Eq. (5) is given in Eq. (10). As a feasible solution of the ICA problem, the obtainde **W** will make **WA**=**ΛP** be satisfied or approximately satisfied to a certain extent, where **ΛP**=diag$[\lambda_1, \lambda_2, \cdots, \lambda_n]$ with each $\lambda_i \neq 0$, and P is a permutation matrix,

$$\mathbf{WA} = \begin{bmatrix} 0.0142 & 0.0034 & 1.4198 & 0.0380 & 0.0146 \\ 0.6805 & 0.0313 & -0.0020 & 0.0243 & -0.0008 \\ 0.0079 & 0.7439 & 0.0132 & -0.0147 & -0.0322 \\ 0.0037 & -0.0051 & -0.0027 & -0.0430 & 0.8376 \\ 0.0052 & -0.0041 & 0.0309 & 1.4245 & 0.0040 \end{bmatrix}, \quad (9)$$

$$\mathbf{WA} = \begin{bmatrix} 0.0141 & 0.0048 & 1.4198 & 0.0379 & 0.0145 \\ 0.6805 & 0.0312 & -0.0019 & 0.0243 & -0.0008 \\ 0.0080 & 0.7439 & 0.0125 & -0.0159 & -0.0315 \\ 0.0037 & -0.0058 & -0.0027 & -0.0431 & 0.8376 \\ 0.0053 & -0.0020 & 0.0311 & 1.4244 & 0.0042 \end{bmatrix}. \quad (10)$$

From Eq. (9), we can observe that the natural gradient learning algorithm with the alternative switching criterion can really solve the noiseless ICA problem with both super- and sub-Gaussian sources since these degenerated elements in **WA** are rather small. In comparison with the results listed in Eq. (10), we can also find that the two switching criteria lead to almost the same solution. However, comparing Eq. (5) with Eq. (7), we can observe that sech(y_i) and tanh(y_i) lead to certain additional computation in the original switching criterion and thus the computation cost of the alternative switching criterion is considerably less than that of the original one, which was indeed demonstrated by the experiments with the fact that the running time of the natural gradient learnign algorithm with the alternative switching criterion is only about 0.5702 that of the natural gradient learning algorithm with the original switching criterion.

Furthermore, we conducted many other experiments on the different noiseless ICA problems with both super- and sub-Gaussian sources in which the natural gradient learning algorithm with the alternative switching criterion always arrives at a satisfactory result as above. Moreover, it was even found that as the number of sources increases, the alternative switching criterion gets a better solution of the ICA problem than the original switching criterion does.

4. Conclusions

An alternative switching criterion for the natural gradient learning algorithm to solve the noiseless ICA problem with both super- and sub-Gaussian sources is constructed with the help of the one-bit-matching conjecture. The experiments show that this alternative switching criterion is as good as the original one, but the computation cost is reduced considerably.

References

[1] S. I. Amari, A. Cichocki, H. Yang, A new learning algorithm for blind separation of sources, Adv. Neural Inf. Process. 8(1998) 757—763.

[2] S. I. Amari, A. Cichocki, H. H. Yang, Blind signal separation and extracting: neural and information theoretical approaches, in: S. Haykin (Ed.), Unsupervised Adaptive, Filtering Blind Source Separation, vol. I, Wiley, New York, 2000, pp. 63—138.

[3] A. Bell, T. Sejnowski, An information-maximization approach to blind separation and blind deconvolution, Neural Comput. 7(6)(1995)1129—1159.

[4] J. F. Cardoso, B. Laheld, Equivalent adaptive source separation, IEEE Trans. Signal Process. 44(12)(1996)3017—3030.

[5] A. Cichocki, W. Kasprzak, S. I. Amari, Adaptive approach to blind source separation with cancellation of additive and convolution noise, in: Proceedings of the Third International Conference on Signal Processing (ICSP'96), vol. I, Beijing, China, 1996, pp. 412—415.

[6] P. Comon, Independent component analysis——a new concept?, Signal Process. 36(3)(1994)287—314.

[7] M. Girolami, C. Fyfe, Extraction of independent signal sources using a deflationary exporatory projection pursuit network with lateral inhibition, IEEE Proc. Vision Image Signal Process. 144(5)(1997)299—306.

[8] T. W. Lee, Independent Component Analysis: Theory and Applications, Kluwer Academic Publishers, Dordrecht, 1998.

[9] T. W. Lee, M. Girolami, T. J. Sejnowski, Independent component analysis using an extended infomax algorithm for mixed sub-gaussian and supergaussian sources, Neural Comput. 11(2)(1999)417—441.

[10] Z. Y. Liu, K. C. Chiu, L. Xu, One-bit-matching conjecture for independent component analysis, Neural Comput. 16(2)(2004)383—399.
[11] J. Ma, Z. Liu, L. Xu, A further result on the ICA one-bit-matching conjecture, Neural Comput. 17(2)(2005)331—334.
[12] L. Xu, C. C. Cheung, S. I. Amari, Learned parametric mixture based ica algorithm, Neurocomputing 22(1—3)(1998)69—80.

原文载于 Neurocomputing 68(2005)267—272.

Independent Particle Filters

Lin Ming T.[1], Zhang Junni L.[2], Cheng Qiansheng[3], Chen Rong[2,4]

1. School of Mathematical Sciences;
2. Department of Business Statistics and Econometrices, Guanghua School of Management, Peking University;
3. School of Mathematical Sciences, Peking University;
4. Department of Information and Decision Sciences, University of Illinois, Chicago

Abstract: Sequential Monte Carlo methods, especially the particle filter(PF) and its various modifications, have been used effectively in dealing with stochastic dynamic systems. The standard PF samples the current state through the underlying state dynamics, then uses the current observation to evaluate the sample's importance weight. However, there is a set of problems in which the current observation provides significant information about the current state but the state dynamics are weak. and thus sampling using the current observation often produces more efficient samples than sampling using the state dynamics. In this article we propose a new variant of the PF, the independent particle filter(IPF), to deal with these problems. The IPF generates exchangeable samples of the current state from a sampling distribution that is conditionally independent of the previous states, a special case of which uses only the current observation. Each sample can then be matched with multiple samples of the previous states in evaluating the importance weight. We present some theoretical results showing that this strategy improves efficiency of estimation as well as reduces resampling frequency. We also discuss some extensions of the IPF, and use several synthetic examples to demonstrate the effectiveness of the method.

Keywords: discharging; nonlinear filtering; particle filter; sequential Monte Carlo; state-space model; target tracking

1. Introduction

Sequential Monte Carlo (SMC) methods provide a general framework for tackling stochastic dynamic systems, which often arise in engineering, bioinformatics, economics, and other fields. They have been applied successfully in various real problems, including computer vision and target tracking (Gordon, Salmond, and Smith 1993; Avitzour 1995; Isard and Blake 1996; Blake, Bascle, Isard, and MacCormick 1998; Salmond and Gordon 2001; McGinnity and Irwin 2001; Hue, Le Cadre, and Perez 2002), economic time series analysis (Hendry and Richard 1991; Durbin and Koopman 1997; Pitt and Shephard 1999; Durham and Gallant 2002), protein structure simulation and energy minimization (Vasquez and Scheraga 1985; Grassberger 1997; Zhang and Liu 2002; Liang, Chen, and Zhang 2002; Zhang, Chen, Tang, and Liang 2003), communications and signal processing (Chen, Wang, and Liu 2000; Wang, Chen, and Guo 2002; Fong, Godsill, Doucet, and West 2002). Also see Liu 2001; Doucet, Gordon, and Krishnamurthy 2001, and references therein. Specially, Liu and Chen (1998) provided a general framework for SMC and unified various sequential simulation algorithms.

We focus our discussion herein on the discrete-time state-space model

$$\text{state equation:} \quad \mathbf{x}_t \sim q_t(\cdot | \mathbf{x}_t - 1); \quad (1)$$

$$\text{observation equation:} \quad \mathbf{y}_t \sim f_t(\cdot | \mathbf{x}_t), \quad (2)$$

where $\mathbf{x}_t$ is the latent state and $\mathbf{y}_t$ is the observation at time t. Let $\mathbf{X}_t = (\mathbf{x}_0, \mathbf{x}_1, \cdots, \mathbf{x}_t)$ and $\mathbf{Y}_t = (\mathbf{y}_1, \cdots, \mathbf{y}_t)$. Statistical inference about $\mathbf{X}_t$ can often be formulated as an estimation of $E_{\pi_t}[h(\mathbf{X}_t)]$, the expectation of $h(\mathbf{X}_t)$ with respect to the posterior distribution

$$\pi_t(\mathbf{X}_t | \mathbf{Y}_t) \propto q_0(\mathbf{x}_0) q_1(\mathbf{x}_1 | \mathbf{x}_0) f_1(\mathbf{y}_1 | \mathbf{x}_1) \cdots q_t(\mathbf{x}_t | \mathbf{x}_{t-1}) f_t(\mathbf{y}_t | \mathbf{x}_t),$$

where, $h(\cdot)$ is a square-integrable function. The SMC algorithm, in the framework of Liu and Chen (1998), draws samples of $\mathbf{X}_t$ from

$$g_t^*(\mathbf{X}_t) = g_0(\mathbf{x}_0) g_1(\mathbf{x}_1 | \mathbf{x}_0, \mathbf{y}_1) \cdots g_t(\mathbf{x}_t | \mathbf{X}_{t-1}, \mathbf{Y}_t),$$

where g_t is the conditional sampling distribution of $\mathbf{x}_t$ given $\mathbf{X}_{t-1}$ and $\mathbf{Y}_t$, and g_t^* is the joint sampling distribution of $\mathbf{X}_t$. Each draw of $\mathbf{X}_t$ needs to be weighted by

$$w_t(\mathbf{X}_t) = \frac{\pi_t(\mathbf{X}_t | \mathbf{Y}_t)}{g_t^*(\mathbf{X}_t)}.$$

It can be shown that such a sample, $\{(\mathbf{X}_t^{(j)}, w_t^{(j)}), j = 1, \cdots, m\}$, is *properly weighted* with respect to the distribution π_t. That is, for any square-integrable function $h(\cdot)$,

$$\frac{\sum_{j=1}^m w_t^{(j)} h(\mathbf{X}_t^{(j)})}{\sum_{j=1}^m w_t^{(j)}} \to E_{\pi_t}[h(\mathbf{X}_t)] \quad \text{as } m \to \infty.$$

In all of the discussions that follow, we assume that the sampling distribution g^* selected and the function h satisfy $\text{var}[w(\mathbf{X}_t)] < \infty$ and $\text{var}[h(\mathbf{X}_t)w(\mathbf{X}_t)] < \infty$.

We consider the situation when h is a function of $\mathbf{X}_t$ only. When SMC is used for filtering purposes and the inference is made directly using the samples (particles), it is also called the *particle filter* (PF) (Kitagawa 1996; Carpenter, Clifford, and Fearnhead 1999; Pitt and Shephard 1999). Note that SMC (PF) can be implemented in a sequential way; at stage t, for $j=1,\cdots,m$:

1. Generate $\mathbf{x}_t^{(j)}$ from $g_t(\mathbf{x}_t^{(j)} | \mathbf{X}_{t-1}^{(j)}, \mathbf{Y}_t)$, and let $\mathbf{X}_t^{(j)} = (\mathbf{X}_{t-1}^{(j)}, \mathbf{x}_t^{(j)})$.

2. compute incremental weight

$$u_t^{(j)} = \frac{\pi_t(\mathbf{X}_{t-1}^{(j)}, \mathbf{x}_t^{(j)} | \mathbf{Y}_t)}{\pi_{t-1}(\mathbf{X}_{t-1}^{(j)} | \mathbf{Y}_{t-1}) g_t(\mathbf{x}_t^{(j)} | \mathbf{X}_{t-1}^{(j)}, \mathbf{Y}_t)}$$

$$\propto \frac{q_t(\mathbf{x}_t^{(j)} | \mathbf{X}_{t-1}^{(j)}) f_t(\mathbf{y}_t | \mathbf{x}_t^{(j)})}{g_t(\mathbf{x}_t^{(j)} | \mathbf{X}_{t-1}^{(j)}, \mathbf{Y}_t)},$$

and let $w_t^{(j)} = w_{t-1}^{(j)} u_t^{(j)}$.

Sample generation and importance weight calculation are two key factors determining the efficiency of SMC methods. The bootstrap filter (Gordon et al. 1993) uses the state dynamics in (1) as the sampling distribution, that is, $g_t(\mathbf{x}_t | \mathbf{X}_{t-1}, \mathbf{Y}_t) = q_t(\mathbf{x}_t | \mathbf{x}_{t-1})$. Fox et al. (2001), Torma and Szepesvafi (2003), Isard and Blake (2004), and Rossi (2004), in several special applications, proposed using a sampling distribution depending only on the current observation,

$$g_t(\mathbf{x}_t | \mathbf{X}_{t-1}, \mathbf{Y}_t) \propto f_t(\mathbf{y}_t | \mathbf{x}_t). \quad (3)$$

Kong, Liu, and Wong (1994) and Liu and Chen (1995, 1998) proposed using a sampling distribution that combines information from both the state dynamics and the current observation, such as

$$g_t(\mathbf{x}_t | \mathbf{X}_{t-1}, \mathbf{Y}_t) \propto q_t(\mathbf{x}_t | \mathbf{x}_{t-1}) f_t(\mathbf{y}_t | \mathbf{x}_t). \quad (4)$$

Various approximation methods have been developed for use when it is difficult to sample from (4), including auxiliary PFs (Pitt and Shephard 1999), unscented PFs (van der Merwe, Doucet, de Freitas, and Wan 2002), mixture Kalman filters (Chen and Liu 2000), and Gaussian sum PFs (Kotecha and Djuric 2003).

In this article we focus on a set of problems for which the current observation provides significant information about the current state but the state dynamics are weak. For example, in visual tracking and robot localization, the observations in the form of images provide significant information, whereas the motion dynamics are not stable (Fox et al. 2001; Blake et al. 1998; Torma and Szepesvari 2003). Such problems also arise in target tracking with fast and unstable maneuvering (Bar-Shalom and Fortmann 1988), fast flat-fading channels (Chen et al. 2000), and population biology. Also, more and more applications in communications are under a high signal-to-noise ratio environment.

Because of the weak state dynamics, the particles from the bootstrap filter do not follow the target distribution closely, resulting in excessive weight variation that will greatly reduce the accuracy of statistical inference (Kong et al. 1994). In contrast, the sampling distribution in (3) uses information from the current observation in the sampling distribution and presumably can work well. In this article, we extend this idea and propose using sampling distributions in the form of

$$g_t(\mathbf{x}_t | \mathbf{X}_{t-1}, \mathbf{Y}_t) \propto g_t(\mathbf{x}_t | \mathbf{Y}_t). \tag{5}$$

We call a PF using (5) as the sampling distribution an *independent particle filter* (IPF), because each draw of $\mathbf{x}_t$ is conditionally independent of individual past particles and of each other. In this article we formulate a general framework of the IPF and study its theoretical and empirical properties. The most intriguing feature of IPF, as also noted by Torma and Szepesvari (2003) and Isard and Blake (2004), is that it allows the draws of $\mathbf{x}_t$ from the sampling distribution (5) to be matched arbitrarily with the past particles of $\mathbf{X}_{t-1}$. We show that for the set of problems on which we focus, an IPF can outperform a PF using full information, that is, using the sampling distribution in (4).

We proceed as follows. Section 2 formally presents IPF and a multiple-matching scheme that uses the independence feature. It also discusses some theoretical properties of the IPF with multiple matching. Section 3 discusses extensions of the IPF and practical guidance for designing IPFs. Section 4 presents several examples. The technical proofs of all theorems are presented in the Appendix.

2. The Independent Particle Filter

2.1 The Basic Form

Formally, the IPF can be described as follows: At stage t, for $j=1,\cdots,m$,

1. Generate $\mathbf{x}_t^{(j)}$ from $g_t(\mathbf{x}_t^{(j)}|\mathbf{X}_{t-1}^{(j)},\mathbf{Y}_t)=g_t(\mathbf{x}_t^{(j)}|\mathbf{Y}_t)$, and let $\mathbf{X}_t^{(j)}=(\mathbf{X}_{t-1}^{(j)},\mathbf{x}_t^{(j)})$.

2. Compute incremental weight

$$u_t^{(j)} \propto \frac{q_t(\mathbf{x}_t^{(j)}|\mathbf{x}_{t-1}^{(j)})f_t(\mathbf{y}_t|\mathbf{x}_t^{(j)})}{g_t(\mathbf{x}_t^{(j)}|\mathbf{Y}_t)}$$

and let $w_t^{(j)}=w_{t-1}^{(j)}u_t^{(j)}$.

If $g_t(\mathbf{x}_t|\mathbf{Y}_t)\propto f_t(\mathbf{y}_t|\mathbf{x}_t)$ is used, then the incremental weight becomes $u_t^{(j)}\propto q_t(\mathbf{x}^{(j)}{}_t|\mathbf{x}_{t-1}^{(j)})$. This would require that $\int f_t(\mathbf{y}_t|\mathbf{x}_t)d\mathbf{x}_t<\infty$; that is, $f_t(\mathbf{y}_t|\mathbf{x}_t)$ as a distribution of $\mathbf{x}_t$ is proper.

2.2 Independent Particle Filter with Multiple Matching

Throughout this article, we let $S_t=\{(\mathbf{x}_t^{(j)},w_t^{(j)}), j=1,\cdots,m\}$ denote a set of weighted random samples properly weighted with respect to $\pi_t(\mathbf{x}_t|\mathbf{Y}_t)$. Because the draws $\{\mathbf{x}_t^{(j)}, j=1,\cdots,m\}$ in the IPF are not coupled with any particular particles in S_{t-1}, they can be matched with the samples in S_{t-1} in any order. Specifically, if $\mathbf{x}_{t-1}^{(i)}$ is matched with $\mathbf{x}_t^{(j)}$, then the importance weight is

$$\lambda_t^{(i,j)}=w_{t-1}^{(i)}u_t^{(i,j)}, \tag{6}$$

$$u_t^{(i,j)}\propto \frac{q_t(\mathbf{x}_t^{(j)}|\mathbf{x}_{t-1}^{(i)})f_t(\mathbf{y}_t|\mathbf{x}_t^{(j)})}{g_t(\mathbf{x}_t^{(j)}|\mathbf{Y}_t)}.$$

Each different match will result in different weights for $\{\mathbf{x}_t^{(j)}, j=1,\cdots,m\}$, and thus different estimates of $E_{\pi_t}[h(\mathbf{x}_t)]$. It is natural to combine these estimates to produce more efficient ones.

Consider L different permutations of $(1,\cdots,m)$: $\mathbf{K}_l\doteq(k_{l,1},\cdots,k_{l,m}), l=1,\cdots,L$. For each permutation, the past particles $\{\mathbf{x}_{t-1}^{(k_{l,j})}, j=1,\cdots,m\}$ are matched with the current particles $\{\mathbf{x}_t^{(j)}, j=1,\cdots,m\}$, and the importance weights are

$$w_{t,l}^{(j)}\equiv\lambda_t^{(k_{l,j},j)}. \tag{7}$$

The resulting set of weighted random samples, $\{(\mathbf{x}_t^{(j)},w_{t,l}^{(j)}), j=1,\cdots,m\}$, is properly weighted with respect to $\pi(\mathbf{x}_t|\mathbf{Y}_t)$. Combining different permutations, we can also construct a new weight, $w_t^{(j)}$, for each $\mathbf{x}_t^{(j)}$,

$$w_t^{(j)}=\frac{\sum_{l=1}^{L}w_{t,l}^{(j)}}{L}, \tag{8}$$

and estimate $E_{\pi_t}[h(\mathbf{x}_t)]$ with

$$\hat{H}_{t,L} = \frac{m^{-1} \sum_{j=1}^{m} w_{t,l}^{(j)} h(\mathbf{x}_t^{(j)})}{m^{-1} \sum_{j=1}^{m} w_t^{(j)}}$$

$$= \frac{(Lm)^{-1} \sum_{j=1}^{m} \sum_{l=1}^{L} w_{t,l}^{(j)} h(\mathbf{x}_t^{(j)})}{(Lm)^{-1} \sum_{j=1}^{m} \sum_{l=1}^{L} w_{t,l}^{(j)}} \equiv \frac{A_{t,L}}{B_{t,L}}. \quad (9)$$

Because $\lambda_t^{(i,j)} h(\mathbf{x}_t^{(j)})$ and $\lambda_t^{(i,j)}$ are exchangeable for different i and j, we can define two constants, $\mathscr{A}_t \equiv E[\lambda_t^{(i,j)} h(\mathbf{x}_t^{(j)})]$ and $\mathscr{B}_t \equiv E(\lambda_t^{(i,j)}) > 0$. We then have the following results.

Proposition 1. For any L permutations and any $1 \leqslant l \leqslant L$,

$$E(A_{t,L}) = \mathscr{A}_t, \quad E(B_{t,L}) = \mathscr{B}_t, \quad E\pi_t[h(\mathbf{x}_t)] = \frac{\mathscr{A}_t}{\mathscr{B}_t}. \quad (10)$$

Proposition 2. Under the assumption that $\{(\mathbf{x}_{t-1}^{(j)}, w_{t-1}^{(j)}), j=1, \cdots, m\}$ are exchangeable, (a) for any fixed $2 \leqslant L \leqslant m$, $\text{var}(A_{t,L})$ and $\text{var}(B_{t,L})$ achieve minimum when the L permutations are mutually exclusive, that is, for any $l_1, l_2 (1 \leqslant l_1 \neq l_2 \leqslant L)$, $k_{l_1, j} \neq k_{l_2, j}, j=1, \cdots, m$; and (b) with mutually exclusive permutations, $\text{var}(A_{t,L})$ and $\text{var}(B_{t,L})$ decrease as L increases ($L \leqslant m$).

Proposition 3. When $\{(\mathbf{x}_{t-1}^{(j)}, w_{t-1}^{(j)}), j=1, \cdots, m\}$ are exchangeable, if $A_{t,m}$ and $B_{t,m}$ are obtained with m mutually exclusive permutations, then for any $L > m$, $\text{var}(A_{t,L}) \geqslant \text{var}(A_{t,m})$ and $\text{var}(B_{t,L}) \geqslant \text{var}(B_{t,m})$.

Remark 1. According to (9), (10), and Proposition 2, to reduce the variance of $\hat{H}_{t,L}$, we should choose mutually exclusive permutations. The largest set containing such permutations is of size $L=m$. Proposition 3 says that using $L>m$ nonexclusive permutations does not help. The size of L controls a tradeoff between estimation efficiency and computational cost.

Remark 2. There is no obvious way to select the "good" matches without first calculating the weights. Torma and Szepesvari (2003) and Isard and Blake (2004) proposed calculating all the weights in complete matching ($L=m$ mutually exclusive permutations) and then use these weights to sample a "good" match that is used for estimation. In our approach, all weights that have been calculated are used in estimation, in partial matching ($L<m$) or complete matching.

Remark 3. Torma and Szepesvari (2003) and Isard and Blake (2004) used complete matching for certain special applications. Although complete matching has the maximum benefit, it is often computationally intensive. As shown in the examples in Section 4, partial matching with $L=5$ or 10 often performs better than complete matching, given the same amount of computation.

2.3 Multiple Matching as a Discharging Tool for the Independent Particle Filter

2.3.1 Discharging in Standard Importance Sampling.

Suppose that the target distribution is $\pi(\mathbf{x})$. To draw a set of random samples from a trial distribution $g(\mathbf{x})$, we can generate $\{(\mathbf{x}^{(j)}, \mathbf{z}^{(j)}), j=1,\cdots,m\}$ from $g^*(\mathbf{x}, \mathbf{z})$ with $\int g^*(\mathbf{x},\mathbf{z})d\mathbf{z} = g(\mathbf{x})$ and consider only $\{\mathbf{x}^{(j)}, j=1,\cdots,m\}$. The importance weight can be calculated in two ways: through the joint trial distribution, $w_j^* = \frac{\pi(\mathbf{x}^{(j)}, \mathbf{z}^{(j)})}{g^*(\mathbf{x}^{(j)}, \mathbf{z}^{(j)})}$, where $\pi(\mathbf{x},\mathbf{z})$ is any distribution satisfying $\int \pi(\mathbf{x},\mathbf{z})d\mathbf{z} = \pi(\mathbf{x})$, and by direct calculation through the marginal distribution $w_j = \frac{\pi(\mathbf{x}^{(j)})}{g(\mathbf{x}^{(j)})}$. Both $\{(\mathbf{x}^{(j)}, w_j^*), j=1,\cdots,m\}$ and $\{(\mathbf{x}^{(j)}, w_j), j=1,\cdots,m\}$ are properly weighted with respect to $\pi(\mathbf{x})$, which can be seen from

$$\iint h(\mathbf{x}) \frac{\pi(\mathbf{x},\mathbf{z})}{g^*(\mathbf{x},\mathbf{z})} g^*(\mathbf{x},\mathbf{z}) d\mathbf{z} d\mathbf{x} = \int h(\mathbf{x}) \frac{\pi(\mathbf{x})}{g(\mathbf{x})} g(\mathbf{x}) d\mathbf{x}$$
$$= \int h(\mathbf{x})\pi(\mathbf{x})d\mathbf{x}. \qquad (11)$$

Proposition 4. With the foregoing w_j^* and w_j, we have

$$\mathrm{var}_g\left[\frac{1}{m}\sum_{j=1}^{m} h(\mathbf{x}^{(j)})w_j\right] \leqslant \mathrm{var}_{g^*}\left[\frac{1}{m}\sum_{j=1}^{m} h(\mathbf{x}^{(j)})w_j^*\right].$$

Hence the weight based on the marginal distribution is more efficient than the weight based on the joint distribution, even though the sample was actually generated using the joint distribution. We call such a procedure *discharging*. It removes the inefficiency due to the additional sampling of $\mathbf{z}$.

2.3.2 Discharging in the Independent Particle Filter.

We demonstrate that multiple matching serves as an approximation of discharging for the IPF. Suppose that we perform complete matching (i.e., using $L=m$ mutually exclusive permutations) in the IPF; then, according to (6)-(8),

$$w_t^{(j)} = \frac{1}{m}\sum_{i=1}^{m} w_{t-1}^{(i)} \frac{\pi_t(\mathbf{X}_{t-1}^{(i)}, \mathbf{x}_t^{(j)} \mid \mathbf{Y}_t)}{\pi_{t-1}(\mathbf{X}_{t-1}^{(i)} \mid \mathbf{Y}_{t-1}) g_t(\mathbf{x}_t^{(j)} \mid \mathbf{Y}_t)}.$$

Because $\{(\mathbf{X}_{t-1}^{(i)}, w_{t-1}^{(i)})\}_{i=1}^{m}$ is properly weighted with respect to $\pi_{t-1}(\mathbf{X}_{t-1} | \mathbf{Y}_{t-1})$, we have

$$w_t^{(j)} \approx \int \frac{\pi_t(\mathbf{X}_{t-1}, \mathbf{x}_t^{(j)} | \mathbf{Y}_t)}{\pi_{t-1}(\mathbf{X}_{t-1} | \mathbf{Y}_{t-1}) g_t(\mathbf{x}_t^{(j)} | \mathbf{Y}_t)} \pi_{t-1}(\mathbf{X}_{t-1} | \mathbf{Y}_{t-1}) d\mathbf{X}_{t-1}$$

$$= \frac{\pi_t(\mathbf{x}_t^{(j)} | \mathbf{Y}_t)}{g_t(\mathbf{x}_t^{(j)} | \mathbf{Y}_t)}, \tag{12}$$

which is the weight as if we had performed discharging. With partial matching ($L<m$), the foregoing calculation still holds, although the approximation in (12) is less accurate.

Remark 4. It should be clear that the IPF with discharging is not necessarily better than the PF without discharging, due to the use of different sampling distributions. However, discharging provides a justification that the IPF with multiple matching performs better than the IPF without multiple matching. In some cases, this improvement can be significant enough so that the IPF outperforms the PF using full information in the sampling distribution (4), as shown by the examples in Section 4.

2.4 Multiple Matching and Resampling

Resampling is an important step in all SMC algorithms, because the importance weight w_t can be increasingly skewed, resulting in many unrepresentative samples of $\mathbf{x}_t$. Suppose that we have obtained S_t. By drawing a new set of samples from $\{\mathbf{X}_t^{(j)}, j=1,\cdots,m\}$ with probabilities proportional to $w_t^{(j)}$ and reassigning the weights to 1, we can effectively discard the sample with small weights and duplicate the more important samples. Liu and Chen (1998) proposed engaging the resampling step when the effective sample size (Kong et al. 1994), $ESS = \frac{m}{1 + C^2(w)}$, is less than a certain threshold, where $C(w)$ is the coefficient of variation of the weights.

A shortcoming of resampling is that it reduces sample diversity. The following proposition shows that multiple matching in the IPF reduces the coefficient of variation of the weights with increasing L, and thus reduces the frequency of resampling, maintaining sample diversity.

Proposition 5. $E(w_t^{(j)}) = \mathscr{B}_t$ is constant for different L, while $\text{var}(w_t^{(j)})$ decreases as L increases.

The following proposition further shows that if complete matching is to be performed at time t, then resampling should not be performed at time $t-1$, because it cannot improve the efficiency of statistical inference about $\mathbf{x}_t$.

Proposition 6. Suppose that at time $t-1$ we have obtained S_{t-1}, where (without loss of generality) the weights are standardized such that $\sum_{j=1}^{m} w_{t-1}^{(j)} = m$. Let $A_{t,m}$ and $B_{t,m}$ be obtained with m mutually exclusive permutation. Suppose that we have performed a resampling step and obtained $\{(\mathbf{x}_{t-1}^{*(j)}, w_{t-1}^{*(j)}=1), j=1,\cdots,m\}$ at time $t-1$. Let $A_{t,m}^*$ and $B_{t,m}^*$ be obtained with m mutually exclusive permutations using the resampled samples and weights. Then

$$E(A_{t,m}^*)=E(A_{t,m}), \quad E(B_{t,m}^*)=E(B_{t,m}),$$
$$\text{var}(A_{t,m}^*) \geq \text{var}(A_{t,m}), \quad \text{var}(B_{t,m}^*) \geq \text{var}(B_{t,m}).$$

3. Extensions of Independent Particle Filters

3.1 Using the Information from Past Particles

The efficiency of the IPF could be improved if the information from past particles is also used. To achieve this and at the same time obtain exchangeable samples of $\mathbf{x}_t$ as required by the IPF, we can use the *general* information from all samples of $\mathbf{x}_{t-1}$. More formally, suppose that we have obtained S_{t-1}. Through the state equation (1), we can construct a distribution $\xi(\mathbf{x}_t | \mathbf{Y}_{t-1})$ for $\mathbf{x}_t$ using the overall information carried by S_{t-1}. Then the sampling distribution for $\mathbf{x}_t$ becomes

$$g_t(\mathbf{x}_t | \mathbf{Y}_t) \propto f_t(\mathbf{y}_t | \mathbf{x}_t) \xi(\mathbf{x}_t | \mathbf{Y}_{t-1}).$$

In particular, suppose that the state equation is in the form of $\mathbf{x}_t = \mathbf{Q}(\mathbf{x}_{t-1}) + \boldsymbol{\varepsilon}_t$, where $\boldsymbol{\varepsilon}_t \sim N(\mathbf{0}, \sigma^2 \mathbf{I})$. Then we can use $\xi(\mathbf{x}_t | \mathbf{Y}_{t-1}) \sim N(\hat{\boldsymbol{\mu}}, \hat{\boldsymbol{\Sigma}})$, where

$$\hat{\boldsymbol{\mu}} = \frac{\sum_{j=1}^{m} w_{t-1}^{(j)} \mathbf{Q}(\mathbf{x}_{t-1}^{(j)})}{\sum_{j=1}^{m} w_{t-1}^{(j)}}, \quad \text{and}$$

$$\hat{\boldsymbol{\Sigma}} = \frac{\sum_{j=1}^{m} w_{t-1}^{(j)} \mathbf{Q}(\mathbf{x}_{t-1}^{(j)}) (\mathbf{Q}(\mathbf{x}_{t-1}^{(j)}))^T}{\sum_{j=1}^{m} w_{t-1}^{(j)}} - \hat{\boldsymbol{\mu}} \hat{\boldsymbol{\mu}}^T + \sigma^2 \mathbf{I}.$$

(13)

Suppose further that $f_t(\mathbf{y}_t | \mathbf{x}_t)$ can be well approximated by $\hat{f}_t(\mathbf{y}_t | \mathbf{x}_t)$, a Gaussian (or mixture Gaussian) approximation of $f_t(\mathbf{y}_t | \mathbf{x}_t)$ as a distribution of

$\mathbf{x}_t$, we form the sampling distribution $g_t(\mathbf{x}_t|\mathbf{Y}_t) \propto \hat{f}_t(\mathbf{y}_t|\mathbf{x}_t)\xi(\mathbf{x}_t|\mathbf{Y}_{t-1})$, which is a Gaussian (or mixture Gaussian) distribution. A similar approach has been used in other cases (e.g., Chopin 2002).

Such a sampling distribution needs to be constructed only once at each stage t, hence bearing minimal additional computational cost. The draws of $\mathbf{x}_t$ are exchangeable, so partial or complete matching can still be conducted, and the theoretical results in Section 2 will hold.

3.2 When y_t is Directly Related to Only Part of x_t

In many applications, the observation is related directly only to part of the state vector $\mathbf{x}_t$. For example, in target tracking problems the state vector includes speed and acceleration, which are not directly related to the observations. In state-space models, if there are some unknown parameters that are involved only in the state equation but not in the observation equation, then they are often included as part of the state vector and are not directly related to the observations.

In these cases, the state space-models can be written as

$$\text{state-equation:} \quad \mathbf{x}_t = (x_{t,1}, x_{t,2})$$
$$\sim q_t(\cdot|\mathbf{x}_{t-1}) = q_{t,1}(x_{t,1}|\mathbf{x}_{t-1})$$
$$\times q_{t,2}(x_{t,2}|\mathbf{x}_{t-1}, x_{t,1});$$

observation equation: $\mathbf{y}_t \sim f_t(\cdot|x_{t,1})$.

The first component, $x_{t,1}$, can be generated using a sampling distribution, $g_{t,1}(x_{t,1}|\mathbf{Y}_t)$, constructed as before. These samples can be matched arbitrarily with multiple previous particles of $\mathbf{x}_{t-1}$ and properly weighted. When $x_{t,1}^{(i)}$ is matched with $\mathbf{x}_{t-1}^{(i)}$, the partial incremental weight is

$$u_{t,1}^{(i,j)} \propto \frac{f_t(\mathbf{y}_t|x_{t,1}^{(j)})q_{t,1}(x_{t,1}^{(j)}|\mathbf{x}_{t-1}^{(i)})}{g_{t,1}(x_{t,1}^{j}|\mathbf{Y}_t)}.$$

Generating the second component, $x_{t,2}$, based on a second sampling distribution, $g_{t,2}(x_{t,2}|\mathbf{x}_{t-1}, x_{t,1})$, will require a one-to-one match of $x_{t,1}$ and $\mathbf{x}_{t-1}$. Given a match between $\mathbf{x}_{t-1}^{(i)}$ and $x_{t,1}^{(j)}$, we can generate $x_{t,2}^{(i,j)} \sim g_{t,2}(x_{t,2}|\mathbf{x}_{t-1}^{(i)}, x_{t,1}^{(j)})$. The corresponding partial incremental weight for this stage is

$$u_{t,2}^{(i,j)} \propto \frac{q_{t,2}(x_{t,2}^{(i,j)}|\mathbf{x}_{t-1}^{(i)}, x_{t,1}^{(j)})}{g_{t,2}(x_{t,2}^{(i,j)}|\mathbf{x}_{t-1}^{(i)}, x_{t,1}^{(j)})}.$$

Under this setting, we modify the IPF as follows.

3.2.1 MIPF-1.

Suppose that at stage $t-1$, we have obtained S_{t-1}; then, at stage t:

1. Generate $x_{t,1}^{(j)}$, $j=1,\cdots,m$, from $g_{t,1}(x_{t,1}|\mathbf{Y}_t)$.

2. For each permutation $\mathbf{K}_l = (k_{l,1},\cdots,k_{l,m})$, $l=1,\cdots,L$, match $\mathbf{x}_{t-1}^{(k_{l,j})}$ with $x_{t,1}^{(j)}$:

- Calculate the partial incremental weight $u_{t,1}^{k_{l,j},j}$, $j=1,\cdots,m$.
- Generate $x_{t,2}^{(k_{l,j},j)} \sim g_{t,2}(x_{t,2}|\mathbf{x}_{t-1}^{(k_{l,j})}, x_{t,1}^{(j)})$, and calculate the partial incremental weight $u_{t,2}^{(k_{l,j},j)}$; let $w_{t,l}^{(j)} = w_{t-1}^{(k_{l,j})} u_{t,1}^{(k_{l,j},j)} u_{t,2}^{(k_{l,j},j)}$.

3. Combine all of the weighted samples of $\mathbf{x}_t$, that is, $\{((x_{t,1}^{(j)}, x_{t,2}^{(k_{l,j})}), w_{t,l}^{(j)}), j=1,\cdots,m, l=1,\cdots,L\}$, to estimate $E_{\pi_t}[h(\mathbf{x}_t)]$,

$$\hat{H}_{t,L} = \frac{\sum_{l=1}^{L}\sum_{j=1}^{m} w_{t,l}^{(j)} h(x_{t,1}^{(j)}, x_{t,2}^{(k_{l,j},j)})}{\sum_{l=1}^{L}\sum_{j=1}^{m} w_{t,l}^{(j)}}. \quad (14)$$

4. Construct the set of weighted random samples, $S_t = \{(\mathbf{x}_t^{(j)}, w_t^{(j)}), j=1,\cdots, m\}$, for propagation to stage $t+1$:

- For each $x_{t,1}^{(j)}$, select a final match $x_{t,2}^{(j)}$ from the set $\{x_{t,2}^{(k_{l,j},j)}, l=1,\cdots,L\}$ with probabilities proportional to $w_{t,l}^{(j)}$.
- Set the weight for each $\mathbf{x}_t^{(j)} = (x_{t,1}^{(j)}, x_{t,2}^{(j)})$ as $w_t^{(j)} = \sum_{l=1}^{L} w_{t,l}^{(j)}/L$.

Proposition 7. For MIPF-1, Propositions 1−3, 5, and 6 hold.

Remark 5. Although estimation can also be done after step 4, the estimator (14) has smaller variation under the Rao-Blackwellization principle (Liu and Chen 1998). If h involves only $x_{t,1}$, then the estimation should be done using

$$\hat{H}_{t,L} = \frac{\sum_{l=1}^{L}\sum_{j=1}^{m} w_{t-1}^{(k_{l,j})} u_{t,1}^{(k_{l,j},j)} h(x_{t,1}^{(j)})}{\sum_{l=1}^{L}\sum_{j=1}^{m} w_{t-1}^{(k_{l,j})} u_{t,1}^{(k_{l,j},j)}}.$$

This estimator further reduces variation under the Rao-Blackwellization principle.

In MIPF-1, for each $\mathbf{x}_t^{(j)}$, we need to generate $x_{t,2}$ and calculate $u_{t,2}$ L times, increasing the computational cost. Next we present another modification of the IPF that has lower computational cost but may be less statistically efficient that MIPF-1.

3.2.2 MIPF-2.

Suppose that at stage $t-1$, we have obtained S_{t-1}; then at stage t:

1. Generate $x_{t,1}^{(j)}, j=1,\cdots,m$, from $g_{t,1}(x_{t,1}|\mathbf{Y}_t)$.

2. For each permutation $\mathbf{K}_l = (k_{l,1},\cdots,k_{l,m}), l=1,\cdots,L$, match $\mathbf{x}_{t-1}^{(k_{l,j})}$ with $x_{t,1}^{(j)}$, and calculate the partial incremental weight $u_{t,1}^{(k_{l,j},j)}$.

3. Construct the set of weighted random samples, $S_t = \{(\mathbf{x}_t^{(j)}, w_t^{(j)}), j=1,\cdots, m\}$, to be used for estimation of $E_{\pi_t}[h(\mathbf{x}_t)]$ and for propagation to stage $t+1$.

- For each $x_{t,1}^{(j)}$, select a final match $\mathbf{x}_{t-1}^{(s_j)}$ from the set $\{\mathbf{x}_{t-1}^{(k_{l,j},j)}, l=1,\cdots, L\}$ with probabilities proportional to $w_{t-1}^{(k_{l,j})} u_{t,1}^{(k_{l,j},j)}$.

- Generate $x_{t,2}^{(j)}$ from $x_{t,2}^{(j)} \sim g_{t,2}(x_{t,2}|\mathbf{x}_{t-1}^{(s_j)}, x_{t,1}^{(j)})$, and calculate $u_{t,2}^{(s_j,j)}$.

- Calculate the weight for each $\mathbf{x}_t^{(j)} = (x_{t,1}^{(j)}, x_{t,2}^{(j)})$ as $w_t^{(j)} = (\sum_{l=1}^{L} w_{t-1}^{(k_{l,j})} u_{t,1}^{(k_{l,j},j)}/L) u_{t,2}^{(s_j,j)}$.

- The estimate of $E_{\pi_t}[h(\mathbf{x}_t)]$ is then $\hat{H}_{t,L} = (\sum_{j=1}^{m} w_t^{(j)} \times h(\mathbf{x}_t^{(j)}))/\sum_{j=1}^{m} w_t^{(j)}$.

Remark 6. If the second sampling distribution, $g_{t,2}$, is chosen as the state dynamics, $q_{t,2}(x_{t,2}|\mathbf{x}_{t-1}, x_{t,1})$, then the matching selection probabilities in step 4 of MIPF-1 and those in step 3 of MIPF-2 become the same, because $u_{t,2}^{(i,j)}$ is constant.

4. Synthetic Examples

4.1 Example 1: Target Tracking with Random Acceleration

Consider a two-dimensional target tracking model with random acceleration,

$$\text{state equation:} \quad \begin{pmatrix} \mathbf{z}_t \\ \mathbf{v}_t \end{pmatrix} = \begin{bmatrix} \mathbf{I}_2 & T_0\mathbf{I}_2 \\ 0 & \mathbf{I}_2 \end{bmatrix} \begin{pmatrix} \mathbf{z}_{t-1} \\ \mathbf{v}_{t-1} \end{pmatrix} + \begin{pmatrix} .5T_0^2\boldsymbol{\varepsilon}_t \\ T_0\boldsymbol{\varepsilon}_t \end{pmatrix}; \quad (15)$$

$$\text{observation equation:} \quad \mathbf{y}_t = \mathbf{z}_t + \boldsymbol{\eta}_t, \quad (16)$$

where $\boldsymbol{\varepsilon}_t \sim N(\mathbf{0}, \sigma^2 \mathbf{I}_2)$ and $\boldsymbol{\eta}_t \sim N(\mathbf{0}, \delta^2 \mathbf{I}_2)$. Here $\mathbf{z}_t, \mathbf{v}_t$, and $\boldsymbol{\varepsilon}_t$ are the position, velocity, and random acceleration vectors for a target moving on a two-dimensional plane. $\mathbf{I}_2$ is the 2×2 indentity matrix, and T_0 is the time interval between observations. The state vector is $\mathbf{x}_t = (\mathbf{z}_t, \mathbf{v}_t)^T$, in which $\mathbf{z}_t$ is observed with noise $\boldsymbol{\eta}_t$. The variances σ^2 and δ^2 specify the strength of information for the current state in the state dynamics and the current observation.

Because the model in (15) and (16) is linear and Gaussian, a Kalman filter can be used to obtain the exact value of $E(\mathbf{z}_t | \mathbf{Y}_t)$, and thus this example is used purely for illustration. To compare different PF methods, we use the root mean squared error (RMSE), defined as

$$RMSE = \left[\frac{1}{T} \sum_{t=1}^{T} \| \hat{\mathbf{z}}_t - E(\mathbf{z}_t | \mathbf{Y}_t) \|_2^2 \right]^{1/2}, \quad (17)$$

where $\hat{\mathbf{z}}_t$ is the estimate of $E(\mathbf{z}_t | \mathbf{Y}_t)$ and T is the total number of observations.

We compared the following PF methods. PF1 is the bootstrap filter, using $g_t(\mathbf{x}_t | \mathbf{X}_{t-1}, \mathbf{Y}_t) \propto q_t(\mathbf{x}_t | \mathbf{x}_{t-1})$. PF2 uses the sampling distribution (4), which combines information from both the state dynamics and the current observation. APF is the auxiliary PF (Pitt and Shephard 1999). Specifically, at each step t, we first resample from the set $\{(\mathbf{x}_{t-1}^{(j)}, w_{t-1}^{(j)}), j=1, \cdots, m\}$ with probabilities proportional to $\tilde{w}_{t-1}^{(j)} = w_{t-1}^{(j)} f_t(\mathbf{y}_t | E(\mathbf{x}_t | \mathbf{x}_{t-1}^{(j)}))$ then draw $\mathbf{x}_t^{(j)}$ from $q_t(\mathbf{x}_t | \mathbf{x}_{t-1}^{(j)})$ for $j=1, \cdots, m$. So, through an approximation, APF also uses information from both $\mathbf{x}_{t-1}$ and $\mathbf{y}_t$. MIPF is the MIPF-2 algorithm described in Section 3.2. According to the state equation, $\mathbf{v}_t$ can be determined exactly by $\mathbf{z}_{t-1}, \mathbf{v}_{t-1}$, and $\mathbf{z}_t$ as $\mathbf{v}_t = \mathbf{v}_{t-1} + 2(\mathbf{z}_t - \mathbf{z}_{t-1} - T_0 \mathbf{v}_{t-1})/T_0$, so the state dynamic, $q_{t,2}(\mathbf{v}_t | \mathbf{x}_{t-1}, \mathbf{z}_t)$, and the sampling distribution, $g_{t,2}(\mathbf{v}_t | \mathbf{x}_{t-1}, \mathbf{z}_t)$, in MIPF-2 are degenerate.

In the simulation that follows, we set the initial position vector and velocity vector as $\mathbf{z}_0 = (0,0)'$ and $\mathbf{v}_0 = (1,0)'$, and let $T_0 = 5$ and $T = 100$. A dynamic resampling schedule is used in which a resampling step is engaged if the effective sample size is smaller than $.1m$. The experiment is repeated 100 times, and the average RMSEs are reported.

We fix $\sigma^2 = .5^2$ and let δ^2 vary, so that δ controls the relative importance of the current observation and the state dynamics. When the same number of particles, $m = 500$, are used, we observed (detailed results not shown) that as L increases, MIPF performs better (confirming Prop. 2) with longer computational time, and that as δ decreases or as the imformation from the observations becomes stronger relative to that from the state dynamices, MIPF performs better than PF1, PF2, and APF. We also noticed that as L increases, the number of resampling steps in MIPF decreases (confirming Prop. 5); also, PF1 requires a significant amount of resampling, due partially to the fact that samples of $\mathbf{x}_t$ are generated with weak state dynamics.

For a fairer comparison, Table 1 reports the average RMSEs of the PF methods when the numbers of particles were chosen so that each method used approximately the same CPU time. It is seen that when the state dynamics is strong (i. e. , large δ), the standard PF is very efficient, because of the ease of computation while enjoying the sufficient information contained in the state dynamics. In contrast, when the current observation contains more information (i. e. , small δ), MIPF performs better than all of the other methods considered. In addition, we see that with the same computation time, L should not be too small or too large to achieve small RMSE. Therefore, in practice, L should be chosen considering the trade-off between estimation efficiency and computational effort, as suggested in Remark 1 in Section 2.2.

Table 1 For Different Values of δ in Example 1, Average RMSEs of PF1, PF2, APF, and MIPF With Different Values of L, and CUP Time

	δ					
	1	2	4	8	16	CPU time (sec.)
PF1 ($m=8,000$)	.2669	.2823	.4138	.7983	1.3860	1.5470
PF2 ($m=2,000$)	.1384	.2126	.5136	1.3229	2.6052	1.3400
APF ($m=6,000$)	.6757	.8715	.9015	.5768	.6500	1.5470
MIPF ($L=1, m=4,000$)	.0460	.1189	.4027	1.4302	5.3018	1.3290
MIPF ($L=5, m=2,100$)	.0464	.1085	.3377	1.1560	4.0246	1.3260
MIPF ($L=10, m=1,400$)	.0474	.1112	.3407	1.1422	3.8801	1.4060
MIPF ($L=20, m=900$)	.0512	.1161	.3541	1.1689	3.9932	1.5470
MIPF ($L=50, m=400$)	.0739	.1653	.4596	1.4347	4.4516	1.5780
MIPF ($L=100, m=200$)	.1038	.2279	.6271	1.9387	5.4115	1.8910
MIPF ($L=150, m=150$)	.1185	.2665	.7297	2.1086	5.8297	1.8250

Note: The numbers of particles were chosen so that each method used approximately the same CPU time.

4.2 Example 2: Target Tracking with Maneuvering

Consider a two-dimensional maneuvering mobility model used by Ikoma et al. (2001),

state equation:
$$\begin{pmatrix} \mathbf{z}_t \\ \mathbf{v}_t \\ \mathbf{r}_t \end{pmatrix} = \mathbf{F} \begin{pmatrix} \mathbf{z}_{t-1} \\ \mathbf{v}_{t-1} \\ \mathbf{r}_{t-1} \end{pmatrix} + \mathbf{G}\boldsymbol{\varepsilon}_t;$$

observation equation:
$$\begin{pmatrix} y_{t,1} \\ y_{t,2} \end{pmatrix} = \begin{pmatrix} \tan^{-1}(z_{t,1}/z_{t,2}) \\ \sqrt{z_{t,1}^2 + z_{t,2}^2} \end{pmatrix} + \delta\boldsymbol{\eta}_t,$$

where $\mathbf{z}_t$, $\mathbf{v}_t$, and $\mathbf{r}_t$ are the position, velocity, and acceleration of the target moving on a two-dimensional plane; $\boldsymbol{\varepsilon}_t$ is the random change of acceleration vector, and $\mathbf{F}$ and $\mathbf{G}$ are two known matrices (see Ikoma, Ichimura, Higuchi, and Maeda 2001) with $\alpha = 1,000$ (Zaidi and Mark 2003).

The state equation is a discretization of a continuous-time mobility model. Because the acceleration of the target may change abruptly with maneuvering, $\varepsilon_{t,i}$ ($i=1,2$) are assumed to follow Cauchy distributions independently,

$$p(\varepsilon_{t,i}) = \frac{q}{\pi(\varepsilon_{t,i}^2 + q^2)}.$$

The observation vector consists of the angle, $y_{t,1}$, and the radius, $y_{t,2}$, of the target, with the observation noise following a Gaussian distribution,

$$\begin{pmatrix} \eta_{t,1} \\ \eta_{t,2} \end{pmatrix} \sim N(\mathbf{0}, \mathbf{R}), \quad \mathbf{R} = \begin{pmatrix} \delta_1^2 & 0 \\ 0 & \delta_2^2 \end{pmatrix}.$$

In comparing the PF methods, the RMSE in (17) is used, with $E(\mathbf{z}_t | \mathbf{Y}_t)$ estimated by a bootstrap filter with 100,000 particles. In the simulation, we set $T_0 = 3.75$, $T = 100$, and $q = 5$. Following Ikoma et al. (2001), we set $\delta_1^2 = 10^{-10}$ and $\delta_2^2 = 10^{-2}$. The prior distribution for the initial state of the target is $N((50, 000, -5,000, 0, 10, 0, 0)', \mathbf{I}_6)$. A dynamic resampling schedule is used, as in Example 1. The experiment is repeated 100 times.

Note that δ in the observation equation again controls the relative importance between the current observation and the state dynamic. For different values of δ, Table 2 reports the average RMSEs of the bootstrap filter (PF), APF, and MIPF-2 with different values of L, using approximately the same computation time. It reveals the same information as in Example 1. Also note that when $L = m$ in the MIPF, resampling actually increases the average RMSEs, which confirms Proposition 6.

Table 2 For Different Values of δ in Example 2, Average RMSE of PF, APF, and MIPF With Different Values of L, and CUP Time

	δ					
	.625	1.25	2.5	5	10	CPU time (sec.)
PF ($m=5,000$)	.6528	.5467	.6424	1.5377	2.4073	1.1090
APF ($m=3,000$)	.2324	.3396	.4785	1.0696	1.4373	1.0780
MIPF ($L=1, m=2,400$)	.1068	.3076	.8389	2.1268	4.8148	1.0320
MIPF ($L=5, m=1,600$)	.0900	.2531	.6797	1.7708	4.4829	1.0310
MIPF ($L=10, m=1,200$)	.0952	.2580	.6361	1.7559	4.1170	1.1250
MIPF ($L=20, m=800$)	.0918	.2568	.6949	1.6992	3.8907	1.1410
MIPF ($L=50, m=400$)	.0877	.2413	.5957	1.5327	3.6006	.1250
MIPF ($L=100, m=200$)	.0784	.2398	.6304	1.6985	3.7970	1.1190
MIPF ($L=150, m=150$, r)	.0998	.2543	.6203	1.5282	3.5762	1.0940
MIPF ($L=150, m=150$, nr)	.0922	.2385	.6210	1.4980	3.4641	1.0780

Note: The last two rows are for MIPF with resampling and without resampling. The numbers of particles were chosen so that each method used approximately the same CUP time.

4.3 Example 3: Nonlinear Filtering

Consider the following nonlinear state-space model (Gordon et al. 1993):

$$\text{state equation:} \quad x_t = .5x_{t-1} + \frac{25x_{t-1}}{1+x_{t-1}^2} + 8\cos(1.2(t-1)) + \varepsilon_t; \quad (18)$$

observation equation: $y_t = x_t^2/20 + \eta_t$,

where $\varepsilon_t \sim N(0, \sigma^2)$, and $\eta_t \sim N(0, \delta^2)$ are Gaussian white noise. Because it is difficult to draw x_t directly from

$$f_t(y_t | x_t) \propto \exp\left\{-\frac{1}{2\delta^2}\left(\frac{x_t^2}{20} - y_t\right)^2\right\}, \quad (19)$$

we sample x_t from a trial distribution that is close to (19) but easier to sample from. Specifically, we linearize $x_t^2/20$ at the points of x_t that maximize the likelihood $f_t(y_t | x_t)$ and set the sampling distribution as a mixture Gaussian distribution. When $y_t > 0$, the points of x_t that maximize $f_t(y_t | x_t)$ are $\bar{x}_t^{(k)} = \pm\sqrt{20y_t}$ ($k=1,2$) and the first derivative of $x_t^2/20$ at these points are $d_k = \bar{x}_t^{(k)}/10 = \pm\sqrt{y_t/5}$ ($k=1,2$); thus we have

$$\frac{x_t^2}{20} \approx \frac{(\bar{x}_t^{(k)})^2}{20} + (x_t - \bar{x}_t^{(k)}) d_k = y_t + (x_t - \bar{x}_t^{(k)}) d_k.$$

Then

$$f_t(y_t | x_t) \approx \hat{f}_t(y_t | x_t)$$

$$\propto .5 \exp\left\{-\frac{1}{2\delta^2}((x_t - \bar{x}_t^{(1)}) d_1)^2\right\} +$$

$$.5 \exp\left\{-\frac{1}{2\delta^2}((x_t - \bar{x}_t^{(2)}) d_2)^2\right\}.$$

Consequently, we set the sampling distribution for x_t as

$$g_t^{(1)}(x_t) \sim .5 N(\bar{x}_t^{(1)}, \delta^2/d_1^2) + .5 N(\bar{x}_t^{(2)}, \delta^2/d_2^2)$$

$$= .5 N(\sqrt{20 y_t}, 5\delta^2/y_t) + .5 N(-\sqrt{20 y_t}, 5\delta^2/y_t),$$

in which the maximum variance is set at $25\delta^2$ to avoid very large variances when y_t is close to 0. When $y_t \leq 0$, we linearize the observation equation at $\bar{x}_t = 0$ and set the variance at $25\delta^2$. In summary, we have

$$g_t^{(1)}(x_t) \sim \begin{cases} .5 N(c, s^2) + .5 N(-c, s^2), & y_t > 0, \\ N(0, 25\delta^2), & y_t \leq 0, \end{cases}$$

where $c = \sqrt{20 y_t}$ and $s^2 = \min(5\delta^2/y_t, 25\delta^2)$.

As described in Section 3.1, we can incorporate information from the past particles to improve our sampling distribution. Specifically, we first let the past particles $\{x_{t-1}^{(j)}, j=1,\cdots,m\}$ propagate to x_t, using the mean state dynamic in (18) without the noise. Then we try to "summarize" the propagated particles into a continuous distribution for x_t.

There are two different cases based on the past observation y_{t-1}. When $y_{t-1} > 0$, the past particles x_{t-1} were generated from a mixture distribution; as a result, they tend to propagate into a mixture distribution. Hence we attempt to summarize the propagated particles using a mixture distribution. Specifically, let

$$\xi(x_t) \sim p_1 N(\mu_1, \tau_1^2) + p_2 N(\mu_2, \tau_2^2),$$

where p_1 and p_2 are estimated mixing proportions that satisfy $p_1 + p_2 = 1$. Because every particle at time $t-1$ is generated from one of the two mixture components, (13) is used separately for the two components to estimate μ_1, τ_1^2, and μ_2, τ_2^2. When $y_{t-1} \leq 0$, x_{t-1} was generated from a Gaussian distribution, and, correspondingly, $\xi(x_t)$ is a Gaussian distribution: $\xi(x_t) \sim N(\mu, \tau^2)$, using (13) to estimate μ and τ^2. After $\xi(x_t)$ is obtained, the sampling distribution at time t can be set as

$$g_t^{(2)}(x_t) \propto g_t^{(1)}(x_t)\xi(x_t),$$

which is a Gaussian distribution or a mixture Gaussian distribution.

In the simulation, we use the prior distribution $p(x_0)=N(0,2)$ and set $T=50$ and $\sigma=\sqrt{10}$. A dynamic resampling schedule is used as in the previous examples, and the experiment is repeated 100 times. Again, δ controls the relative importance between the current observation and the state dynamic. For different values of δ, Table 3 reports the average RMSEs of the bootstrap filter (PF1), a PF using the sampling distribution $g_t(x_t|X_{t-1},Y_t) \propto q_t(x_t|x_{t-1})g_t^{(1)}(x_t)$ to approximate (4)(PF2), an IPF with $g_t^{(1)}(x_t)$ as the sampling distribution (IPF1), and an IPF with $g_t^{(2)}(x_t)$ as the sampling distribution (IPF2), using approximately the same computational time. For each IPF method, Table 3 shows similar results as the previous examples. In addition, it is also seen that when L becomes large and/or δ becomes large, IPF2 performs slightly better than IPF1, due to the incorporation of information from x_{t-1}.

Table 3 For Different Values of δ in Example 3, Average RMSE for PF1, PF2, and IPF1 and IPF2 With Different Values of L, and CPU Time

	δ				
	1/8	1/4	1/2	1	CPU time (sec.)
PF1($m=5,000$)	.6676	.4767	.3263	.3062	.3170
PF2($m=1,200$)	.3672	.3407	.3401	.3371	.3180
IPF1 ($L=1,m=2,600$)	.2879	.2848	.3193	.3921	.3130
IPF1 ($L=5,m=1,300$)	.2697	.2784	.3093	.3713	.3120
IPF1 ($L=10,m=850$)	.3077	.3134	.3554	.3852	.3440
IPF1 ($L=20,m=500$)	.3639	.3441	.3741	.4467	.3290
IPF1 ($L=100,m=100$)	.6595	.6957	.7644	.9021	.3120
IPF2 ($L=1,m=2,000$)	.3130	.3281	.2996	.3856	.3280
IPF2($L=5,m=1,100$)	.2880	.3009	.3006	.3794	.3120
IPF2 ($L=10,m=750$)	.3342	.3068	.3117	.3692	.3280
IPF2 ($L=20,m=500$)	.3814	.3510	.3638	.4199	.3440
IPF2 ($L=100,m=100$)	.6381	.6359	.6661	.8270	.3280

Note: The number of particles were chosen so that each method used approximately the same CPU time.

Appendix: Proofs

Proof of Proposition 1

The proof is trivial.

Proof of Proposition 2

Because of exchangeability, we have the following constants:

$$C_1 = \mathrm{var}[\lambda_t^{(i,j)} h(\mathbf{x}_t^{(j)})];$$
$$C_2 = \mathrm{cov}[\lambda_t^{(i_1,j)} h(\mathbf{x}_t^{(j)}), \lambda_t^{(i_2,j)} h(\mathbf{x}_t^{(j)})], \quad i_1 \neq i_2;$$
$$C_3 = \mathrm{cov}[\lambda_t^{(i,j_1)} h(\mathbf{x}_t^{j_1}), \lambda_t^{(i,j_2)} h(\mathbf{x}_t^{(j_2)})], \quad j_1 \neq j_2;$$

and

$$C_4 = \mathrm{cov}[\lambda_t^{(i_1,j_1)} h(\mathbf{x}_t^{(j_1)}), \lambda_t^{(i_2,j_2)} h(\mathbf{x}_t^{(j_2)})], \quad i_1 \neq i_2, j_1 \neq j_2.$$

For $i_1 \neq i_2, j_1 \neq j_2$, we have

$$\mathrm{var}\left[\frac{(\lambda_t^{(i_1,j_1)} + \lambda_t^{(i_2,j_1)}) h(\mathbf{x}_t^{(j_1)}) + (\lambda_t^{(i_1,j_2)} + \lambda_t^{(i_2,j_2)}) h(\mathbf{x}_t^{(j_2)})}{2}\right]$$
$$\leq \mathrm{var}[\lambda_t^{(i_1,j_1)} h(\mathbf{x}_t^{(j_1)}) + \lambda_t^{(i_2,j_2)} h(\mathbf{x}_t^{(j_2)})].$$

This gives $C_1 + C_2 + C_3 + C_4 \leq 2C_1 + 2C_4$, hence

$$C_1 + C_4 - C_2 - C_3 \geq 0. \tag{A.1}$$

Define $a_{t,l} = m^{-1} \sum_{j=1}^{m} w_{t,l}^{(j)} h(\mathbf{x}_t^{(j)})$. Consider any pair of permutations, $\mathbf{K}_{l_1}$ and $\mathbf{K}_{l_2}$ ($l_1 \neq l_2$). Suppose that there are m_1 j's that satisfy $k_{l_1,j} = k_{l_2,j}$ and m_2 j's that satisfy $k_{l_1,j} \neq k_{l_2,j}$ ($m_1 + m_2 = m$). Using (A.1), we can easily show that when $m_1 = 0$, $\mathrm{cov}(a_{t,l_1}, a_{t,l_2})$ is minimized, and

$$\mathrm{cov}(a_{t,l_1}, a_{t,l_2}) = \frac{C_2 + C_3 + (m-2)C_4}{m}. \tag{A.2}$$

At the same time, for $l = 1, \cdots, L$, $\mathrm{var}(a_{t,l}) = m^{-1}[C_1 + (m-1)C_4]$. Because $A_{t,L} = \sum_{l=1}^{L} a_{t,l}/L$, $\mathrm{var}(A_{t,L})$ is minimized when for any $1 \leq l_1 \neq l_2 \leq L$, $k_{l_1,j} \neq k_{l_2,j}$, $j = 1, \cdots, m$. This proves part (a).

With mutually exclusive permutations, by plugging (A.2) into the formula of $\mathrm{var}(A_{t,L})$ and using (A.1), we get that $\mathrm{var}(A_{t,L})$ decreases as L increases. This proves part (b).

By setting $h(\mathbf{x}_t) = 1$, we can prove similar conclusions for $\mathrm{var}(B_{t,L})$.

Proof of Proposition 3

According to (9), for any $L>m$ permutations,

$$A_{t,L} = \frac{1}{m}\sum_{l=1}^{L}\sum_{j=1}^{m}\frac{1}{L}\lambda_t^{(k_{l,j},j)}h(\mathbf{x}_t^{(j)}) = \frac{1}{m}\sum_{i=1}^{m}\sum_{j=1}^{m}d_{i,j}\lambda_t^{(i,j)}h(\mathbf{x}_t^{(j)}),$$

where $d_{i,j}$ satisfy

$$\sum_{i=1}^{m}\sum_{j=1}^{m}d_{i,j} = m, \quad d_{i,j} \geqslant 0. \tag{A.3}$$

The Lagrangian expression of minimizing $\text{var}(A_{t,L})$ is

$$J = \text{var}(A_{t,L}) + \mu\left(\sum_{i=1}^{m}\sum_{j=1}^{m}d_{i,j} - m\right)$$

$$= \frac{1}{m^2}\Bigg\{C_1\sum_{i=1}^{m}\sum_{j=1}^{m}d_{i,j}^2 + C_2\sum_{1\leqslant i_1\neq i_2\leqslant m}\sum_{j=1}^{m}d_{i_1,j}d_{i_2,j} +$$

$$C_3\sum_{i=1}^{m}\sum_{1\leqslant j_1\neq j_2\leqslant m}d_{i,j_1}d_{i,j_2} + C_4\sum_{1\leqslant i_1\neq i_2\leqslant m}\sum_{1\leqslant j_1\neq j_2\leqslant m}d_{i_1,j_1}d_{i_2,j_2}\Bigg\} +$$

$$\mu\left(\sum_{i=1}^{m}\sum_{j=1}^{m}d_{i,j} - m\right),$$

where C_1, C_2, C_3 and C_4 are as defined in the proof of Proposition 2. Setting the derivative of J with respect to a particular d_{i_0,j_0} to 0, we get, for any i_0 and j_0,

$$(C_1 - C_2 - C_3 + C_4)d_{i_0,j_0} + (C_2 - C_4)\sum_i d_{i,j_0}$$

$$+ (C_3 - C_4)\sum_j d_{i_0,j} + C_4 m = -.5\mu m^2. \tag{A.4}$$

By summing (A.4) over i_0, j_0 and (i_0, j_0) jointly, we find that d_{i_0,j_0} is constant for different i_0 and j_0. By (A.3), $\text{var}(A_{t,L})$ achieves its minimum when $d_{i_0,j_0} = 1/m$, and thus $A_{t,L}$ equals $A_{t,m}$ for m mutually exclusive permutations. Because when $h(\mathbf{x}_t)=1$, $A_{t,L}=B_{t,L}$, similar conclusion holds for $\text{var}(B_{t,L})$.

Proof of Proposition 4

It can be seen by

$$\text{var}_{g^*}\left[h(\mathbf{x})\frac{\pi(\mathbf{x},\mathbf{z})}{g^*(\mathbf{x},\mathbf{z})}\right] \geqslant \text{var}_g\left[E_{g^*}\left\{h(\mathbf{x})\frac{\pi(\mathbf{x},\mathbf{z})}{g^*(\mathbf{x},\mathbf{z})}\bigg|x\right\}\right]$$

$$= \text{var}_g\left[h(\mathbf{x})\int\frac{\pi(\mathbf{x},\mathbf{z})}{g^*(\mathbf{x},\mathbf{z})}\frac{g^*(\mathbf{x},\mathbf{z})}{g(\mathbf{x})}d\mathbf{z}\right]$$

$$= \text{var}_g\left[h(\mathbf{x})\frac{\pi(\mathbf{x})}{g(\mathbf{x})}\right].$$

Proof of Proposition 5

According to (8), $E(w_t^{(i)}) = \mathscr{B}_t$ is a constant for different L's. Because of exchangeability, we have two constants, $D_1 = \text{var}(\lambda_t^{(i,j)})$ and $D_2 = \text{cov}(\lambda_t^{(i_1,j)}, \lambda_t^{(i_2,j)})$, $i_1 \neq i_2$:

$$\text{var}(\lambda_t^{(i_1,j)} - \lambda_t^{(i_2,j)}) = 2D_1 - 2D_2 \geq 0 \Rightarrow D_1 \geq D_2$$

and

$$\text{var}(w_t^{(j)}) = \text{var}\left[\frac{\sum_{l=1}^L \lambda_t^{(k_l,j)}}{L}\right] = \frac{LD_1 + L(L-1)D_2}{L^2} = \frac{D_1 - D_2}{L} + D_2,$$

which decrease as L increases.

Proof of Proposition 6

A complete matching ($L = m$) between $\{\mathbf{x}_t^{(j)}, j = 1, \cdots, m\}$ and the resampled set $\{\mathbf{x}_{t-1}^{*(i)}, i = 1, \cdots, m\}$ results in the new weight (under $w_{t-1}^{*(i)} = 1$),

$$w_t^{*(j)} = \frac{\sum_{l=1}^m w_{t,l}^{*(j)}}{m} = \frac{\sum_{i=1}^m \lambda_t^{*(i,j)}}{m}$$

$$= \frac{1}{m}\sum_{i=1}^m \frac{q_t(\mathbf{x}_t^{(j)} \mid \mathbf{x}_{t-1}^{*(i)}) f_t(\mathbf{y}_t \mid \mathbf{x}_t^{(j)})}{g_t(\mathbf{x}_t^{(j)} \mid \mathbf{Y}_t)}. \quad (A.5)$$

Recall that

$$A_{t,m}^* = \frac{1}{m}\sum_{j=1}^m w_t^{*(j)} h(\mathbf{x}_t^{(j)}) \quad \text{and} \quad B_{t,m}^* = \frac{1}{m}\sum_{j=1}^m w_t^{*(j)}. \quad (A.6)$$

Because the resampling probability for $\mathbf{x}_{t-1}^{(k)}$ is $w_{t-1}^{(k)}/m$, we have

$$E[\lambda_t^{*(i,j)} \mid \{\mathbf{x}_{t-1}^{(k)}, w_{t-1}^{(k)}\}_{k=1}^m, \mathbf{x}_t^{(j)}] = \sum_{k'=1}^m \left[\frac{w_{t-1}^{(k')}}{m} \frac{q_t(\mathbf{x}_t^{(j)} \mid \mathbf{x}_{t-1}^{(k')}) f_t(\mathbf{y}_t, \mathbf{x}_t^{(j)})}{g_t(\mathbf{x}_t^{(j)} \mid \mathbf{Y}_t)}\right]$$

$$= \frac{\sum_{k'=1}^m \lambda_t^{(k',j)}}{m} = w_t^{(j)}. \quad (A.7)$$

From (A.5) and (A.7), we have

$$E[w_t^{*(j)} \mid \{\mathbf{x}_{t-1}^{(k)}, w_{t-1}^{(k)}\}_{k=1}^m, \mathbf{x}_t^{(j)}] = w_t^{(j)}. \quad (A.8)$$

According to (A.6) and (A.8),

$$E(A_{t,m}^* \mid \{\mathbf{x}_{t-1}^{(j)}, w_{t-1}^{(j)}, \mathbf{x}_t^{(j)}\}_{j=1}^m) = A_{t,m} \quad \text{and} \quad E(B_{t,m}^* \mid \{\mathbf{x}_{t-1}^{(j)}, w_{t-1}^{(j)}, \mathbf{x}_t^{(j)}\}_{j=1}^m) = B_{t,m}.$$

So we have

$$E(A_{t,m}^*) = E(A_{t,m}), \quad E(B_{t,m}^*) = E(B_{t,m}),$$
$$\mathrm{var}(A_{t,m}^*) \geqslant \mathrm{var}(A_{t,m}), \quad \mathrm{var}(B_{t,m}^*) \geqslant \mathrm{var}(B_{t,m}).$$

Proof of Proposition 7

The proof is similar to the proofs for Propositions 1—3,5 and 6.

References

Avitzour,D. (1995),"A Stochastic Simulation Bayesian Approach to Multitarget Tracking," *IEE Proceedings on Radar*, *Sonar and Navigation*,142,41—44.

Bar-Shalom,Y. ,and Fortmann,T. (1988), *Tacking and Data Association*, Boston: Academic Press.

Blake,A. , Bascle, B. , Isard, M. , and MacCormick, J. (1998), "Statistical Models of Visual Shape and Motion," *Philosophical Transactions of the Royal Society*, Ser. A, 356, 1283—1302.

Carpenter,J. ,Clifford,P. ,and Fearnhead,P. (1999),"Improved Particle Filter for Nonlinear Problems,"*IEE Proceedings on Radar*,*Sonar*,*and Navigation*,146,2—7.

Chen,R. ,and Liu,J. S. (2000),"Mixture Kalman Filters,"*Journal of the Royal Statistical Society*,Ser. B,62,493—508.

Chen,R. ,Wang,X. ,and Liu,J. (2000),"Adaptive Joint Detection and Decoding in Flat-Fading Channels via Mixture Kalman Filtering."*IEEE Transactions on Information Theory*,46,2079—2094.

Chopin,N. ,(2002),"A Sequential Particle Filter Method for Static Models,"*Biometrica*,89, 539—552.

Doucet,A. ,Gordon,N. J. ,and Krishnamurthy,V. (2001),"Particle Filters for State Estimation of Jump Markov Linear Systems," *IEEE Transactions on Signal Processing*,49, 613—624.

Durbin,J. ,and Koopman,S. (1997),"Monte Carlo Maximum Likelihood Estimation for Non-Gaussian State-Space Models,"*Biometrika*,84,669—684.

Durham,G. ,and Gallant,A. R. (2002),"Numerical Techniques for Maximum Likelihood Estimation of Continuous-Time Diffusion,"*Journal of Business & Economic Statistic*,20, 297—338.

Fong,W. ,Godsill,S. ,Doucet,A. ,and West,M. (2002),"Monte Carlo Smoothing With Application to Audio Signal Enhancement,"*IEEE Transactions on Signal Processing*,50, 438—449.

Fox,D. , Thrun, S. , Burgar, W. , and Dellaert, F. (2001), "Particle Filters for Mobile Robot

Localization," in *Sequential Monte Carlo Methods in Practice*, eds. A. Doucet, N. de Freitas, and N. Gordon, New York: Springer-Verlag, pp. 401—428.

Gordon, N. J., Salmond, D. J., and Smith, A. F. M. (1993), "Novel Approach to Nonlinear/Non-Gaussian Bayesian State Estimation," *IEE Proceedings on Radar and Signal Processing*, 140, 107—113.

Grassberger, P. (1997), "Prunded-Enriched Rosenbluth Method: Simulation of θ Polymers of Chain Length up to 1,000,000," *Physics Review*, Ser. E, 56, 3682—3693.

Hendry, D., and Richard, J.-F. (1991), "Likelihood Evaluation for Dynamic Latent Variable Models," in *Computational Economics and Econometrics*, eds. H. Amman, D. Belsley, and L. Pau, Dordrech: Kluwer, pp. 3—17.

Hue, C., Le Cadre, J.-P., and Perez, P. (2002), "Sequential Monte Carlo Methods for Multiple Target Tracking and Data Fusion," *IEEE Transactions on Signal Processing*, 50, 309—315.

Ikoma, N., Ichimura, N., Higuchi, T., and Maeda, H. (2001), "Maneuvering Target Tracking by Using Particle Filter," *Joint 9th IFSA World Congress and 20th NAFIPS International Conference*, 4, 2223—2228.

Isard, M., and Blake, A. (1996), "Contour Tracking by Stochastic Propagation of Conditional Density," in *Computer Vision-ECCV' 96*, eds. B. Buxton and R. Cipolla, New York: Springer-Verlag, pp. 343—356.

—— (2004), "ICONDENSATION: Unifying Low-Level and High-Level Tracking in a Framework," technical report, Oxford University, available at *http://robots.ox.ac.uk/~ab*.

Kitagawa, G. (1996), "Monte Carlo Filter and Smoother for Non-Gaussian Nonlinear State-Space Models," *Journal of Computational and Graphical Statistics*, 5, 1—25.

Kong, A., Liu, J. S., and Wong, W. H. (1994), "Sequential Imputations and Bayesian Missing Data Problems," *Journal of the American Statistical Association*, 90, 567—576.

Kotecha, J., and Djuric, P. (2003), "Gaussian Sum Particle Filters," *IEEE Transactions on Signal Processing*, 51, 2602—2612.

Liang, J., Chen, R., and Zhang, J. (2002), "On Statistical Geometry of Packing Defects of Lattice Chain Polymer," *Journal of Chemical Physics*, 117, 3511—3521.

Liu, J. S. (2001), *Monte Carlo Strategies in Scientific Computing*, New York: Springer-Verlag.

Liu, J. S., and Chen, R. (1995), "Blind Deconvolution via Sequential Imputations," *Journal of the American Statistical Association*, 90, 567—576.

—— (1998), "Sequential Monte Carlo Methods for Dynamic Systems," *Journal of the American Statistical Association*, 93, 1032—1044.

McGinnity, S., and Irwin, G. (2001), "Manoeuvring Target Tracking Using a Multiple-Model Bootstrap Filter," in *Sequential Monte Carlo in Practice*, eds. A. Doucet, J. F. G. de Freitas, and N. J. Gordon, New York: Springer-Verlag, pp. 479—498.

Pitt, M. K., and Shephard, N. (1999), "Filtering via Simulation: Auxiliary Particle Filters," *Journal of the American Statistical Association*, 94, 590—599.

Rossi, G. D. (2004), "The Two-Factor Cox-Ingersoll-Ross Model as a Self-Organizing State Space," working paper.

Salmond, D., and Gordon, N. (2001), "Particles and Mixtures for Tracking and Guidance," in *Sequential Monte Carlo in Practice*, eds. A. Doucet, J. F. G. de Freitas, and N. J. Gordon, New York: Springer-Verlag, pp. 517—532.

Torma, P., and Szepesvari, P. (2003), "Sequential Importance Sampling for Visual Tracking Reconsidered," in *Proceedings of the AI and Statistics*, pp. 198—205.

van der Merwe, R., Doucet, A., de Freitas, N., and Wan, E. (2002), "The Unscented Particle Filter," in *Advances in Neural Information Processing Systems (NIPS13)*, eds. T. K. Leen, T. G. Dietterich, and V. Tresp, Cambridge, MA: MIT Press.

Vasquez, M., and Scheraga, H. (1985), "Use of Build-up and Energy-Minimization Procedures to Compute Low-Energy Structures of the Backbone of Enkephalin," *Biopolymers*, 24, 1437—1447.

Wang, X., Chen, R., and Gou, D. (2002), "Delayed Pilot Sampling for Mixture Kalman Filter With Application in Fading Channels," *IEEE Transactions on Signal Processing*, 50, 241—264.

Zaidi, Z. R., and Mark, B. L. (2003), "A Mobility Tracking Model for Wireless ad hoc Networks," *IEEE Wireless Communication and Networking Conference (WCNC' 2003)*, 3, 1790—1795.

Zhang, J., Chen, R., Tang, C., and Liang, J. (2003), "Origin of Scaling Behavior of Protein Packing Density: A Sequential Monte Carlo Study of Compact Long-Chain Polymer," *Journal of Chemical Physics*, 118, 6102—6109.

Zhang, J. L., and Liu, J. S. (2002), "A New Sequential Importance Sampling Method With Application to the 2D Hydrophobic-Hydrophilic Model," *Journal of Chemical Physics*, 117, 3492—3498.

原文载于 Journal of the American Statistical Association, Vol. 100, No. 472, December 2005, 1412—1421.

Particle Filters for Maneuvering Target Tracking Problem

Yu Yihua[1], Cheng Qiansheng[2]

1. School of Information Engineering, Beijing University of Posts and Telecommunications;
2. Department of Information Science, School of Mathematical, Sciences, Peking University

Abstract: In this paper, we address the target tracking problem for the case of maneuvering target, including single target and multiple target tracking. We propose a suitable model to characterize the maneuvering acceleration and develop a state space model to describe the maneuvering target tracking problem. Algorithms based on particle filters are developed. Simulations are given to demonstrate the performance of the procedures.

Keywords: target tracking; particle filter; data association

1. Introduction

Mobility tracking is an important issue for cellular radio network [1, 2]. It enables efficient network control and provides the ability to offer additional services. Similar issues appear in many other civilian and military applications [3, 4]. All these problems are related in that they can be described by state space models. Particle filters [5, 6], which can be applied to estimate non-linear and non-Gaussian dynamic process, have received much attention in recent years. Many solutions for target tracking based on particle filters have been proposed [7, 4].

Multitarget tracking deals with the state estimation of a number of moving targets. The extension of the particle filters to multitarget tracking has progressively received attention only in the last 10 years [3, 8, 9].

But these methods are mainly devoted to the random-acceleration target tracking and little work is related to the case of maneuvering target. Since the acceleration cannot be observed directly, maneuvering acceleration is usually difficult to track. In this paper, we focus on the maneuvering target tracking problem, including single target and multiple target tracking. We propose a suitable model to characterize the maneuvering acceleration and develop a state space model to describe the maneuvering target tracking problem. Particle filtering approach is applied and tracking algorithms are derived.

The rest of this paper is organized as follows. In Section 2, we describe our maneuvering acceleration model and motion models of target. In Sections 3 and 4, we present our algorithms for single target tracking and multiple target tracking, respectively. Simulation results are provided in Section 5 and a brief conclusion is given in Section 6.

2. Models

Target tracking on a two-dimensional plane can be modelled as follows: let $s_t = (s_{t1}, s_{t2})^T$ be position vector, $v_t = (v_{t1}, v_{t1})^T$ be velocity vector and $a_t = (a_{t1}, a_{t2})^T$ be acceleration vector, the target is supposed to evolve in the following way:

$$\begin{pmatrix} s_t \\ v_t \end{pmatrix} = \begin{pmatrix} I_2 & \Delta t \cdot I_2 \\ 0 & I_2 \end{pmatrix} \begin{pmatrix} s_{t-1} \\ v_{t-1} \end{pmatrix} + \begin{pmatrix} \frac{\Delta t^2}{2} \cdot I_2 \\ \Delta t \cdot I_2 \end{pmatrix} a_t, \tag{1}$$

where Δt is the sample period, and I_2 is 2×2 identity matrix.

If no maneuver exists, a_t is random acceleration and usually considered as the state noise. When some unknown maneuver exists, a_t is maneuvering acceleration. In this case, we characterize a_t by the following way:

$$a_t = a_{t-1} + \Delta a_t, \tag{2}$$

where Δa_t denotes the change of acceleration between time $t-1$ and t. We model Δa_t as

$$\Delta a_t = u_t + \varepsilon_t, \tag{3}$$

where ε_t is a zero mean continuous random variable and u_t is a discrete random variable. The choice of ε_t and u_t is dependent on different applications. For example, suppose the change of acceleration between two successive time points is not too large, we can choose $\varepsilon_t \sim N(\mathbf{0}, 0.1I_2)$ and

$$u_t \in \mathbf{\Omega} \equiv \{(0,0)^T, (1,0)^T (1,1)^T, (0,1)^T, (-1,1)^T, \\ (-1,0)^T, (-1,-1)^T, (0,-1)^T, (1,-1)^T\}.$$

The distribution of u_t on $\mathbf{\Omega}$ is determined by some prior information.

From (1)—(3), we obtain the motion equation of maneuvering target as

$$\begin{pmatrix} s_t \\ v_t \\ a_t \end{pmatrix} = \begin{bmatrix} I_2 & \Delta t \cdot I_2 & \frac{\Delta t^2}{2} \cdot I_2 \\ 0 & I_2 & \Delta t \cdot I_2 \\ 0 & 0 & I_2 \end{bmatrix} \begin{pmatrix} s_{t-1} \\ v_{t-1} \\ a_{t-1} \end{pmatrix} + \begin{bmatrix} \frac{\Delta t^2}{2} \cdot I_2 \\ \Delta t \cdot I_2 \\ I_2 \end{bmatrix} u_t + \begin{bmatrix} \frac{\Delta t^2}{2} \cdot I^2 \\ \Delta t \cdot I_2 \\ I_2 \end{bmatrix} \varepsilon_t. \quad (4)$$

Basically, the observations that are related to the target can be described as

$$y_t = h(s_t) + \eta_t, \quad (5)$$

where the observation noise η_t are characterized by its density and $h(\cdot)$ is determined by different applications, such as the distance [4] relative to a reference point $p = (p_1, p_2)^T$,

$$h(s_t) = \| s_t - p \|_2,$$

and the direction of the target [7, 4] relative to the reference point

$$h(s_t) = \tan^{-1}\left(\frac{s_{t2} - p_2}{s_{t1} - p_1}\right).$$

3. Single target tracking

3.1 Description of the algorithm

The state equation (4) and the observation equation (5) form a dynamic system for the maneuvering target tracking problem. Our goal is to estimate s_t, v_t, a_t based on all available observations $y_{1,t} = (y_1, \cdots, y_t)$ up to time t. In this section, we develop an algorithm based on auxiliary particle filter [10] to this problem.

From the Bayes rule,

$$p(s_t, v_t, a_t | y_{1,t}) = \frac{p(y_t | s_t, v_t, a_t) p(s_t, a_t, v_t | y_{1,t-1})}{p(y_t | y_{1,t-1})} \quad (6)$$

$$= \frac{p(y_t | s_t) p(s_t, a_t, v_t | y_{1,t-1})}{p(y_t | y_{1,t-1})}.$$

Moreover,

$$p(s_t, v_t, a_t | y_{1,t-1})$$

$$= \int p(s_t, v_t, a_t, s_{t-1}, v_{t-1}, a_{t-1}, u_t | y_{1,t-1}) ds_{t-1} dv_{t-1} da_{t-1} du_t$$

$$= \int p(s_t, v_t, a_t | s_{t-1}, v_{t-1}, a_{t-1}, u_t, y_{1,t-1})$$

$$\times p(u_t \mid s_{t-1}, v_{t-1}, a_{t-1}, y_{1,t-1})$$
$$\times p(s_{t-1}, v_{t-1}, a_{t-1} \mid y_{1,t-1}) \mathrm{d}s_{t-1} \mathrm{d}v_{t-1} \mathrm{d}a_{t-1} \mathrm{d}u_t$$
$$= \int p(s_t, v_t, a_t \mid s_{t-1}, v_{t-1}, a_{t-1}, u_t) p(u_t)$$
$$\times p(s_{t-1}, v_{t-1}, a_{t-1} \mid y_{1,t-1})$$
$$\times \mathrm{d}s_{t-1} \mathrm{d}v_{t-1} \mathrm{d}a_{t-1} \mathrm{d}u_t, \tag{7}$$

where we utilize the Markovian structure of the state space model, and u_t is independent on $s_{t-1}, v_{t-1}, a_{t-1}, y_{1,t-1}$.

Suppose a set of weighted random samples $\{(s_{t-1}^{(i)}, v_{t-1}^{(i)}, a_{t-1}^{(i)}), w_{t-1}^{(i)}\}_{i=1}^N$ from $p(s_{t-1}, v_{t-1}, a_{t-1} \mid y_{1,t-1})$ and $\{u_t^{(j)}\}_{j=1}^M$ from $p(u_t)$ are available, then (7) can be expressed as

$$p(s_t, v_t, a_t \mid y_{1,t-1})$$
$$\approx \frac{1}{M} \sum_{i=1}^N \sum_{j=1}^M p(s_t, v_t, a_t \mid s_{t-1}^{(i)}, v_{t-1}^{(i)}, a_{t-1}^{(i)}, u_t^{(j)}) w_{t-1}^{(i)}$$
$$\triangleq \frac{1}{M} \sum_{k=1}^{MN} p(s_t, v_t, a_t \mid s_{t-1}^{(k)}, v_{t-1}^{(k)}, a_{t-1}^{(k)}, u_t^{(k)}) w_{t-1}^{(k)}, \tag{8}$$

where we reset the indices of the summation.

From (6) and (8), an approximation of $p(s_t, v_t, a_t \mid y_{1,t})$ is given by

$$\hat{p}(s_t, v_t, a_t \mid y_{1,t}) \propto p(y_t \mid s_t) \sum_{k=1}^{MN} p(s_t, v_t, a_t \mid s_{t-1}^{(k)}, v_{t-1}^{(k)}, a_{t-1}^{(k)}, u_t^{(k)}) w_{t-1}^{(k)}. \tag{9}$$

Next we develop an algorithm based on auxiliary particle filter to generate random samples of $p(s_t, v_t, a_t \mid y_{1,t})$ from (9).

We add an auxiliary random variable k in (9) and get a new density

$$p(s_t, v_t, a_t, k \mid y_{1,t}) \propto p(y_t \mid s_t) p(s_t, v_t, a_t \mid s_{t-1}^{(k)}, v_{t-1}^{(k)}, a_{t-1}^{(k)}, u_t^{(k)}) w_{t-1}^{(k)}, \tag{10}$$

where the value set of k is $\{1, \cdots, MN\}$. Obviously, $\hat{p}(s_t, v_t, a_t \mid y_{1,t})$ is the marginal density of $p(s_t, v_t, a_t, k \mid y_{1,t})$.

In order to generate random samples from $p(s_t, v_t, a_t, k \mid y_{1,t})$, we choose the following importance sample density:

$$q(s_t, v_t, a_t, k \mid y_{1,t}) \propto p(y_t \mid \mu_t^{(k)}) p(s_t, v_t, a_t \mid s_{t-1}^{(k)}, v_{t-1}^{(k)}, a_{t-1}^{(k)}, u_t^{(k)}) w_{t-1}^{(k)}. \tag{11}$$

Here, we let $\mu_t^{(k)} = E\{s_t \mid s_{t-1}^{(k)}, v_{t-1}^{(k)}, a_{t-1}^{(k)}, u_t^{(k)}\}$. From (4), we have

$$\mu_t^{(k)} = s_{t-1}^{(k)} + \Delta t \cdot v_{t-1}^{(k)} + \frac{\Delta t^2}{2} a_{t-1}^{(k)} + \frac{\Delta t^2}{2} u_t^{(k)}.$$

From (11),

$$q(k \mid y_{1,t}) \propto \int p(y_t \mid \mu_t^{(k)}) p(s_t, v_t, a_t \mid s_{t-1}^{(k)}, v_{t-1}^{(k)}, a_{t-1}^{(k)}, u_t^{(k)}) w_{t-1}^{(k)} ds_t dv_t da_t$$
$$= p(y_t \mid \mu_t^{(k)}) w_{t-1}^{(k)},$$

then

$$q(s_t, v_t, a_t, k \mid y_{1,t}) \propto p(s_t, v_t, a_t \mid s_{t-1}^{(k)}, v_{t-1}^{(k)}, a_{t-1}^{(k)}, u_t^{(k)}) q(k \mid y_{1,t}).$$

The procedure to generate random samples from $q(s_t, v_t, a_t, k \mid y_{1,t})$ can be implemented by the following way:

(1) Generate N random samples $\{k^{(j)}\}_{j=1}^{N}$ from $q(k \mid y_{1,t})$.

(2) For $j=1,\cdots,N$, generate a random sample $(s_t^{(j)}, v_t^{(j)}, a_t^{(j)})$ according to the density $p(s_t, v_t, a_t \mid s_{t-1}^{(k^{(j)})}, v_{t-1}^{(k^{(j)})}, a_{t-1}^{(k^{(j)})}, u_t^{(k^{(j)})})$.

If each random sample $(s_t^{(j)}, v_t^{(j)}, a_t^{(j)}, k^{(j)})$ is assigned a weight

$$w_t^{(j)} \propto \frac{p(s_t^{(j)}, v_t^{(j)}, a_t^{(j)}, k^{(j)} \mid y_{1,t})}{q(s_t^{(j)}, v_t^{(j)}, a_t^{(j)}, k^{(j)} \mid y_{1,t})} = \frac{p(y_t \mid s_t^{(j)})}{p(y_t \mid \mu_t^{(k^{(j)})})},$$

then we get the weighted random samples $\{(s_t^{(j)}, v_t^{(j)}, a_t^{(j)}, k^{(j)}), w_t^{(j)})\}_{j=1}^{N}$ which are distributed from $p(s_t, v_t, a_t, k \mid y_{1,t})$ and $\{(s_t^{(j)}, v_t^{(j)}, a_t^{(j)}), w_t^{(j)}\}_{j=1}^{N}$ which are distributed from $p(s_t, v_t, a_t \mid y_{1,t})$.

Algorithm 1 (Single target tracking algorithm).

$$[\{(s_t^{(i)}, v_t^{(i)}, a_t^{(i)}), w_t^{(i)}\}_{i=1}^{N}] = \text{STT}[\{(s_{t-1}^{(i)}, v_{t-1}^{(i)}, a_{t-1}^{(i)}), w_{t-1}^{(i)}\}_{i=1}^{N}].$$

- For each sample $(s_{t-1}^{(i)}, v_{t-1}^{(i)}, a_{t-1}^{(i)}), i=1,\cdots,N$, generate M random samples $\{u_t^{(j)}\}_{j=1}^{M}$ from $p(u_t)$, and denote
$$\{(s_{t-1}^{(k)}, v_{t-1}^{(k)}, a_{t-1}^{(k)}, u_t^{(k)}), w_{t-1}^{(k)}\}_{k=1}^{NM} \triangleq \{(s_{t-1}^{(i)}, v_{t-1}^{(i)}, a_{t-1}^{(i)}, u_t^{(j)}), w_{t-1}^{(i)}\}_{i=1,j=1}^{N,M}.$$

- For $k=1,\cdots,NM$, compute $\mu_t^{(k)}$ and $q(k \mid y_{1,t})$.

- Generate N random samples $\{k^{(j)}\}_{j=1}^{N}$ from the set $\{1,\cdots,NM\}$ with the probability proportional to $q(k \mid y_{1,t})$.

- For $j=1,\cdots,N$, generate one random sample $(s_t^{(j)}, v_t^{(j)}, a_t^{(j)})$ from the density $p(s_t, v_t, a_t \mid s_{t-1}^{(k^{(j)})}, v_{t-1}^{(k^{(j)})}, a_{t-1}^{(k^{(j)})}, u_t^{(k^{(j)})})$ and assign weight $w_t^{(j)} = p(y_t \mid s_t^{(j)}) / p(y_t \mid \mu_t^{(k^{(j)})})$.

3.2 Delayed-Weight Estimation

The delayed-weight estimation method [11] is a simple extension of the particle filters. Since the state space model is highly correlated, the future observations often contain information about the current state. Hence, a delayed estimate is usually more accurate than the concurrent estimate.

The basic idea of the delayed-weight estimation is as follows. Suppose at time $t+D, D>0$, the samples $\{x_{1,t+D}^{(j)}\}_{j=1}^{N}$ and weights $\{w_{t+D}^{(j)}\}_{j=1}^{N}$, where $x_{1,t+D}=(x_1,\cdots,x_{t+D})$, are properly generated from $p(x_{1,t+D}|y_{1,t+D})$, then we have the samples $\{x_t^{(j)}\}_{j=1}^{N}$ and weights $\{w_{t+D}^{(j)}\}_{j=1}^{N}$ which are properly distributed from $p(x_t|y_{1,t+D})$. With these samples and weights, we obtain a delayed estimate of the quantity of interest at time t as

$$E\{h(x_t)\mid y_{1,t+D}\} \approx \sum_{j=1}^{N} h(x_t^{(j)}) w_{t+D}^{(j)} \Big/ \sum_{j=1}^{N} w_{t+D}^{(j)}.$$

The delayed-weight estimation method incurs no additional computational cost, but it requires some extra memory for storing $\{(x_{t+1}^{(j)},\cdots,x_{t+D}^{(j)})\}_{j=1}^{N}$.

4. Multiple Target Tracking

4.1 Multiple Target Tracking Model

Consider a system of M targets. Let $x_t^m = (s_t^m, v_t^m, a_t^m), m=1,\cdots,M$, be the state of the M targets at time t. The state of each target evolves in the following way:

$$s_t^m = s_{t-1}^m + \Delta t \cdot v_{t-1}^m + \frac{\Delta t^2}{2} \cdot a_t^m, \tag{12}$$

$$v_t^m = v_{t-1}^m + \Delta t \cdot a_t^m, \tag{13}$$

$$a_t^m = a_{t-1}^m + \Delta a_t^m. \tag{14}$$

We model Δa_t^m in the same way as in Section 2, i.e., $\Delta a_t^m = u_t^m + \varepsilon_t^m$. Let $x_t = (x_t^1,\cdots,x_t^M)$ and $u_t = (u_t^1,\cdots,u_t^M)$.

The observaion vector collected at time t is denoted by $y_t = (y_t^1,\cdots,y_t^{N_t})$, where N_t is the number of observations collected at time t. Because of the effect of observation noise and false alarm, the origin of each observation is not known. It may come from one target or from false alarm. The false alarms are assumed to be uniformly distributed in the observation area. Their number is assumed to arise from Poisson density of parameter λV, where V is the volume of the observation area, and λ is the number of false alarms per unit volume.

The most common traditional approaches for multitarget tracking consist of two sequential steps: (1) data association, and (2) target state estimation.

4.2 Data Association

The most commonly used method for data association is probably the joint probability data association filter (JPDAF) [12,13]. In this paper, we take advantage of JPDAF for data association.

As we do not know the origin of each observation, one has to introduce the vector A_t to describe the associations between the observations and the targets. Each component $A_t^n, n=1,\cdots,N_t$, is a random variable that takes its values among $\{0,\cdots,M\}$. $A_t^n=0$ indicates that y_t^n is associated with the false alarm and $A_t^n=m (m\neq 0)$ indicates that y_t^n is associated with the mth target. In the case of $A_t^n=m(m\neq 0)$, y_t^n is a realization of the following observation equation:

$$y_t^n = h(x_t^m) + \eta_t^m.$$

The JPDAF begins with the gating of the observations. Only the observations that are inside an ellipsoid around the predicted state of one target are kept. Then the probabilities of each data association

$$\beta_t^{nm} = p(A_t^n = m), \quad n=1,\cdots,N_t, \quad m=0,\cdots,M,$$

are estimated. For more details of JPDAF, we refer to [13,14].

4.3 State Estimation

With the output of data association, we estimate the state of each target. Our goal is to track M posterior density $p(x_t^m | y_{1:t}), m=1,\cdots,M$.

From the Bayes rule,

$$p(x_t^m | y_{1:t}) = \frac{p(y_t | x_t^m) p(x_t^m | y_{1:t-1})}{p(y_t | y_{1:t-1})},$$

and

$$\begin{aligned}
p(x_t^m | y_{1:t-1}) &= \int p(x_t^m, x_{t-1}, u_t | y_{1:t-1}) dx_{t-1} du_t \\
&= \int p(x_t^m | x_{t-1}, u_t) p(x_{t-1}, u_t | y_{1:t-1}) dx_{t-1} du_t \\
&= \int p(x_t^m | x_{t-1}^m, u_t^m) p(x_{t-1}^m, u_t^m | y_{1:t-1}) dx_{t-1}^m du_t^m \\
&= \int p(x_t^m | x_{t-1}^m, u_t^m) p(x_{t-1}^m, y_{1:t-1}) p(u_t^m) dx_{t-1}^m du_t^m.
\end{aligned}$$

Here, we utilize the fact the x_t^m is only related to x_{t-1}^m and u_t^m and independent on other elements in x_{t-1} and u_t from Eqs. (12)—(14).

Suppose a set of random samples and weights $\{x_{t-1}^{m,(i)}, w_{t-1}^{m,(i)}\}_{i=1}^N$ from $p(x_{t-1}^m | y_{1:t-1})$ is available at time $t-1$, and $\{u_t^{m,(j)}\}_{j=1}^R$ are generated from $p(u_t^m)$, then

$$p(x_t^m | y_{1:t-1}) \approx \frac{1}{R} \sum_{i=1}^N \sum_{j=1}^R p(x_t^m | x_{t-1}^{m,(i)}, u_t^{m,(j)}) w_{t-1}^{m,(i)}, \qquad (15)$$

and an approximation of $p(x_t^m | y_{1:t})$ is given by

$$\hat{p}(x_t^m | y_{1:t}) \propto \frac{1}{R} \sum_{i=1}^N \sum_{j=1}^R p(y_t | x_t^m) p(x_t^m | x_{t-1}^{m,(i)}, u_t^{m,(j)}) w_{t-1}^{m,(i)}$$
$$\triangleq \frac{1}{R} p(y_t | x_t^m) \sum_{k=1}^{NR} p(x_t^m | x_{t-1}^{m,(k)}, u_t^{m,(k)}) w_{t-1}^{m,(k)}, \qquad (16)$$

where we reset the indices of the summation.

The following procedure can be used to generate the weighted random samples $\{x_t^{m,(i)}, w_t^{m,(i)}\}_{i=1}^N$ from (16) which are distributed from $p(x_t^m | y_{1:t})$, $m=1,\cdots,M$.

Algorithm 2 (Multitarget tracking algorithm).

$$[\{x_t^{m,(i)}, w_t^{m,(i)}\}_{i=1\ m=1}^{N\ M}] = \text{MTT}[\{x_{t-1}^{m,(i)}, w_{t-1}^{m,(i)}\}_{i=1\ m=1}^{N\ M}].$$

- For $m=1\cdots,M, i=1,\cdots,N$, generate R random samples $\{u_t^{m,(j)}\}_{j=1}^R$ from $p(u_t^m)$ and denote
$$\{(x_{t-1}^{m,(k)}, u_t^{m,(k)}), w_{t-1}^{m,(k)}\}_{R=1}^{NR} \triangleq \{(x_{t-1}^{m,(i)}, u_t^{m,(j)}), w_{t-1}^{m,(i)}\}_{i=1\ j=1}^{N\ R}.$$

- For $m=1,\cdots,M, k=1,\cdots,NR$, generate one random sample $x_t^{m,(k)}$ from the density $p(x_t^m | x_{t-1}^{m,(k)}, u_t^{m,(k)})$.

- For each $x_t^{m,(k)}$, assign a weight
$$w_t^{m,(k)} = w_{t-1}^{m,(k)} p(y_t | x_t^{m,(k)}), \quad m=1,\cdots,M, \quad k=1,\cdots,NR. \qquad (17)$$

- For $m=1,\cdots,M$, generate N random samples from the set $\{x_t^{m,(k)}\}_{k=1}^{NR}$ according to the density $p(x_t^{m,(k)}) \propto w_t^{m,(k)}$, and denote the results as $\{x_t^{m,(i)}, w_t^{m,(i)}\}_{i=1}^N$, where each weight $w_t^{m,(i)}$ is reset as $1/N$.

The computation of the weight (17) in Algorithm 2 needs the output of the data association. From the total probability theorem with the event ($A_t^n = m$), $n=1,\cdots,N_t$,

$$p(y_t \mid x_t^m) = \sum_{n=1}^{N_t} p(y_t \mid x_t^m, A_t^n = m) p(A_t^n = m)$$

$$= \sum_{n=1}^{N_t} p(y_t^n \mid x_t^m, A_t^n = m) \beta_t^{nm}.$$

5. Experiments

5.1 Single Target Tracking

With the sample period $\Delta t = 1$s in (1), we obtain the motion model of the target. The observation includes two elements: the relative distance and the relative direction from the target to a reference point $p = (p_1, p_2)^T$,

$$y_t = \begin{pmatrix} \|s_t - p\|_2 \\ \tan^{-1}\left(\dfrac{s_{t2} - p_2}{s_{t1} - p_1}\right) \end{pmatrix} + \eta_t. \tag{18}$$

In this example, we choose $p = (-1, -1)^T$ and $\eta_t \sim N(\binom{0}{0}, \binom{(2.0)^2 \ 0}{0 \ (0.05)^2})$.

Before the simulation, we artificially give the maneuvering accelerations, the actual initial position $s_0 = (0,0)^T$ and the actual initial velocity $v_0 = (0,0)^T$ of the target. With these settings, we get the simulated position s_t and velocity v_t from (1) and the simulated observation data y_t from (18), where $t = 1, \cdots, L$ and $L = 100$ is taken.

With the simulated data $y_{1:L} = \{y_1, \cdots, y_L\}$, we apply Algorithm 1 to estimate the simulated position, velocity and acceleration of the target, with $N = 500$. The distribution of $p(u_t)$ is set to $p(u_t) = \dfrac{4}{5}$ if $u_t = (0,0)^T$, and $p(u_t) = \dfrac{1}{5 \times 8}$ otherwise. In Algorithm 1, $M = 5$ samples are generated from $p(u_t)$ every time.

The results of the estimation are shown in Fig. 1 and the results of the delayed-weight estimation with $D = 10$ are shown in Fig. 2.

Since the position of target can be directly computed from the noisy observations. Fig. 1(f) presents the results of direct computation. But the velocity and acceleration cannot be directly computed.

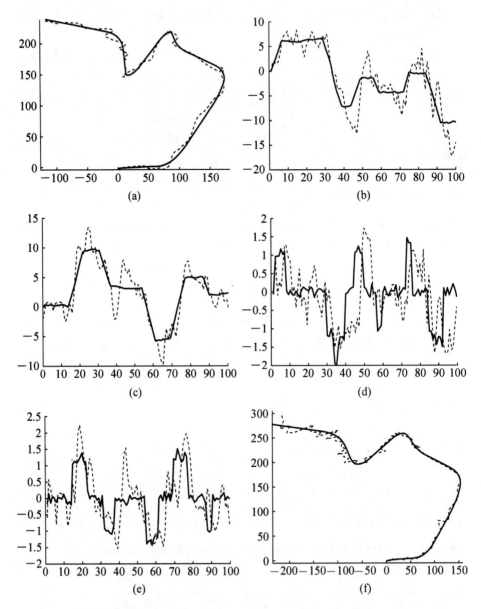

Fig. 1 The simulated data (solid line) and estimated data by Algorithm 1 (dashed line): (a) target position; (b) x-coordinate of target velocity; (c) y-coordinate of target velocity; (d) x-coordinate of target acceleration; (e) y-coordinate of target acceleration; (f) the result of direct computation

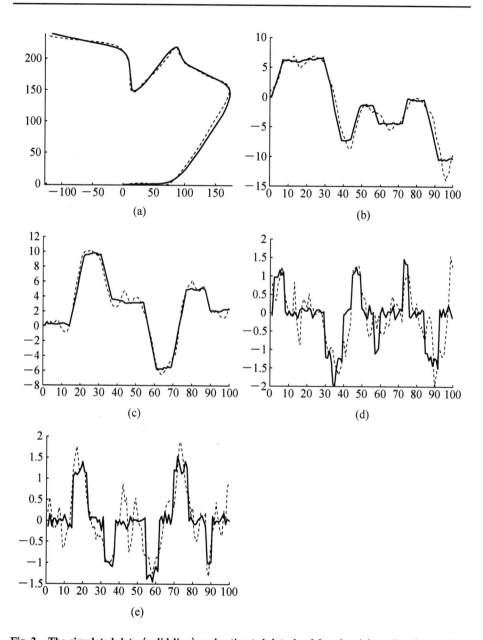

Fig. 2 The simulated data (solid line) and estimated data by delayed-weight estimation method (dashed line): (a) target position; (b) x-coordinate of target velocity; (c) y-coordinate of target velocity; (d) x-coordinate of target acceleration; (e) y-coordinate of target acceleration

We perform $G=100$ runs of the simulations. All the computations are performed on Pentium 4 1.7GHz PC with Matlab 6.0. The performance measure is root mean squared error (RMS) computed as follows [15]:

$$(\text{RMS})^2 = \frac{1}{L}\sum_{t=1}^{L}\left\{\frac{1}{G}\sum_{g=1}^{G}[(s_{t1}(g)-\hat{s}_{t1}(g))^2+(s_{t2}(g)-\hat{s}_{t2}(g))^2]\right\}^{1/2},$$

where $(s_{t1}(g),s_{t2}(g))^T$ is the simulated target position and $(\hat{s}_{t1}(g),\hat{s}_{t2}(t))^T$ is the estimated target position at time t of the gth simulation. The results of performance are RMS=2.3977 for Algorithm 1, RMS=1.8737 for the delayed-weight estimation method, and RMS=2.7536 for the direct computation, respectively.

5.2 Multitarget Tracking

In this simulation study, the proposed MTT algorithm is implemented with five targets and $\Delta t=1s$. An observation produced by the mth target is generated according to

$$y_t = \begin{pmatrix} \|s_t^m - p\|_2 \\ \tan^{-1}\left(\frac{s_{t2}^m-p_2}{s_{t1}^m-p_1}\right) \end{pmatrix} + \eta_t,$$

where $p=(-1,-1)^T$ and $\eta_t \sim N(\binom{0}{0},\binom{(1.0)^2\ 0}{0\ 0.04})^2)$.
The number of false alarm follows the Poisson distribution with mean $\lambda V=2$. These false alarms are assumed to be independent and uniformly distributed within the observation volume V.

With these settings, the states of the five targets and the observations are simulated with $t=1,\cdots,L$ and $L=100$. The trajectories of the targets are represented in Fig. 3, where each "o" indicates the starting position and each "*" indicates the ending position of one target.

From the simulated observations, the proposed MTT algorithm is implemented to estimate the states of the targets. The results are presented in Fig. 3(a). The delayed-weight estimation method is also applied here and the results are presented in Fig. 3(b).

If two targets are in the approximately same position in the two-dimensional plane at the same time, it is usually difficult to distinguish them. Since our algorithm estimates the velocity of target, we can take advantage of the velocity information to distinguish them. Moreover, if the velocities of these two targets are also approximate, we can take advantage of the acceleration information. But if the positions, velocities and accelerations of these two targets are all approximate, we cannot distinguish them any more.

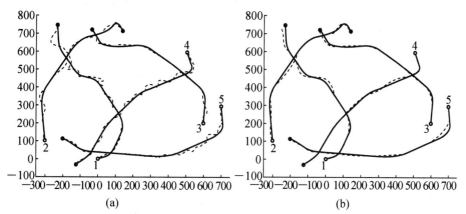

Fig. 3 The simulated positions (solid line) and the estimated positions (dashed line) of the multitarget tracking with five targets by (a) Algorithm 2 and (b) delayed-weight estimation method

6. Conclusions

In this paper, algorithms are presented for the target tracking problem for the case of the maneuvering target. A suitable model is proposed to characterize the maneuvering acceleration. Algorithms based on particle filters are developed. Finally, two simulations are given to demonstrate the performance of the procedures.

References

[1] M. Hellebrandt, R. Mathar, Location tracking of mobiles in cellular radio networks, IEEE Trans. Veh. Technol. 48 (5) (September 1999) 1158—1162.

[2] T. Liu, P. Bahl, I. Chlamtac, Mobility modeling, location tracking, and trajectory prediction in wireless ATM networks, IEEE J. Select. Areas Commun. 16 (6) (August 1998) 922—936.

[3] D. Avitzour, Stochastic simulation Bayesian approach to multitarget tracking, IEE Proc. -F 142 (2) (April 1995) 41—44.

[4] F. Gustafsson, F. Gunnarsson, N. Bergman, U. Forssell, J. Jansson, R. Karlsson, P. Nord-

lund, Particle filters for positioning, navigation and tracking, IEEE Trans. Signal Process. 50 (2) (February 2002) 425—437.

[5] A. Doucet, S. J. Godsill, C. Andrieu, On sequential Monte Carlo sampling methods for Bayesian filtering, Statist. Comput. 10 (3) (July 2000) 197—208.

[6] J. S. Liu, R. Chen, Sequential Monte Carlo methods for dynamic systems, J. Amer. Statist. Assoc. 93 (433) (September 1998) 1032—1044.

[7] N. J. Gordon, D. J. Salmond, A. F. M. Smith, Novel approach to nonlinear/non-Gaussian Bayesian state estimation, IEE Proc. -F 140 (2) (April 1993) 107—113.

[8] N. Gordon, A hybrid bootstrap filter for target tracking in clutter, IEEE Trans. Aerosp. Electron. Syst. 33 (1) (January 1997) 353—358.

[9] C. Hue, J. L. Cadre, P. Perez, Sequential Monte Carlo methods for multiple target tracking and data fusion, IEEE Trans. Signal Process. 50 (2) (February 2002) 309—325.

[10] M. K. Pitt, N. Shephard, Filtering via simulation: auxiliary particle filters, J. Amer. Statist. Assoc. 94 (446) (June 1999) 590—599.

[11] R. Chen, X. D. Wang, J. S. Liu, Adaptive joint detection and decoding in flat-fading channels via mixture Kalman filtering, IEEE Trans. Inform. Theory 46 (6) (September 2000) 2079—2094.

[12] Y. Bar-Shalom, T. E. Fortmann, Tracking and Data Association, Academic Press, New York, 1988.

[13] T. E. Fortmann, Y. Bar-Shalom, M. Scheffe, Sonar tracking of multiple targets using joint probabilistic data association, IEEE J. Oceanic Eng. OE-8 (July 1983) 173—184.

[14] B. Zhou, N. K. Bose, Multitarget tracking in clutter: fast algorithm for data association, IEEE Trans. Aerosp. Electron. Syst. 29 (2) (April 1993) 352—363.

[15] A. Doucet, N. J. Gordon, V. Krishnamurthy, Particle filters for state estimation of jump Markov linear systems, IEEE Trans. Signal Process. 49 (3) (March 2001) 613—624.

原文载于 Elsevier Computer Science, Signal Processing, 86(2006) 195—203.

Nonlinear Discriminant Analysis on Embedded Manifold

Yan Shuicheng[1], *Member, IEEE*, Hu Yuxiao[1], Xu Dong[2],
Zhang Hongjiang[3], *Fellow, IEEE*, Zhang Benyu[3], Cheng Qiansheng[4]

1. Beckman Institute, University of Illinois, Urbana-Champaign;
2. Department of Electrical Engineering, Columbia University;
3. Microsoft Research Asia, Beijing;
4. School of Mathematical Science, Peking University

Abstract: Traditional manifold learning algorithms, such as ISOMAP, LLE, and Laplacian Eigenmap, mainly focus on uncovering the latent low-dimensional geometry structure of the training samples in an unsupervised manner where useful class information is ignored. Therefore, the derived low-dimensional representations are not necessarily optimal in discriminative capability. In this paper, we study the discriminant analysis problem by considering the nonlinear manifold structure of data space. To this end, firstly, a new clustering algorithm, called *Intra-Cluster Balanced K-Means* (ICBKM), is proposed to partition the samples into multiple clusters while ensure that there are balanced samples for the classes within each cluster; approximately, each cluster can be considered as a local patch on the embedded manifold. Then, the local discriminative projections for different clusters are simultaneously calculated by optimizing the *global Fisher Criterion* based on the cluster weighted data representation. Compared with traditional linear/kernel discriminant analysis (KDA) algorithms, our proposed algorithm has the following characteristics: 1) it essentially is a KDA algorithm with specific geometry-adaptive-kernel tailored to the specific data structure, in contrast to traditional KDA in which the kernel is fixed and independent to the data set; 2) it is approximately a locally linear while globally nonlinear discriminant analyzer; 3) it does not need to store the original samples for computing the low-dimensional representation of a new data; and 4) it is computationally efficient compared with traditional KDA when the sample number is large. The toy problem on artificial data demonstrates the effectiveness of our proposed algorithm in deriving discriminative representations for problems with nonlinear classification hyperplane. The face recognition experiments on YALE and CMU PIE databases show that our proposed algorithm significantly outperforms linear discriminant analysis (LDA) as well as Mixture LDA, and has higher accuracy than KDA with traditional kernels.

Index Terms: kernel design; kernel machine; kernel selection; linear discriminant analysis (LDA); manifold learning; principal component analysis (PCA); subspace learning

I . Introduction

Previous works on manifold learning [6],[10],[9],[12],[14] focus on uncovering the compact, low-dimensional representations of the observed high-dimensional unorganized data that lie on or nearly on a manifold in an unsupervised manner. According to the property of the mapping functions, these algorithms can be roughly classified into two types. One type is the algorithms without explicit mapping functions, such as ISOMPA [26], LLE [21], and Laplacian Eigenmap [3]. For this type, the low-dimensional representations are available often only for the sample data and by preserving certain local or global properties of the manifold structure. Another type is the algorithms with explicit mapping functions for the whole data space. Roweis *et al*. [25] proposed an algorithm that automatically aligns a mixture of local dimensionality reducers into a single global one. Brand [7] presented a similar work to merge local representations and construct a global nonlinear mapping function for the whole data space. He *et al*. [13] proposed the Locality Preserving Projections algorithm to linearly approximate the Laplacian Eigenmap algorithm. Bengio *et al*. [5] proposed the kernel explanations of ISOMAP, LLE, Laplacian Eigenmap and other spectral analysis algorithms to compute the low-dimensional representation for the out-of-sample data. All these algorithms are unsupervised and most of them are merely evaluated on simple toy problems. For real-world classification problems, such as face recognition, the unsupervised learning algorithms are unnecessarily optimal since they ignore the useful class label information of the sample data, which can be effectively applied to further improve classification performance.

In this paper, we study the problem to utilize the class label information of the training data for discriminant analysis by exploring the underlying nonlinear manifold structure of data space, and propose a novel algorithm for nonlinear discriminant analysis. This algorithm is motivated from the following observations. First, previous works on manifold learning focus on exploring the low-dimensional representations that best preserve some characteristics of a manifold, while the best representative features are not always the best discriminating ones for general classification task. Secondly, if we directly perform supervised learning based on the low-dimensional representations derived from previous manifold learning algorithms, there may be some important features for classification lost in the unsupervised dimensionality reduction step, which may be very strong in discriminative capability although weak in representative capability; and consequently it may degrade the performance of posterior discriminant analysis. Finally, linear discriminant analysis (LDA) can only well handle the linearly separable problem and is conducted directly on the Euclidean feature space without considering the possibly nonlinear manifold structure of the data space. Kernel discriminant analysis (KDA) [4],[16],[17],[29] utilizes the kernel trick to extend the LDA for handing linearly inseparable classification problems; and owing to the capability to extract nonlinear discriminant features instead of linear ones, it has been widely studied and used in many applications. In the past years, many researches have been devoted to presenting more robust solutions to KDA and alleviating the *ill-posed* problem encountered by KDA. Mika *et al.* [19] proposed to add a small scalar matrix to the denominator matrix of objective function, consequently avoiding the singularity issue when using generalized eigenvalue decomposition method for computing the solution. To resolve the same issue, Baudat and Anouar [1] utilized the QR decomposition technique and Yang [30] proposed to use the principal component analysis (PCA) plus LDA strategy as in Fisher-faces [2] instead. When the sample number is large, KDA often suffers from the high computation cost and it requires to store all the training samples for computing the low-dimensional representation of a new data. Moreover, despite of the wide study of KDA, most works study KDA independent to the kernel itself, and the mapping function from the input feature space to the higher Hilbert space does not consider the manifold structure of the data, either. Following the above analysis, it is desired to propose an efficient algorithm for discriminant analysis by explicitly considering the possibly nonlinear manifold structure of data space.

In this work, the sample data of multiple classes are assumed lying on or nearly on a curved low-dimensional manifold; whereas it is often the case that the globally linearly inseparable manifold may be easily separable locally. The intuition of this work is to reside multiple local linear discriminant analyzers on the curved manifold, then merge these local analyzers into a global discriminant analzer by optimizing the *global Fisher Criterion*. In summary, there are two subproblems to be solved: one is how to place the local discriminant analyzers on the curved manifold; the other is how to merge these local discriminant analyzers into s single global discriminant analyzer.

For the first subproblem, traditional methods like mixture factor analysis (MFA) [25] and K-means [11],[18] cannot be directly applied, since they cannot guarantee that there are balanced samples for all the classes within a cluster and the performance of the local discriminant analyzers will be degraded in the cases with unbalanced samples of even no sample for some classes. In this work, we formulate this task as a special supervised clustering problem and present a novel clustering approach that ensures there are balanced samples for the classes in a cluster, thus, called *Intra-Cluster Balanced K-Means* (ICBKM). Two extra terms penalizing the unbalance between clusters and within each cluster are imposed to the objective function of ICBKM; in addition to the *reassignment optimization* method utilized in K-means, a novel method called *exchange optimization* is proposed to minimize the sum of the intracluster variances, while keeps the penalizing terms constant.

Taking the advantage of the clustering results of ICBKM, the sample data are reset as multiple clusters, and local discriminant analysis can be conducted within each cluster. The traditional way to classify a new data using these local analyzers is to conduct the classification using the nearest local analyzer; therefore, these local analyzers are independent in both learning and inferring steps, which is not optimal since it is obvious that different analyzers can collaborate to improve the classification performance. To solve the second subproblem, we propose a novel approach in which the local analyzers are mutually dependent in both learning and inferring stages, and the optimal discriminative features for each cluster are computed simultaneously to optimize the global Fisher Criterion. Frist, PCA is conducted within each cluster; then the posterior probability of each cluster for a given data, i. e. , $p(c|x)$ can be obtained. The optimal discriminative features for each cluster are computed by maximizing the global Fisher criterion, i. e. , maximizing the ratio of the weighted global interclass and intra-class scatters, where the scatters are computed based on the $p(c|x)$ weighted representations for the samples. In the inferring stage, the low-dimensional representation for new data is derived as the $p(c|x)$ weighted sum of the projections from different clusters and the classification can be conducted using the Nearest Neighbor (NN) algorithm based on the derived low-dimensional representations. This algorithm can be understood from two different perspectives: 1) naively, it automatically merges local linear discriminant analyzers and conducts dimensionality reduction and classification in a single coordinate farmework; and 2) in theory, it is essentially a special KDA algorithm with geometry-adaptive-kernel tailored to the special nonlinear manifold structure of sample data, in contrast to traditional KDA in which the kernel is predefined and independent to sample data. Recently, Kim *et al.* [15] proposed a similar yet independently developed algorithm for locally LDA, which can be considered as the K-means-Daemon algorithm as discussed in our experiment part.

It is worthwhile to highlight some characteristics of our proposed algorithm here.

1) A supervised clustering algorithm is proposed to ensure that there are balanced samples for all classes in each cluster, which facilitates the following local discriminant analysis and successfully makes the local analyzers escape from the unbalanced problem existed in other clustering algorithm such as K-means.

2) The local discriminant analyzers collaborate to optimize the global Fisher Criterion in the learning stage, and the final classification is also based on the fused results from different local discriminant analyzers, which is superior to the Mixture LDAs that are learned and performed independently.

3) Our proposed algorithm is supervised and considers the nonlinear manifold stucture of sample data; thus, it is natural to be superior to other unsupervised manifold learning algorithms, such as ISOMAP, LLE, Laplacian Eigenmap, and the recent proposed approach for out-of-sample extensions.

4) The kernel justification proves that our proposed algorithm is sound in theory. Moreover, the geometry-adaptive-kernel is tailored to the special nonlinear geometry structure of the sample data and superior to other fixed kernels in terms of classification performance.

The rest of the paper is structured as follow. The ICBKM clustering method and the locally linear while global nonlinear discriminant analysis algorithm based on the clustering result are introduced in Section II. In Section III, we present the kernel justification for the proposed algorithm. The toy problem on the artifical data and the real world face recognition experimental results compared with LDA, Mixture LDA and KDA on the YALE and CMU PIE database are illustrated in Section IV. Finally, we give the conclusion remarks in Sectio V.

II. Nonlinear Discriminant Analysis on Embedded Manifold

Suppose $X = \{x_1, x_2, \cdots, x_N\}$ be a set of sample points that lie on or nearly on a low-dimensional manifold embedded in the high-dimensional observed space. For each sample $x_i \in \mathbb{R}^D$, a class label is given as $l_i \in \{1, 2, \cdots, L\}$. Most previous works on manifold learning are unsupervised; in this section, we show how to utilize the class information for nonliner discriminant analysis by considering the nonlinear manifold structure of sample data. A continuous manifold may be considered as a combined set of a series of open sets, and when specific to the discrete sample data on it, they are the combination of a series of clusters. Furthermore, the globally linearly inseparable manifold may be easily separable within these local clusters. The above analysis motivates us to conduct local discriminant analysis within each local cluster, and then merge these local analyzers into a global discriminant analyzer. Following this idea, we first segment the sample data into multiple clusters. The traditional clustering algorithms like K-means [18] and Normalized Cut [23] cannot be directly applied to the problem discussed here since there may be unbalanced samples for the classes in a cluster and even only a single class in some clusters, which makes the local discriminant analysis difficult or even impossible. To address this problem, we propose a novel clustering algorithm called ICBKM to ensure that the sample numbers for the classes in a cluster are balanced. Second, we search for local optimal features in each cluster by following the global Fisher Criterion in which we maximize the ratio of the cluster weighted inter- and intra-class scatters. In the following subsections, we will introduce the ICBKM and the global Fisher Criterion in detail, respectively.

A. Intra-Cluster Balanced K-Means Clustering (ICBKM)

K-Means clustering algorithm aims at putting more similar samples in the same cluster. It is unsupervised, thus, cannot guarantee that there are balanced samples in a cluster as desired. Compared to the traditional clustering algorithms, the clustering problem we concern here may have the following characteristics: 1) the class label for each sample is available, thus, the clustering process can be conducted in a supervised manner; 2) its purpose is not only to put the similar samples in the same cluster, but also to ensure that the samples for the classes in each cluster are balanced such that the local discriminant analysis can be conducted within each cluster. Cheung et al. [8] proposed a variation K-Means approach called cluster balance K-Means (CBKM), in which the concept *cluster balance* was proposed. However, CBKM only ensures that the sample number in each cluster is balanced and does not take into account the class label information and does not require the class balance. To provide a solution to this special clustering problem, we propose a novel supervised clustering approach, namely ICBKM here. ICBKM satisfies the requirement that there are balanced samples for classes in each cluster by adding an extra regularization term to constrain the sample number variation for the classes in each cluster. The cluster variance and the class variation constraints for the samples in each cluster collaborate to derive the clustering result with both cluster compactness and class balance.

Formally, the objective function of ICBKM is represented as

$$\arg \min_{K_i \in 1,2,\cdots,K} \sum_{i}^{N} \frac{|x_i - \bar{x}^{K_i}|^2}{\delta^2} + \alpha \sum_{k=1}^{K} |N^k - \bar{N}|^2 + \beta \sum_{k=1}^{K} \sum_{c=1}^{c_k} |N_c^k - \bar{N}^k|^2$$
$$\text{subject to}: c_k \geq 2 \quad (k=1,2,\cdots,K), \tag{1}$$

where K_i is the cluster index for the sample x_i; $\bar{x}^k$ is the average of the samples in cluster k; N^k is the sample number in cluster k; $\bar{N}$ is the average sample number for each cluster; N_c^k is the sample number of the c-th class in cluster k; $\bar{N}^k$ is the average sample number for each class in cluster k; c_k is the class number in cluster k; δ is the standard deviation of the sample data; and α and β are the weighting coefficients for the last two terms.

In the objective function of ICBKM, the first term minimizes cluster variance so as to make clusters compact; the second term makes sample numbers of different clusters balanced; and the third term is to ensure that each class has similar number of samples in a given cluster. The objective function is not trivial and we cannot obtain the closed form solution directly. Here, we apply an iterative procedure as traditional K-Means does to optimize the objective function. The pseudo-code is listed in Fig. 1.

ICBKM: Given the class label set $S_i = \{1, 2, \cdots, L\}$, the data set χ, the class label l_i for each sample x_i in χ and the final cluster number K.

1. **Initialization**: Compute the standard deviation δ of the data set χ; randomly select $\bar{x}_1, \bar{x}_2, \cdots, \bar{x}_K$ as the initial cluster centers, then assign each x_i to the cluster whose center is the nearest to x_i. For each cluster with only one class of samples, randomly select another one of different class to this cluster.

2. **Reset Cluster Centers**: For each cluster C^k, reset the center as the average of all the samples assigned to cluster C^k.

3. **Assignment Optimization**: For each $x_i \in \chi$, assign it to the cluster that makes the objective function minimal and the result satisfies the constraint in (1).

4. **Exchange Optimization**: For each cluster, exchange the cluster labels for the sample in C^k that is the farthest to cluster center and the sample of the same class $\notin C^k$ that is the nearest to cluster center. If no improvement or the requirment of multiple classes for each cluster unsatisfied, keep the previous lables.

5. **Evaluation**: If current step has no improvement, return the final clustering results $\{C^1, C^2, \cdots, C^K\}$; else, go step 2.

Fig. 1 Procedure for ICBKM

In this procedure, the first optimization step is called *Assignment Optimization*, which is similar to K-Means. For each sample, we first check whether there exist at least two classes of other samples within the cluster to which this sample belongs. If yes, this sample is reassigned to each of the other clusters, and is finally put to the one resulting in the minimal object function value. We can see for each reassignment, only the terms related to the original cluster and the assigned one are required to be updated, hence the process to find the target cluster for each sample is fast.

Unlike K-Means, an extra step called *exchange optimization* is proposed in ICBKM. It is designed to solve the problem that traditional assignment optimization approach cannot well handle. That is, when a sample is reassigned to another cluster, the in-cluster variance is reduced but the other two terms increase; In the exchange optimization step, the first term is optimized while the last two terms remain constant, thus, the optimization conflict between the first term and last two terms is avoided. In the exchange optimization step, the exchange resulting in the issue of only one class of samples within a cluster will be canceled; and to avoid the heavy computational cost, the exchange pair is constrained to the ones near the cluster margins as denoted in the Fig. 1. These two steps are iteratively to minimize the objective function until a local optimum is obtained.

1) *Discussion*: The proposed ICBKM is based on the assumption that it is possible to have relatively balanced numbers of samples for the classes within a cluster, so it may be improper for certain extreme cases, such as the two-class problem with extremely unbalanced training samples, and the multi-class problem in which the sample number of one class is much larger than that of all other classes. For most applications such as face recognition, the training sample numbers of different classes are relatively balanced; hence ICBKM can be well applied in these applications.

B. Global Discriminant Analysis by Merging Local Analyzers

Taking the advantage of the proposed ICBKM approach, the sample data are separated into multiple clusters with balanced samples for different classes. The traditional way to utilize these clustering results is to conduct discriminant analysis within each cluster, then determine the class label of a new date according to the classification result from its nearest discriminant analyzer. In this way, the local analyzers are independent and the final classification uses only part of the available information. We propose to utilize the global Fisher criterion to combine the local discriminant analyzers into a globally nonlinear discriminant analyzer. The global Fisher criterion maximizes the ratio of the class weighed inter- and intra-cluster scatters. The algorithm has three steps and they are introduced in detail as follows.

1) PCA Projections: In each cluster, PCA is conducted for dimensionality reduction; moreover, like in Fisher-face [2], PCA step can prevent the algorithm from suffering from the singular problem when the sample number is less than the feature number. In our experiments, we retain 98% of the energy in the sense of reconstruction error. Thus, in each cluster, each data $x_i \in \chi$ projected into a low-dimensionality feature space as

$$z^k(x_i) = (W_{pca}^k)^T (x_i - \bar{x}^k), \quad k=1,\cdots,K, \tag{2}$$

where $W_{pca}^k \in \mathbb{R}^{D \times n_k}$ is the n_k leading eigenvectors of the covariance matrix from the sample data belonging to cluster k, and $z^k(x_i) \in \mathbb{R}^{n_k}$. The conditional probability of cluster k for a given data x, $p(C^k|x)$ also simplified as $p^k(x)$, can be obtained using a simple formulation as in [22]

$$p^k(x) = p(C^k \mid x) = \frac{p(x,C^k)}{\sum_{j=1}^{K} p(x,C^j)}, \tag{3}$$

where $p(x,C^k) = \exp\{-a^k(x)\}$ and $a^k(x)$ is the *activity signal* of the data for cluster k. In our experiments, $a^k(x)$ is set as the Mahalanobis Distance [20] of the data in the PCA space of cluster k.

2) Nonlinear Dimensionality Reduction by Optimizing Global Fisher Criterion: LDA algorithm cannot well handle the nonlinear classification problem; and the KDA may suffer from high computational cost in the classification stage when the sample number is too large, it is desirable to propose a novel efficient discriminant analysis algorithm to conduct nonlinear discriminant analysis in consideration of nonlinear manifold structure of the sample data. As previously described, each sample $x_i \in \chi$ can be represented as a low-dimensional vector $z^k(x_i)$ in cluster k. Denote the optimal feature directions for dimensionality reduction within cluster k as $W_f^k \in \mathbb{R}^{n_k \times n}$ and the translations as $W_0^k \times \mathbb{R}^{1 \times n}$ in cluster k, and $W^k = ((W_f^k)^T, (W_0^k)^T)^T \in \mathbb{R}^{(n_k+1) \times n}$, then the optimal low-dimensional representation of x_i can be represented as the weighted sum of the projections from different clusters as

$$\begin{aligned}\Gamma(x_i) &= \sum_{k=1}^{K} p(C^k \mid x_i)(W_f^{kT} W_{pca}^{kT}(x_i - \bar{x}^k) + W_0^k) \\ &= \sum_{k=1}^{K} p^k(x_i)(W_f^{kT} z^k(x_i) + W_0^k) = W^T z(x_i),\end{aligned} \tag{4}$$

where
$$z^T(x_i) = (p^1(x_i)z^1(x_i)^T, p^1(x_i), \cdots, p^K(x_i)z^K(x_i)^T, p^K(x_i)) \in \mathbb{R}^{\sum_{k=1}^{K}(n_k+1)}$$
and matrix $W = ((W^1)^T, (W^2)^T, \cdots, (W^K)^T)^T \in \mathbb{R}^{\sum_{k=1}^{K}(n_k+1)\times n}$. The global intra-class and inter-class scatters can be represented as

$$S_w = \left| \sum_{i=1}^{N} (\Gamma(x_i) - \overline{\Gamma}^{l_i})(\Gamma(x_i) - \overline{\Gamma}^{l_i})^T \right|$$

$$= \left| W^T \sum_{i=1}^{N} (z(x_i) - \bar{z}^{l_i})(z(x_i) - \bar{z}^{l_i})^T W \right|$$

$$= | W^T M_w W |,$$

$$S_b = \left| \sum_{l=1}^{L} N_l (\overline{\Gamma}^l - \overline{\Gamma})(\overline{\Gamma}^l - \overline{\Gamma})^T \right|$$

$$= \left| W^T \sum_{l=1}^{L} N_l (\bar{z}^l - \bar{z})(\bar{z}^l - \bar{z})^T W \right|$$

$$= | W^T M_b W |, \qquad (5)$$

where

$$M_w = \sum_{i=1}^{N} (z(x_i) - \bar{z}^{l_i})(z(x_i) - \bar{z}^{l_i})^T,$$

$$M_b = \sum_{l=1}^{L} N_l (\bar{z}^l - \bar{z})(\bar{z}^l - \bar{z})^T \qquad (6)$$

and $\overline{\Gamma}^l$ is the mean of $\Gamma(x_i)$ where x_i belongs to class l and

$$\overline{\Gamma} = (1/N) \sum_{i=1}^{N} \Gamma(x_i);$$

$\bar{z}^l$ is the mean of $z(x_i)$ where x_i belongs to class l and $\bar{z} = (1/N) \sum_{i=1}^{N} z(x_i)$. The global Fisher Criterion is to maximize the cluster weighted inter-class scatter while minimize the cluster weighted intra-class scatter with respect to W, i.e.,

$$W^* = \arg \max_{W} \frac{|W^T M_b W|}{|W^T M_w W|}. \qquad (7)$$

It has closed form solution and can be directly computed out by solving the generalized eigenvalue decomposition problem [9] as

$$M_b w_i = \lambda_i M_w w_i, \quad W = [w_1, w_2, \cdots, w_n]. \qquad (8)$$

3) Nonlinear Dimensionality Reduction for Classification: For a new data x, the posterior probabilities for each cluster can be computed according to (3) and its low-dimensional representation is obtained via the following nonlinear mapping function in term of the derived local optimal discriminant features for each cluster

$$M(x) = \sum_{k=1}^{K} p(C^k \mid x)(W_f^{*kT} W_{pca}^{kT}(x - \bar{x}^k) + W_0^{*k}). \tag{9}$$

It is an explicit nonlinear mapping function from the original data space to the low-dimensional space. The consequent classification can be conducted based on these low-dimensional representations by using the traditional approaches like NN or Nearest Feature Line (NFL). In all our experiments, we used the NN method for final classification for simplicity.

III. Kernel Justification and Discussions

The two approaches described above are integrated into an algorithm for conducting **di**scriminant **a**nalysis on **em**bedded ma**ni**fold (Daemon). That is, Daemon consists of two steps: 1) separate the sample set into a set of class balanced clusters; and 2) merge the local discriminant analyzers into a single global nonlinear discriminant analyzer by following the global Fisher Criterion. It supervises the local analyzers and automatically decides the responsibility of each analyzer, which is somewhat like the background procedure named *Daemon* in UNIX system, thus, called *Daemon*. As described above, the intuition of Daemon is to merge the local discriminant analyzers into a unified coordinate framework. In this section, we analyze Daemon from a different point of view and justify that Daemon is a specific KDA algorithm, in which the kernel is data dependent and geometry adaptive, unlike traditional kernel machines that are independent to the data set to be analyzed. Then, we will discuss the relationship between Daemon and Mixture LDA as well as LLE variant [25].

A. Kernel Justification

Daemon follows the global Fisher Criterion in the learning stage and essentially is a discriminant analysis algorithm [1]; yet, as shown in Section II, it is nonlinear and is adaptive to the geometry structure of the data set. As shown in (4), Daemon can be considered a process in which the training sample, denoted as x, is mapped into another data space, denoted as $z(x)$, then, LDA are conducted on this new feature space by considering $z(x)$ as the new object to be analzed. Therefore, Daemon can be considered a special KDA algorithm with the kernel defined as

$$k(x,y)=\phi(x) \cdot \phi(y), \qquad (10)$$

where

$$\phi(x)=z(x)=(p^1(x)z^1(x)^T, p^1(x), \cdots, p^K(x)z^K(x)^T, p^K(x))^T$$

as in (4). This kernel has the following characteristics: 1) it has explicit mapping function from the input space to another feature space as the polynomial kernel does; and 2) it is dependent on the training samples and adaptive to the geometry structure of the sample data, which directly leads to its superiority over traditional kernels that are fixed as independent to the data set. Moreover, it is a special Marginalized kernel as introduced below.

Marginalized Kernel [27] defines a kernel between two visible variables x, x' with an extra hidden variable $h \in \mathscr{H}$, where $\mathscr{H}$ is a finite set. Let $\xi = (x, h)$, the marginalized kernel is derived by taking the expectation with respect to the hidden variable as

$$K(x,x') = \sum_{h \in \mathscr{H}} \sum_{h' \in \mathscr{H}} p(h \mid x) p(h' \mid x') K_\xi(\xi, \xi') \qquad (11)$$

in which the joint kernel $K_\xi(\xi, \xi')$ is designed between two combined variables ξ and ξ'.

In Daemon, the kernel $k(x,y) = \phi(x) \cdot \phi(y)$ can be rewritten as

$$K(x,x') = \phi(x) \cdot \phi(x') = \sum_k p^k(x) p^k(x')(x^k \cdot x'^k + 1)$$

$$= \sum_k p(C^k \mid x) p(C^k \mid x')((x - \bar{x}'^k)^T W_{pca}^k W_{pca}^{kT}(x' - \bar{x}'^k) + 1).$$

Thus, it can be considered as a special marginalized kernel with $h = k, \mathscr{H} = \{1, 2, \cdots, K\}$ and the joint kernels are defined as

$$K_\xi(\xi, \xi') = (x - \bar{x}'^h)^T W_{pca}^h W_{pca}^{hT}(x' - \bar{x}'^h) + 1. \qquad (12)$$

B. Discussions

In the following, we discuss the connections between Daemon and other algorithms, including Mixture LDA, LLE, and its variant [25]. Then, we analyze the computational complexity of Daemon.

1) Relationship With Mixture LDA: A nature extension of LDA to handle nonlinear classification problem is to train specific LDA within each cluster, and we call this strategy as Mixture LDA here. Similar to Daemon, Mixture LDA algorithm also separates the sample data into multiple clusters and conducts local discriminant analysis. However, Daemon is different from Mixture LDA in both learning and inferring steps.

1) In the training stage, for Mixture LDA, the local discriminant features are learned directly from the samples within the corresponding cluster by optimizing the *local* Fisher Criterion; hence, these local discriminant analyzers are independent. For Daemon, the local discriminant features for different clusters are calculated simultaneously by optimizing the *global* Fisher Criterion. Daemon aims at aligning the local discriminant features with a single global coordinate framework.

2) In the inferring stage, Mixture LDA classifies a new data only by the nearest local discriminant analyzer thus, can only use part of the available information for classification. In Daemon, the low-dimensional representation of a new data is the weighted sum of the projections from all the local discriminant analyzers and the classification is based on all the sample data. Fig. 2 demonstrates the difference between Daemon and Mixture LDA, and notice that, for Daemon, the final most similar sample of a new data may not exist in the nearest cluster.

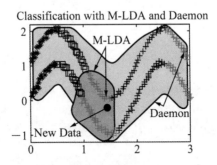

Fig. 2 Classification strategies compared between Mixture-LDA and Daemon. Note that 1) for Mixture-LDA, the small semi-trasnparent area means the sample area that will be used for classification when the new data in figure comes, and we can see only part of the training samples will be used for final classification; and 2) for Daemon, the large semi-transparent area means the sample area that will be used for classification when the new data in figure comes, and we can see that all the samples are used and the samples most similar to the new data may not be in the nearest cluster

2) *Relationship With LLE and its Variant*: LLE [21] is a manifold learning algorithm that maps the input data to a lower dimensional feature space by preserving the relationship among the neighboring points. First, the sparse local reconstruction coefficient matrix M is computed, such that $\sum_{j \in N_k(i)} M_{ij} = 1$ where the set $N_k(i)$ is the index set of the k nearest neighbors of the sample x_i and $\sum_{j \in N_k(i)} \| x_i - M_{ij} x_j \|^2$ is minimized; then, the low-dimensional representation y_i for sample x_i is obtained by minimizing $\sum_i \sum_{j \in N_k(i)} \| y_i - M_{ij} y_j \|^2$.

Roweis [25] proposed a procedure to align disparate local linear presentations into a global coherent coordinate system by preserving the relationship between neighboring points as LLE does. It has some similar ideas as Daemon, but they are intrinsically different in two-folds: 1) Daemon is supervised, but LLE variant [25] is unsupervised, and Daemon presents a novel clustering algorithm to derive desirable clustering results for supervised learning; 2) the merging strategy in LLE variant is similar to that in Daemon, whereas Daemon derives and understands it from the kernel design perspective which presents an insight between the kernel machine and manifold learning. With this in-depth understanding, the strategy to globally merge local analyzers can be used as a general kernel design procedure for other kernel-based algorithms and we are planning to explore this extension in our future work.

3) *Computational Complexity*: In the training stage of Daemon, the ICBKM is required for clustering and, hence, it is often computationally more expensive than traditional KDA. In real application, namely testing stage, Daemon has the superiority over traditional KDA in terms of computational efficinecy. Denote the time complexity for a multiplication operator of two real values as T_*, and the time complexity for an exponential operator is T_e. The complexity to compute a feature for Daemon is $O(2KDT_* + KT_e)$, while for Gaussian Kernel-based KDA, it is $O(NDT_* + NT_e)$, and for polynomial kernel, it is $O(NDT_* + N_p T_*)$ where p is the exponent. We can see that the complexity for the testing of Daemon is independent to the sample number and it is often the case that $K \ll N$. Therefore, it is much more computationally effi-

cient compared with KDA when the sample number is large.

IV. Experiments

In this section, we present three sets of experiments with both artificial data and real-world data to evaluate the effectiveness of our proposed Daemon algorithm. The first toy problem on the artificial data demonstrates the effectiveness of Daemon in deriving discriminative feature for nonlinear classification problem; the face recognition results on YALE and CMU PIE databases show that Daemon significantly outperforms Fisherface [2] and has better accuracy than traditional KDA [19] as well as Mixture LDA. Another toy problem explores the property of the geometry-adaptive-kernel when applied for Kernel PCA (KPCA) [30] algorithm.

A. Toy Problem

The objective of this experiment is two folds. One is to explore the performance of the supervised algorithm for clustering, namely ICBKM, in deriving the clustering results with balanced samples for each class within a cluster. Another is to explicitly introduce the characteristics of the local discriminant analyzers in deriving satisfying projection directions.

As shown in the top-left image of Fig. 3, the original data is composed of two classes of samples and they cannot be separated linearly. Formally, these data are synthesized according to the following distribution:

$$\begin{cases} x_i^k = 0.03 * i + \tilde{\delta}, \\ y_i^k = \sin(\pi x_i) + k + \tilde{\delta}, \end{cases} \quad k = 0, 1, \tilde{\delta} \sim N(0, 0.1). \tag{13}$$

We have systematically compared the clustering results of three K-Means-like algorithms. The original K-Means algorithm produced clustering result that aims at obtaining least sums of intra-cluster variances. As shown in the up-right image of Fig. 3, the sample numbers for the classes in a cluster is not balanced and some clusters have only one class of samples. It makes the consequent local discriminant analysis impossible in these clusters. The cluster-balanced K-Means algorithm produced similar result as that of the original K-Means algorithm; yet, the sample numbers for different clusters are balanced.

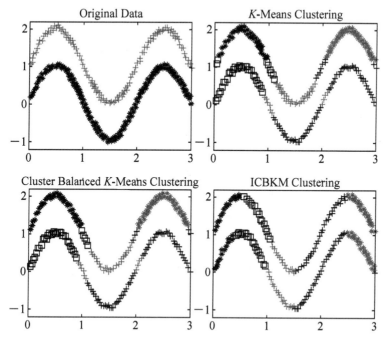

Fig. 3 Toy problem on clustering on the synthesized data ($\alpha=0.15, \beta=0.1$). Note that top-left is the plot of the original samples of two classes, top-right is the plot of the clustering result of K-means, bottom-left is the plot of the clustering result of CBKM and bottom-right is the plot of the clustering result of our proposed ICBKM

As shown in the bottom-right plot of Fig. 3, the clustering result from our proposed ICBKM algorithm has the following properties: 1) the sample numbers for the classes in a cluster are balanced, which is helpful for the local discriminant anylsis in each cluster; and 2) the two classes of samples in each cluster are linearly separable. It is obvious that the proposed ICBKM algorithm has produced more useful clustering result than the other two methods and intra-cluster balanced clustering result presents proper structure representation for the consequent analysis. We have also calculated the optimal local discriminant feature in each cluster to optimize the global Fisher Criterion, and the computed local feature direction in each cluster is illustrated in Fig. 4. It shows that the local feature direction is approximately optimal for the samples in each cluster in the senses of classification capability and they are merged into a global coordinate such that the samples of the same class are close to each other and samples of different classes are far to each other.

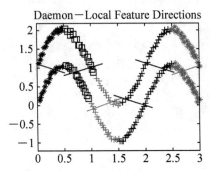

Fig. 4 Derived local feature directions using Daemon. We can see that the plotted directions are approximately optimal for local classification

B. Face Recognition Experiments

The YALE [28] and CMU PIE [24] face databases were used to evaluate the effectiveness of our proposed Daemon algorithm for face recognition problem. In both experiments, each face image is normalized by fixing the positions of two eyes and scaling to size of 64 * 64 pixels. Yale face database was constructed by the Yale Center for Computational Vision and Control and it contains 165 grayscale images of 15 individuals. For each individual, six faces are used for training, and the other five are used for testing. Fig. 5 plots the 11 images of one person in the YALE database. In all our experiments, the reduced dimension number for face recognition using Fisher-face is $L-1$ as in [2]. The kernel used for traditional Kernel-DA algorithm is Gaussian Kernel $K(x,y)=\exp(-\|x-y\|^2/2\eta^2)$, and for a fair comparison, we tested 21 different parameters as $\eta_i=2^{(i-11)/2}\delta$ where δ is the standard variance of the sample data. Moreover, in this experiment, we compare Daemon with Mixture LDA that learns different LDA models for different clusters derived from ICBKM, and we also compare Daemon with K-Means-Daemon algorithm which is similar to Daemon by replacing the ICBKM algorithm with the K-Means for clustering. For Daemon, we set $\alpha=0.15, \beta=0.1$ as in the toy problem; the cluster number K is explored between 2-7 and the best results is reported. Table I illustrates the face recognition results of Fisher-face, KDA, Mixture LDA, K-Means-Daemon, and Daemon. It shows that Daemon significant outperforms LDA, Mixture LDA and *K*-Means-Daemon, and also has better results than traditional KDA with Gaussian Kernel.

Fig. 5   Eleven images of one person in the YALE database and the images are aligned by fixing the positions of the two eyes

Table I   Comparison between Fisher-Face, Kernel-Da, Mixture LDA, *K*-Means-Daemon, and Daemon on Yale

Algorithm	Fisher-face	Kernel-DA	Mixture-LDA	*K*-Means-Daemon	Daemon
Accuracy	80%	85.3%	82.7%	82.7% (K=2)	88% (K=2)

We also conducted the face recognition experiments on the PIE database. Some example images of one person in PIE database are plotted in Fig. 6. The face images of pose 02, 37, 05, 27, 29, and 11 in the illumination directory are used in our experiments and referred to as PIE-1 subdatabase. The results over ten random splits with equal numbers of training and testing samples are averaged and the results listed in Table II show that Daemon outperforms the other two algorithms in the PIE-1 subdatabase as in YALE database. It also demonstrates that Daemon has strong capability to handle nonlinear classification problems and can improve the accuracy in the general classification problems compared with Fisher-face and Kernel-DA. Moreover, we also conducted a relative simple experiment on the frontal images with only 15 types of different illumination conditions in the illumination directory of the PIE database, referred to as PIE-2 subdatabase. The results in Table III show that our proposed Daemon algorithm performs best and Fig. 7 plots the comparative face recognition accuracies on different feature dimensions. Note that the directions from 68—80th are selected as the eigenvalues with larger nonzero eigenvalues produced by computational error.

Fig. 6   Some sample images of one person in the CMU PIE database and the images are aligned by fixing the positions of the two eyes

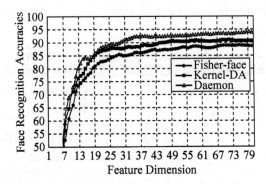

Fig. 7 Face recognition accuracy versus feature dimension compared between Fisher-face, Kernel-DA, and Daemon on PIE-2 subdatabase

Table Ⅱ Comparison between Fisher-Face, Kernel-DA, and Daemon on PIE-1 Database

Algorithm	Fisher-face	Kernel-DA	Daemon($K=5$)
Accuracy	63.6%	68.4%	71.1%

Table Ⅲ Comparison between Fisher-Face, Kernel-DA, and Daemon on PIE-2 Subdatabase

Algorithm	Fisher-face	Kernel-DA	Daemon($K=5$)
Accuracy	88.5%	90.7%	93.2%

C. Property of Geometry-Adaptive-Kernel

We have proved that Daemon is essentially a special KDA algorithm with geometry-adaptive-kernel, which is dynamic and adaptive with the data distribution and label information. In this section, we explore the property of this kernel when applied to KPCA for nonlinear PCA.

KPCA has the potential to capture nonlinear structure of the data set. Hence, we compare the linear PCA with the Kernel-PCA using four different kernels: Gaussian, polynomial, inverse multiquadric kernel and our proposed geometry-adaptive-kernel. The data set are sampled from a structure integrating the symbol "V" and a horizontal line, i.e, with both nonlinear and linear structures. Fig. 8 demonstrates the distributions of the data projections to the first principal component of KPCA and the lines are the contours of the projection values. It is evident that KPCA algorithm with geometry-adaptive-kernel best captures the intrinsic nonlinear structure of the data set. PCA cannot capture the nonlinear structure of the data as the linear property.

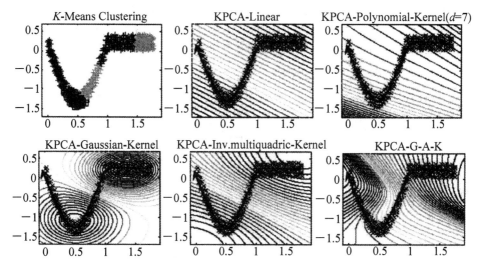

Fig. 8 Toy problem to illustrate the capability of KPCA with geometry-adaptive-kernel to capture the principal curve of the nonlinearly distributed data. From left to right and top to bottom are the contour line image of the projection to the first principal component direction of PCA and KPCA with different types of kernels. Note that contour value at one point is computed by projecting its coordinates to the first principal direction. The curve with arrows arrow means the desired principal curve direction; and for a good result, the contour line should be orthogonal to the curve with arrow and the projection value should change monotonously along the direction of the curve with arrow. Therefore, KPCA with geometry-adaptive-kernel is the best

V. Discussions and Future Work

We have presented a novel algorithm called Daemon for general nonlinear classification problem. Daemon is a nonlinear discriminant analysis algorithm that effectively utilizes the underlying nonlinear manifold structure. In this work, the discrete sample data on a manifold are clustered using the ICBKM algorithm such that the sample numbers for the classes within each cluster are balanced; and then the local optimal discriminant features are simultaneously derived by optimizing the global Fisher Criterion. Deamon can be justified from kernel-machine perspective and is a KDA algorithm with geometry-adaptive-kernel, a special marginalized kernel tailed to the nonlinear manifold structure of the sample data.

To the best of our knowledge, this is the first work to conduct nonlinear discriminant analysis while explicitly considering the embedded geometry structure of the data set. In this work, we have only utilized the basic property of manifold that a manifold can be covered by a series of open sets; how to combine the other topology properties of a manifold with discriminant analysis for general classification problem is the future direction of our work. Moreover, in the clustering task, the initialization issue and how to select the optimal cluster number are still open problems, and we also plan to explore these problems in our future work.

References

[1] G. Baudat and F. Anouar, "Generalized discriminant analysis using a kernel approach," *Neural Computal.*, vol. 12, no. 10, pp. 2385—2404, 2000.

[2] P. Belhumeur, J. Hespanha, and D. Kriegman, "Eigenfaces versus Fisher-faces: Recognition using class specific linear projection," *IEEE Trans. Pattern Anal. Mach. Intell.*, vol. 19, no. 7, pp. 711—720, Jul. 1997.

[3] M. Belkin and P. Niyogi, "Laplacian eigenmaps and spectral techniques for embedding and clustering," in *Proc. Adv. Neural Inf. Process. Syst.* Vancouver, BC, Canada, 2001, vol. 15, pp. 585—591.

[4] S. Billings and K. Lee, "Nonlinear Fisher discriminant analysis using a minimum squared error cost function and the orthogonal least squares algorithm," *Neural Netw.*, vol. 15, no. 2, pp. 263—270, 2002.

[5] Y. Bengio, J. Paiement, and P. Vincent, "Out-of-sample extensions for LLE, Isomap, MDS, eigenmaps, and spectral clustering," in *Adv. Neural Inf. Process. Syst.*, 2003, vol. 15.

[6] C. Bregler and S. M. Omohundro, "Nonlinear manifold learning for visual speech recognition," in *Proc. 5th Int. Conf. Comp. Vis.*, Jun. 1995, pp. 494—499.

[7] M. Brand, "Charting a manifold," in *Proc. Adv. Neural Inf. Process. Syst.*, 2002, vol. 15, pp. 977—984.

[8] D. Cheung, S. Lee, and Y. Xiao, "Effect of data skewness and workload balance in parallel data mining," *IEEE Trans. Knowl. Data Eng.*, vol. 14, no. 3, pp. 498—513, May 2002.

[9] F. Chung, "Spectral graph theory," in *Proc. Regional Conf. Series in Math.*, 1997, vol. 92.

[10] D. Freedman, "Efficient simplicial reconstructions of manifolds from their samples," *IEEE Trans. Pattern Anal. Mach. Intell.*, vol. 24, no. 10, pp. 1349—1357, Oct. 2002.

[11] K. Fukunaga, *Introduction to Statistical Pattern Recognition*, 2nd ed. New York: Academic, 1991.

[12] J. Gomes and A. Mojsilovic, "A variational approach to recovering a manifold from sample points," in *Proc. Eur. Conf. Comput. Vis.*, Copenhagen, May 2002, pp. 3—17.

[13] X. He and P. Niyogi, "Locality Preserving Projections (LPP)," TR-2002-09, 2002.

[14] G. Hinton and S. Roweis, "Stochastic neighbor embedding," in *Proc. Adv. Neural Inf. Process. Syst.*, 2002, vol. 15. pp. 857—864.

[15] T. Kim and J. Kittler, "Locally linear discriminant analysis for multimodally distributed classes for face recognition with a single model image," *IEEE Trans. Pattern Anal. Mach. Intell.*, vol. 27, no. 3, pp. 318—327, Mar. 2005.

[16] Q. Liu, R. Huang, H. Lu, and S. Ma, "Face recognition using kernel based Fisher discriminant analysis," in *Proc. 5th IEEE Int. Conf. Autom. Face Gesture Recognit.*, May 2002, p. 197.

[17] J. Lu, K. Plataniotis, and A. Venetsanopoulos, "Face recognition using kernel direct discriminant analysis algorithms," *IEEE Trans. Neural Netw.*, vol. 14, no. 1. pp. 117—126, 2003.

[18] J. MacQueen, "On convergence of k-means and partitions with minimum average variance," *Ann. Math. Statist.*, vol. 36. p. 1084 (abstract), 1965.

[19] S. Mika, G. Ratsch, J. Weston, B. Scholkopf, and K. Muller, "Fisher discriminant analysis with kernels," in *Proc. IEEE Int. Workshop Neural Netw. Signal Process. IX*, Aug. 1999, pp. 41—48.

[20] B. Moghaddam and A. Pentland, "Probabilistic visual learning for object representation," *IEEE Trans. Pattern Anal. Mach. Intell.*, vol. 19, pp. 696—710, 1997.

[21] S. Roweis and L. K. Saul, "Nonlinear dimensionality reduction by locally linear embedding," *Science*, vol. 290, pp. 2268—2269, Dec. 2000.

[22] S. Roweis, L. Saul, and G. Hinton, "Global coordination of local linear models," in *Proc. Adv. Neural Inf. Process. Syst.*, 2001, vol. 14. pp. 889—896.

[23] J. Shi and J. Malik, "Normalized cust and image segmentation," *IEEE Trans. Pattern Anal. Mach. Intell.*, vol. 22, pp. 888—905, 2000.

[24] T. Sim, S. Baker, and M. Bsat, "The CMU pose, illumination, and expression (PIE) database," in *Proc. IEEE Int. Conf. Autom. Face Gesture Recognit.*, May 2002, pp. 53—56.

[25] Y. Teh and S. Roweis, "Automatic alignment of hidden representations," in *Proc. Adv. Neural Inf. Process. Syst.*, 2002, vol. 15, pp. 841—848.

[26] J. Tenenbaum, V. de Silva, and J. C. Langford, "A global geomertic framework for non-

linear dimensionality reduction," *Science*, vol. 290, pp. 2319—2323, Dec. 2000.

[27] K. Tsuda, T. Kin, and K. Asai, "Marginal kernels for biological sequences," *Proc. BIOINFORMATICS*, vol. 1, no. 1, pp. 1—8, 2002.

[28] *Yale Univ. Face Database*. 2002 [Online]. Available: http://cvc.yale.edu/projects/yalefaces/yalefaces.html.

[29] J. Yang, A. Frangi, J. Yang, D. Zhang, and Z. Jin, "KPCA plus LDA: A complete kernel Fisher discriminant framework for feature extraction and recognition," *IEEE Trans. Pattern Anal. Mach. Intell.*, vol. 27, no. 2, pp. 230—244, Feb. 2005.

[30] M. Yang, "Kernel eigenfaces versus kernel Fisher-faces: Face recognition using kernel methods," in *Proc. 5th IEEE Int. Conf. Autom. Face Gesture Recognit.*, May 2002, pp. 215—220.

原文载于 IEEE Transactions on Circuits and Systems for Video Technology, Vol. 17, No. 4, April 2007, 468—477.

谱估计与分析

谱估计中的最佳高分辨时窗函数

程乾生 谢衷洁
北京大学数学力学系

  无论对随机过程还是对时间序列,截取一段进行谱估计,都要产生时窗函数问题[7].用不同方法可给出不同时窗函数.给出这些时窗函数并没有统一的标准,但是为了比较这些时窗函数的好坏,往往要分析这些时窗函数的频谱,看其频谱主瓣与旁瓣的关系如何,为了提高谱估计的分辨率,我们希望主瓣所占的比重尽可能大,这样就给出了一个标准,就可求最佳高分辨时窗函数[1].这个问题是和信号的不定原理紧密联系在一起的[1,3,6].但是,[1]讨论的是离散情况,[3]讨论的是连续情况,二者之间的关系并不清楚(见文献[1]).本文的目的是:(1) 讨论最佳高分辨离散时窗函数与最佳高分辨连续时窗函数的关系,(2) 对以下问题作一说明:最佳高分辨时窗函数近似算法,具有非负频谱的最佳高分辨时窗函数的求法,加权最佳高分辨时窗函数及某些时窗函数的性质.

一、最佳高分辨连续时窗函数与最佳高分辨离散时窗函数的关系

  设连续时窗函数 $h(t)$ 为实的,并且

$$h(t)=\begin{cases}h(t), & |t|\leqslant T,\text{\textcircled{1}}\\ 0, & |t|>T,\end{cases} \qquad (1)$$
$$\int_{-T}^{T}h^2(t)dt<+\infty,$$

相应的频谱为

$$H(f)=\int_{-T}^{T}h(t)e^{-i2\pi ft}dt. \qquad (2)$$

---

① 这里说的连续时窗函数 $h(t)$,是指自变量 t 连续变化.

考虑能量比
$$Q(h) = \frac{\int_{-\tau}^{\tau} |H(f)|^2 df}{\int_{-\infty}^{\infty} |H(f)|^2 df}. \tag{3}$$

所谓最佳高分辨连续时窗函数 $h_0(t)$，是指 $h_0(t)$ 满足条件(1)，并且有
$$Q(h_0) = \sup Q(h), \tag{4}$$
其中 $h(t)$ 要满足(1)。

在(1)、(3)中出现了 T 和 τ，其中 $2T$ 为时窗长度，2τ 为低频宽度，T 和 τ 都是事先给定的。

把(1)中的 $h(t)$ 离散化。设 N 为正整数($N>1$)，记
$$\left. \begin{array}{l} \Delta_{N-1} = \dfrac{2T}{N-1}, \ \Delta_N = \dfrac{2T}{N}, \ t_n = -T + n\Delta_n, \\ \hat{t}_n = -T + n\Delta_{N-1}, \quad n = 0, 1, \cdots, N-1. \end{array} \right\} \tag{5}$$

于是得到离散时窗函数 $h(\hat{t}_n)$ 和它的频谱 $H_{\Delta_N}(f)$：
$$h(\hat{t}_n), \quad n = 0, 1, \cdots, N-1. \tag{6}$$
$$H_{\Delta_N}(f) = \Delta_N \sum_{n=0}^{N-1} h(\hat{t}_n) e^{-i2\pi i n f}. \tag{7}$$

考虑能量比
$$Q_N(h) = \frac{\int_{-\tau}^{\tau} |H_{\Delta_N}(f)|^2 df}{\dfrac{\Delta_{N-1}}{\Delta_N} \int_{-1/2\Delta_{N-1}}^{1/2\Delta_{N-1}} |H_{\Delta_N}(f)|^2 df}. \tag{8}$$

我们注意，$|H_{\Delta_N}(f)|$ 是以 $1/\Delta_{N-1}$ 为周期的函数，且 $|H_{\Delta_N}(f)| = |H_{\Delta_N}(-f)|$。

所谓最佳高分辨离散时窗函数 $h_1(\hat{t}_n)$，是指 $h_1(\hat{t}_n)$ 在(5)、(6)下有
$$Q_N(h_1) = \sup Q_N(h), \tag{9}$$
其中 h 表示 $h(\hat{t}_n)$，满足(6)。

由积分方程理论知，最佳高分辨连续时窗函数(以下简称最佳连续时窗) $h_0(t)$ 是某一积分方程的最大特征函数(即对应最大特征值的特征函数)，$\sup Q(h)$ 是该积分方程的最大特征值[3,4]①。由普通的求多元函数极值的方法就可以知道，最佳离散时窗 $h_1(t_n)$ 是某一代数方程的最大特征向量，$\sup Q_N(h)$ 是该代数方程的最大特征值[1]。

现在我们要讨论 $\sup Q(h)$ 与 $\sup Q_N(h)$ 的关系，$h_0(t)$ 与 $h_1(t_n)$ 的关系，为

---

① 本文所说的特征值与[4]中的特征值互为倒数，这样做是为了和代数方程特征值的称呼相一致。

此,首先我们把以上有关公式改写成另一形式.

令
$$P(f)=1. \tag{10}$$
$$K(\mu,\nu)=\int_{-\tau}^{\tau}P(f)e^{i2\pi(\mu-\nu)f}df=\int_{-\tau}^{\tau}P(f)\cos2\pi(\mu-\nu)fdf. \tag{11}$$

设 L^2 表示 $[-T,T]$ 上所有实的平方可积函数组成的空间. 若 $h,g\in L^2$, 则内积为
$$(h,g)=\int_{-T}^{T}h(t)g(t)dt, \tag{12}$$

h 的模 $\|h\|$ 为
$$\|h\|=\left[\int_{-T}^{T}h^2(t)dt\right]^{1/2}. \tag{13}$$

若 $g\in L^2$, 则 Kg 定义为
$$Kg=\int_{-T}^{T}K(t,s)g(s)ds. \tag{14}$$

在以上记号下, (3)式的分子可写为
$$\int_{-\tau}^{\tau}|H(f)|^2df=\int_{-\tau}^{\tau}P(f)\left(\int_{-T}^{T}h(\mu)e^{-i2\pi\mu f}d\mu\int_{-T}^{T}h(\nu)e^{i2\pi\nu f}d\nu\right)df$$
$$=\int_{-T}^{T}\int_{-T}^{T}h(\mu)h(\nu)\left(\int_{-\tau}^{\tau}P(f)e^{i2\pi(\mu-\nu)}df\right)d\mu d\nu$$
$$=(Kh,h).$$

按照能量等式, (3)式的分母为
$$\int_{-\infty}^{+\infty}|H(f)|^2df=\int_{-T}^{T}h^2(t)dt=\|h\|^2.$$

因此, (3)式可写为
$$Q(h)=\frac{(Kh,h)}{\|h\|^2}. \tag{15}$$

由此可知,
$$\sup Q(h)=\sup_{\|h\|=1}(Kh,h). \tag{16}$$

根据积分方程理论[4], $\sup Q(h)$ 是积分方程
$$Kh=\lambda h$$

即
$$\int_{-T}^{T}K(t,s)h(s)ds=\lambda h(t) \tag{17}$$

的最大特征值 λ_0, 而最佳连续时窗 $h_0(t)$ 是积分方程(17)的最大特征函数.

现在把离散问题转换成另一形式.

令

$$\chi_n(t) = \begin{cases} 1, & t \in I_n, \\ 0, & \text{其他}, \end{cases} \quad n=0,1,\cdots,N-1. \tag{18}$$

其中

$$I_{N-1} = [t_{N-1}, t_N], \quad I_n = [t_n, t_{n+1}), \quad n=0,1,\cdots,N-2.^{①} \tag{19}$$

$$\tilde{h}(t) = \sum_{n=0}^{N-1} h(\hat{t}_n)\chi_n(t), \tag{20}$$

$$\widetilde{K}(\mu,\nu) = \sum_{m=0}^{N-1}\sum_{n=0}^{N-1} K(\hat{t}_m,\hat{t}_n)\chi_m(\mu)\chi_n(\nu). \tag{21}$$

在以上记号下,(8)式分子为

$$\begin{aligned}
\int_{-\tau}^{\tau} |H_{\Delta_N}(f)|^2 df &= \int_{-\tau}^{\tau} P(f)|H_{\Delta_N}(f)|^2 df \\
&= \sum_{m=0}^{N-1}\sum_{n=0}^{N-1} h(\hat{t}_m)h(\hat{t}_n)\int_{-\tau}^{\tau} P(f)e^{i2\pi(\hat{t}_m-\hat{t}_n)f}df\Delta_N\Delta_N \\
&= \sum_{m=0}^{N-1}\sum_{n=0}^{N-1} h(\hat{t}_m)h(\hat{t}_n)K(\hat{t}_m,\hat{t}_n)\Delta_N\Delta_N \\
&= \int_{-T}^{T}\int_{-T}^{T} \tilde{h}(\mu)\tilde{h}(\nu)\widetilde{K}(\mu,\nu)d\mu d\nu = (\widetilde{K}\tilde{h},\tilde{h}).
\end{aligned} \tag{22}$$

(8)式分母为

$$\frac{\Delta_{N-1}}{\Delta_N}\int_{-1/2\Delta_{N-1}}^{1/2\Delta_{N-1}} |H_{\Delta_N}(f)|^2 df = \sum_{n=0}^{N-1} h^2(\hat{t}_n)\Delta_N = \int_{-T}^{T} \tilde{h}^2(t)dt = \|\tilde{h}\|^2. \tag{23}$$

因此,(8)式可写为

$$Q_N(h) = \frac{(\widetilde{K}\tilde{h},\tilde{h})}{\|\tilde{h}\|^2}. \tag{24}$$

由积分方程理论[4]知道,积分方程

$$\widetilde{K}h = \tilde{\lambda}h$$

即

$$\int_{-T}^{T} \widetilde{K}(t,s)h(s)ds = \tilde{\lambda}h(t) \tag{25}$$

的最大特征值 $\tilde{\lambda}_0 = \sup_{\|h\|=1}(\widetilde{K}h,h)$ 对应的最大特征函数要满足(25)式,而 $\widetilde{K}(t,s)$ 由(21)确定,因此(25)左边为形如(20)的阶梯函数,这说明 $\widetilde{K}$ 的最大特征函数

---

① 我们指出, $\hat{t}_n \in I_n, 0 \leqslant n \leqslant N-1$. 具体有: $\hat{t}_0 = t_0, \hat{t}_{N-1} = t_N$, 而当 $0 < n < N-1$ 时, $t_n < \hat{t}_n < t_{n+1}$.

为形如(20)的阶梯函数,我们记为 $\tilde{h}_0(t)$. 这也说明,
$$\sup_{\|h\|=1}(\widetilde{K}h,h)=\sup Q_N(h), \tag{26}$$
其中 $\sup Q_N(h)$ 的意义见(8),(9).

我们指出,任一个离散时窗函数 $h(\hat{t}_n)$(见(6))和阶梯函数 $\tilde{h}(t)$(见(20))是一一对应的. 因此,我们也可以称 $\tilde{h}_0(t)$ 为最佳阶梯时窗,它与最佳离散时窗是一一对应的.

下面,给出最佳连续时窗与最佳离散时窗的三个定理.

定理 1 对连续时窗函数 $h(t)$(见(1)),$Q(h)$ 由(3)确定,对离散时窗函数 $h(\hat{t}_n)$(见(6)),$Q_N(h)$ 由(8)确定,则
$$|\sup Q(h)-\sup Q_N(h)|\leqslant \varepsilon_N=\sqrt{\frac{14}{3}}\pi\int_{-\tau}^{\tau}P(f)|f|df\frac{(2T)^2}{N}. \tag{27}$$

证明 设 $h(t)\in L^2, \|h\|=1$.

我们先估计 $|Kh-\widetilde{K}h|=|(K-\widetilde{K})h|$.
$$|(K-\widetilde{K})h|^2=\left|\int_{-T}^{T}(K(t,s)-\widetilde{K}(t,s))h(s)ds\right|^2$$
$$\leqslant \int_{-T}^{T}|K(t,s)-\widetilde{K}(t,s)|^2ds\int_{-T}^{T}h^2(s)ds$$
$$=\int_{-T}^{T}|K(t,s)-\widetilde{K}(t,s)|^2ds.$$

所以
$$\|(K-\widetilde{K})h\|^2=\int_{-T}^{T}|(K-\widetilde{K})h|^2dt\leqslant\int_{-T}^{T}\int_{-T}^{T}|K(t,s)-\widetilde{K}(t,s)|^2dsdt. \tag{28}$$

把(28)右边的二重积分分解为许多小区域上的积分,则有
$$\int_{-T}^{T}\int_{-T}^{T}|K(t,s)-\widetilde{K}(t,s)|^2dsdt=\sum_{m=0}^{N-1}\sum_{n=0}^{N-1}\int_{I_n}ds\int_{I_m}dt|K(t,s)-\widetilde{K}(t,s)|^2. \tag{29}$$

当 $(t,s)\in I_m\times I_n$ 时,由(21)和(11)知
$$|K(t,s)-\widetilde{K}(t,s)|=|K(t,s)-K(\hat{t}_m,\hat{t}_n)|$$
$$=\left|\int_{-\tau}^{\tau}P(f)[\cos 2\pi(t-s)f-\cos 2\pi(\hat{t}_m-\hat{t}_n)f]df\right|$$
$$=\left|\int_{-\tau}^{\tau}P(f)2\pi f\sin 2\pi\xi f df((t-s)-(\hat{t}_m-\hat{t}_n))\right|$$
(其中 ξ 在 $t-s$ 与 $\hat{t}_m-\hat{t}$ 之间)
$$\leqslant 2\pi\int_{-\tau}^{\tau}P(f)|f|df\cdot|(t-\hat{t}_m)-(s-\hat{t}_n)|. \tag{29}'$$

令
$$\beta = 2\pi \int_{-\tau}^{\tau} P(f)|f|df, \tag{30}$$

则
$$\int_{I_n} ds \int_{I_m} dt |K(t,s) - \widetilde{K}(t,s)|^2$$
$$\leq \beta^2 \int_{t_n}^{t_{n+1}} ds \int_{t_m}^{t_{m+1}} dt |(t-\hat{t}_m) - (s-\hat{t}_n)|^2$$
$$= \beta^2 \int_{t_n-\hat{t}_n}^{t_{n+1}-\hat{t}_n} dy \int_{t_m-\hat{t}_m}^{t_{m+1}-\hat{t}_m} dx (x-y)^2$$
$$\leq \frac{7}{6}\beta^2 \Delta_N^4. ① \tag{31}$$

由(28)、(29)、(31)得
$$\|(K-\widetilde{K})h\|^2 \leq \frac{7}{6}N^2\beta^2\Delta_N^4,$$

即
$$\|(K-\widetilde{K})h\| \leq \varepsilon_N, \tag{32}$$

其中
$$\varepsilon_N = \sqrt{\frac{14}{3}}\pi \int_{-\tau}^{\tau} P(f)|f|df \frac{(2T)^2}{N}. \tag{33}$$

现在来证明(27).

对于 $\|h\|=1$ 有
$$|(Kh,h) - (\widetilde{K}h,h)| = |((K-\widetilde{K}h),h)| \leq \|(K-\widetilde{K})h\| \|h\| \leq \varepsilon_N, \tag{34}$$

即
$$(\widetilde{K}h,h) - \varepsilon_N \leq (Kh,h) \leq (\widetilde{K}h,h) + \varepsilon_N.$$

由上可得
$$\sup_{\|h\|=1}(\widetilde{K}h,h) - \varepsilon_N \leq \sup_{\|h\|=1}(Kh,h) \leq \sup_{\|h\|=1}(\widetilde{K}h,h) + \varepsilon_N. \tag{35}$$

由(16)、(26)、(35)即可得(27).定理1证毕.

现说明一下定理1的意义. $\sup Q(h)$ 是积分方程(17)的最大特征值，$\sup Q_N(h)$ 是一线性代数方程的最大特征值[1].定理1给出了两者之间的误差估计.由定理1自然得到：当 $N \to +\infty$ 时，$\sup Q_N(h) \to \sup Q(h)$.这就解决了文献[1]中所不能回答的问题.

下面我们进一步给出最佳连续时窗函数 $h_0(t)$ 和最佳离散时窗函数 $h_1(t_n)$

---

① 此不等式可通过直接计算积分然后再由(5)式得到.

的关系.

定理 2 设 $h_0(t)$ 是使 $Q(h)$ (见(3)) 达到极大值的最佳连续时窗函数, 且满足 $\|h_0(t)\|=1$, $h_1(\hat{t}_n)$ 是使 $Q_N(h)$ (见(8)) 达到极大值的最佳离散时窗函数, $\tilde{h}_1(t)$ 是由 $h_1(\hat{t}_0)$ 构造的一个阶梯函数(见(20)), 且要求 $h_1(\hat{t}_n)$ 使 $\|\tilde{h}_1\|=1$, $(h_0,\tilde{h}_1) \geqslant 0$, 则

$$\int_{-T}^{T} |h_0(t)-\tilde{h}_1(t)|^2 dt \leqslant \left(\frac{2\varepsilon_N}{\lambda_0-\lambda_1}\right)^2 + \frac{2\varepsilon_N}{\lambda_0-\lambda_1}, \tag{36}$$

其中 ε_N 由(27)确定, λ_0, λ_1 分别是积分方程(17)的最大和第二大特征值.

证明 由前面讨论知, $h_0(t), \tilde{h}_1(t)$ 分别满足

$$Kh_0 = \lambda_0 h_0, \quad \widetilde{K}\tilde{h}_1 = \tilde{\lambda}_0 \tilde{h}_1, \tag{37}$$

其中 $\lambda_0, \tilde{\lambda}_0$ 分别是方程(17)和(25)的最大特征值. 由(16)和(26)知, $\lambda_0 = \sup Q(h)$, $\tilde{\lambda}_0 = \sup Q_N(h)$. 根据定理1的(27)式有

$$|\lambda_0 - \tilde{\lambda}_0| \leqslant \varepsilon_N. \tag{38}$$

由(37)可知

$$(Kh_0, h_0) = \lambda_0, \quad (\widetilde{K}\tilde{h}_1, \tilde{h}_1) = \tilde{\lambda}_0. \tag{39}$$

由(34)知

$$|(K\tilde{h}_1, \tilde{h}_1) - \tilde{\lambda}_0| = |(K\tilde{h}_1, \tilde{h}_1) - (\widetilde{K}\tilde{h}_1, \tilde{h}_1)| \leqslant \varepsilon_N. \tag{40}$$

再由(38),(40)可得

$$|(K\tilde{h}_1, \tilde{h}_1) - \lambda_0| \leqslant |(K\tilde{h}_1, \tilde{h}_1) - \tilde{\lambda}_0| + |\tilde{\lambda}_0 - \lambda_0| \leqslant 2\varepsilon_N. \tag{41}$$

因此可把 $(K\tilde{h}_1, \tilde{h}_1)$ 写为

$$(K\tilde{h}_1, \tilde{h}_1) = \lambda_0 - \delta, \tag{42}$$

其中

$$\delta \leqslant 2\varepsilon_N. \tag{43}$$

设 $\lambda_j (j=0,1,2,\cdots)$ 是积分方程(17)所有的特征值, 依次有 $\lambda_0 > \lambda_1 > \lambda_2 \cdots$[3], $h_j(t)$ 是对应 λ_j 的特征函数, 要求 $\|h_j\|=1$. 对任何 $h(t) \in L^2$, 只要 $\|h\| \neq 0$, 则 $h(t)$ 的频谱 $H(f)$ 几乎处处不为 0[6], 因此按照(14)式后面的一个式子知

$$(Kh,h) = \int_{-\tau}^{\tau} |H(f)|^2 df > 0.$$

由希尔伯脱-施密特定理[4]就可知 $h_j(t), j=0,1,2,\cdots$, 是 L^2 中完备标准正交系.

因此, $\tilde{h}_1(t)$ 可表为

$$\tilde{h}_1(t) = \alpha_0 h_0(t) + \alpha_1 h_1(t) + \alpha_2 h_2(t) + \cdots. \tag{44}$$

由于 $\|\tilde{h}\|=1$, 所以

$$\sum_{j=0}^{+\infty} \alpha_j^2 = 1. \tag{45}$$

由(44)得

$$K\tilde{h}_1 = \sum_{j=0}^{+\infty} \alpha_j K\lambda_j = \sum_{j=0}^{+\infty} \alpha_j \lambda_j h_j.$$

由上式和(44)有

$$(K\tilde{h}_1, \tilde{h}_1) = \sum_{j=0}^{+\infty} \lambda_j \alpha_j^2.$$

再由(42)就得到

$$\sum_{j=0}^{+\infty} \lambda_j \alpha_j^2 = \lambda_0 - \delta. \tag{46}$$

因为 $\lambda_1 > \lambda_2 > \cdots$,所以由(45)和(46)可得

$$\lambda_0 \alpha_0^2 + \lambda_1 (1 - \alpha_0^2) = \lambda_0 \alpha_0^2 + \lambda_1 \sum_{j=1}^{+\infty} \alpha_j^2 \geqslant \lambda_0 \alpha_0^2 + \sum_{j=1}^{+\infty} \lambda_j \alpha_j^2 = \lambda_0 - \delta.$$

由上式可得

$$0 \leqslant 1 - \alpha_0^2 \leqslant \frac{\delta}{\lambda_0 - \lambda_1}. \tag{47}$$

当 N 充分大时 δ 充分小,因此 α_0^2 就接近1,我们总要求 $\alpha_0 = (\tilde{h}_1, h_0) > 0$,否则用 $-\tilde{h}_1(t)$ 代替 $\tilde{h}_1(t)$ 就可达到这个要求. 在这条件下有

$$1 - \alpha_0 \leqslant (1 - \alpha_0)(1 + \alpha_0) \leqslant 1 - \alpha_0^2 \leqslant \frac{\delta}{\lambda_0 - \lambda_1}. \tag{48}$$

由(44),(47),(48),我们有

$$\begin{aligned}
\|h_0 - \tilde{h}_1\|^2 &= \|(1-\alpha_0)h_0 - \alpha_1 h_1 - \alpha_2 h_2 - \cdots\|^2 \\
&= (1-\alpha_0)^2 + \alpha_1^2 + \alpha_2^2 + \cdots \\
&= (1-\alpha_0)^2 + (1-\alpha_0^2) \\
&\leqslant \frac{\delta^2}{(\lambda_0 - \lambda_1)^2} + \frac{\delta}{(\lambda_0 - \lambda_1)}.
\end{aligned}$$

再由(43)知,定理 2 的(36)式成立,定理 2 证毕.

利用定理 1 和定理 2 还可对最佳连续时窗函数 $h_0(t)$ 和最佳阶梯时窗函数 $\tilde{h}_1(t)$ 的误差 $|h_0(t) - \tilde{h}_1(t)|$ 进行估计. 见下面定理 3.

定理 3 设 $h_0(t), h_1(\hat{t}_n), \tilde{h}_1(t)$ 满足定理 2 的条件,则

$$|h_0(t) - \tilde{h}_1(t)| \leqslant \frac{1}{\lambda_0} \varepsilon_N', \quad t \in [-T, T], \tag{49}$$

特别地有

$$|h_0(t_n)-h_1(\hat{t}_n)|\leqslant \frac{1}{\lambda_0}\varepsilon'_N, \quad n=0,1,\cdots,N-1, \tag{50}$$

其中

$$\varepsilon'_N=\sqrt{2T}\int_{-\tau}^{\tau}P(f)df\left(\frac{\varepsilon_N}{\lambda_0-\varepsilon_N}+\sqrt{\left(\frac{2\varepsilon_N}{\lambda_0-\lambda_1}\right)^2+\frac{2\varepsilon_N}{\lambda_0-\lambda_1}}\right)$$
$$+\sqrt{2T}4\pi\int_{-\tau}^{\tau}P(f)|f|df\frac{2T}{N}, \tag{51}$$

上式中 $\varepsilon_N,\lambda_0,\lambda_1$ 的意义见定理 2.

证明 由(37)知,$\lambda_0 h_0=Kh_0,\tilde{\lambda}_0\tilde{h}_1=\widetilde{K}\tilde{h}_1$,因此

$$\lambda_0 h_0-\tilde{\lambda}_0\tilde{h}_1=Kh_0-\widetilde{K}\tilde{h}_1=K(h_0-\tilde{h}_1)+(K-\widetilde{K})\tilde{h}_1. \tag{52}$$

由(11)知,$|K(t,s)|$ 的最大值 M 为

$$M=\int_{-\tau}^{\tau}P(f)df, \tag{53}$$

所以

$$|K(h_0-\tilde{h}_1)|=\left|\int_{-T}^{T}K(t,s)[h_0(s)-\tilde{h}_1(s)]ds\right|$$
$$\leqslant M\int_{-T}^{T}|h_0(s)-\tilde{h}_1(s)|ds$$
$$\leqslant M\sqrt{2T}\|h_0-\tilde{h}_1\|. \tag{54}$$

由(29)′知,当 $(t,s)\in I_m\times I_n$ 时,

$$|K(t,s)-\widetilde{K}(t,s)|\leqslant 2\pi\int_{-\tau}^{\tau}P(f)|f|df(|t-t_m|+|s-t_n|)$$
$$\leqslant 2\pi\int_{-\tau}^{\tau}P(f)|f|df\cdot 2\Delta_N. \tag{55}$$

记

$$\rho_N=4\pi\int_{-\tau}^{\tau}P(f)|f|df\Delta_N. \tag{56}$$

由(55)知

$$|(K-\widetilde{K})\tilde{h}_1|=\left|\int_{-T}^{T}[K(t,s)-\widetilde{K}(t,s)]\tilde{h}_1(s)ds\right|$$
$$\leqslant \rho_N\int_{-T}^{T}|\tilde{h}_1(s)|ds\leqslant \rho_N\sqrt{2T}. \tag{57}$$

根据(54),(57)和(52),可知

$$|\lambda_0 h_0-\tilde{\lambda}_0\tilde{h}_1|\leqslant \sqrt{2T}M\|h_0-\tilde{h}_1\|+\sqrt{2T}\rho_N. \tag{58}$$

由(37)和(53)知,

$$|\tilde{\tilde{h}}_1| = \frac{1}{\tilde{\lambda}_0}|\widetilde{K}\tilde{h}_1| = \frac{1}{\tilde{\lambda}_0}\left|\int_{-T}^{T}\widetilde{K}(t,s)\tilde{h}_1(s)ds\right|$$

$$\leqslant \frac{1}{\tilde{\lambda}_0}M\int_{-T}^{T}|\tilde{h}_1(s)|ds \leqslant \frac{1}{\tilde{\lambda}_0}M\sqrt{2T},$$

又由(38), $\frac{1}{\tilde{\lambda}_0} \leqslant \frac{1}{\lambda_0 - \varepsilon_N}$, 所以有

$$|\tilde{\tilde{h}}_1| \leqslant \frac{M\sqrt{2T}}{\lambda_0 - \varepsilon_N}. \tag{59}$$

根据(58),(38),(59)可以得到

$$|\lambda_0 h_0 - \lambda_0 \tilde{\tilde{h}}_1| \leqslant |\lambda_0 h_0 - \tilde{\lambda}_0 \tilde{\tilde{h}}_1| + |\tilde{\lambda}_0 \tilde{\tilde{h}}_1 - \lambda_0 \tilde{\tilde{h}}_1|$$

$$\leqslant \sqrt{2T}M\|h_0 - \tilde{h}_1\| + \sqrt{2T}\rho_N + \varepsilon_N \frac{\sqrt{2T}M}{\lambda_0 - \varepsilon_N},$$

再由(53),(36),(56)就得到(49),(51)式. 定理3证毕.

在定理3中出现的 ε'_N, 当 $N \to +\infty$ 时 $\varepsilon'_N \to 0$. 由(49)知,当 $N \to +\infty$ 时, $\tilde{h}_1(t)$(它是依赖 N 的最佳阶梯时窗函数)一致收敛到 $h_0(t)$.

二、关于最佳高分辨时窗函数的某些说明

1. 最佳高分辨时窗函数的近似算法

当 N 较大时,要求最佳离散时窗函数就必须计算一个高阶线性代数方程的最大特征根和最大特征向量[1]. 无论如何,这个计算量是比较大的. 定理1到定理3,在理论上告诉我们,当 N 较大时,最佳离散时窗函数与最佳连续时窗函数很接近. 因此,只要求出最佳连续时窗函数 $h_0(t)$,对较大的 N,用 $h_0(t_n)$ 作为最佳离散时窗函数 $h_1(t_n)$ 的近似就行了.

最佳连续时窗函数 $h_0(t)$ 是积分方程(17)的最大特征函数, λ_0 为最大特征值, $h_0(t)$ 满足关系

$$\lambda_0 h_0(t) = \int_{-T}^{T} K(t,s)h(s)ds. \tag{60}$$

我们知道,函数系

$$\frac{1}{\sqrt{2T}}, \quad \frac{1}{\sqrt{T}}\sin 2\pi \frac{1}{2T}t, \quad \frac{1}{\sqrt{T}}\cos 2\pi \frac{1}{2T}t,$$

$$\frac{1}{\sqrt{T}}\sin 2\pi \frac{2}{2T}t, \quad \frac{1}{\sqrt{T}}\cos 2\pi \frac{2}{2T}t, \cdots$$

是 $[-T, T]$ 上的完备标准正交系. $h_0(t)$ 是偶函数[3],因此 $h_0(t)$ 可表为

$$h_0(t) = \sum_{j=1}^{+\infty} \alpha_j \psi_j(t), \tag{61}$$

其中

$$\psi_0(t) = \frac{1}{\sqrt{2T}}, \quad \psi_1(t) = \frac{1}{\sqrt{T}} \cos 2\pi \frac{j}{2T} t, \quad j \geqslant 1. \tag{62}$$

现在我们要求 $h_0(t)$ 的近似函数 $\hat{h}_0(t)$

$$\hat{h}_0(t) = \sum_{j=0}^{m} \beta_j \psi_j(t), \tag{63}$$

选取 $\beta_0, \beta_1, \cdots, \beta_m$ 使下式

$$q_m(\hat{h}_0) = \frac{(K\hat{h}_0, \hat{h}_0)}{(\hat{h}_0, \hat{h}_0)} \tag{64}$$

达极大值 $q_m = \sup\limits_{\|\hat{h}_0\|=1} (K\hat{h}_0, \hat{h}_0)$.

令

$$A_{ij} = (K\psi_i, \psi_j) = \int_{-T}^{T} \int_{-T}^{T} K(t,s) \psi_i(t) \psi_j(s) dt ds, \tag{65}$$

则(64)可写为

$$q_m(\hat{h}_0) = \frac{\sum\limits_{i=0}^{m} \sum\limits_{j=0}^{m} A_{ij} \beta_i \beta_j}{\sum\limits_{j=0}^{m} \beta_j^2}. \tag{66}$$

用多元函数求极值的方法可知,我们要求的 $\beta_0, \beta_1, \cdots, \beta_m$ 是矩阵

$$[A_{ij}]_{(m+1)\times(m+1)} \tag{67}$$

的最大特征向量.

这样,求近似最佳连续时窗函数 $\hat{h}_0(t)$ 的问题和求最佳离散时窗函数 $h_1(t_n)$ 一样,最后变成一个求矩阵最大特征向量的问题,但是,关键在于矩阵阶数的大小. 求 $h_1(t_n)$ 时,若离散数据有 N 个,就要求 $N\times N$ 阶矩阵的最大特征根. 一般,$N\geqslant 100$,因此计算量是相当大的. 而求 $\hat{h}_0(t)$(见(63),(62))时,常常可取 $m=2,3,4$. 当取 $m=1$ 时,时窗函数

$$\hat{h}_0(t) = \beta_0 \psi_0(t) + \beta_1 \psi_1(t) = \frac{\beta_0}{\sqrt{2T}} + \frac{\beta_1}{\sqrt{T}} \cos 2\pi \frac{t}{2T} = \hat{\beta}_0 + \hat{\beta}_1 \cos 2\pi \frac{t}{2T}.$$

这是 Blackman-Tukey 类型时窗. 现在我们从高分辨角度给出了确定 $\hat{\beta}_0, \hat{\beta}_1$ 的方法.

和上面讨论的求近似最佳连续时窗方法一样,也可求近似最佳离散时窗[2],但这里要求对每个 N 都要求一次 $(m+1)\times(m+1)$ 阶矩阵的最大特征向

量,这就给应用带来不方便之处,如 $N=512$ 和 $N=1024$ 时,就要分别计算两次.

最后我们说明 A_{ij} (65)的计算. 对 $\psi_j(t)$ (62),首先可计算

$$\Phi_j(f)=\int_{-T}^{T}\psi_j(t)e^{-i2\pi ft}dt$$

的解析表达式. 根据(11)式,(65)可表示为

$$A_{ij}=\int_{-\tau}^{\tau}P(f)\Phi_j(f)\Phi_i(f)df=2\int_{0}^{\tau}P(f)\Phi_i(f)\Phi_j(f)df. \quad (68)$$

在上式中被积函数是已知的,因此可选一种计算量少、精度高的积分近似算法计算出 A_{ij}.

在[9]中,Kaiser 用第一类 0 阶贝塞尔函数给出了近似最佳连续时窗.

2. 具有非负频谱的最佳高分辨时窗函数的求法

利用相关函数计算功率谱,如果时窗函数 $h(t)$(见(1))的频谱 $H(f)$(见(2))具有负值,在计算功率谱时就有可能出现负值. 然而,功率谱总是非负的. 因此,往往要求时窗函数的频谱 $H(f)$ 具有非负值. 在这个条件下,要求最佳高分辨时窗达到极大面积比就行了. 具体问题如下:

已知 $h(t)$ 满足条件(1),并且

$$H(f)=\int_{-T}^{T}h(t)e^{-i2\pi ft}dt\geqslant 0. \quad (69)$$

考虑面积比

$$R(h)=\frac{\int_{-\tau}^{\tau}H(f)df}{\int_{-\infty}^{+\infty}H(f)df}. \quad (70)$$

求最佳高分辨时窗函数 $h_0(t)$,它满足(1),(69),并且满足

$$R(h_0)=\sup R(h). \quad (71)$$

如何求 $h_0(t)$?

满足(1),(69)的 $h(t)$,皆存在一个函数 $g(t)^{[6]}$,它满足:

$$g(t)=\begin{cases}g(t), & |t|\leqslant\dfrac{T}{2},\\ 0, & \text{其他}.\end{cases} \quad (72)$$

$$G(f)=\int_{-T/2}^{T/2}g(t)e^{-i2\pi ft}dt, \quad (73)$$

使

$$h(t)=\int_{-T/2}^{T/2}g(t+s)g(s)ds=\begin{cases}\int_{-T/2}^{T/2-t}g(t+s)g(s)ds, & 0\leqslant t\leqslant\dfrac{T}{2},\\ h(-t), & -\dfrac{T}{2}\leqslant t\leqslant 0,\end{cases}$$

(74)

$$H(f)=|G(f)|^2.\tag{75}$$

因此,(70)就可写为

$$Q(g)=\frac{\int_{-\tau}^{\tau}|G(f)|^2 df}{\int_{-\infty}^{+\infty}|G(f)|^2 df}.\tag{76}$$

这样,问题就化为在条件(72)下求 $g_0(t)$ 使(76)达极大值. 这就是在第一节中讨论的问题. 关于它的计算在第二节 1 中也讨论过了,我们只需指出,在求出形如(63)的近似解 $\hat{g}_0(t)$ (这时要把(62)中 $\psi_j(t)$ 中出现的 T 换成 $\dfrac{T}{2}$)之后,通过(74)可以直接求出 $h_0(t)$ 近似解 $\hat{h}_0(t)$ 的解析表达式.

3. 加权最佳高分辨时窗函数

在第一节中,最佳高分辨时窗函数的问题是:在(1),(2)条件下,求 $h_0(t)$ 使(3)式达极大值.(3)可写为

$$Q(h)=\frac{\int_{-\tau}^{\tau}P(f)|H(f)|^2 df}{\int_{-\infty}^{+\infty}|H(f)|^2 df},\tag{77}$$

其中 $P(f)=1$(见(10)).

在(77)中,$P(f)$ 是加权函数. 从提高分辨率角度考虑,如果要求时窗函数的频谱 $H(f)$ 不仅主瓣能量大、旁瓣能量小,而且要求 $H(f)$ 在 $(-\tau,\tau)$ 内呈尖锐状,那么就要求权函数 $P(f)$ 满足:在 $(-\tau,\tau)$ 内 $P(f)$ 为正实偶函数,在 $(0,\tau)$ 内 $P(f)$ 为下降函数. 例如,可取 $P(f)$ 为

$$\left.\begin{aligned}P_1(f)&=1-\frac{1}{\tau}|f|,\\ P_1(f)&=1-\frac{1}{\tau^2}f^2,\\ P_1(f)&=\cos\frac{\pi}{2\tau}f,\\ P_1(f)&=\left(\cos\frac{\pi}{2\tau}f\right)^2,\end{aligned}\right\}\tag{78}$$

或者取 $P(f)$ 为
$$P(f)=aP_1(f)+b, \tag{79}$$
其中 $P_1(f)$ 为(78)中的一种，$a>0, b\geqslant 0$.

在条件(1),(2)之下，求 $h_0(t)$ 使(77)式达极大值(在(77)式中，不要求 $P(f)=1$)，则称 $h_0(t)$ 为加权最佳高分辨时窗函数. $h_0(t)$ 的近似求法，与第二节 1 中的讨论相同.

4. 在不同标准下的最佳时窗函数

对时窗函数 $h(t)$（见(1)），在不同标准下，可求出不同的最佳时窗函数，下面举三个例子说明.

1) 在条件(1),(2)之下，求 $h_0(t)$ 使
$$Q_1(h)=\frac{|H(0)|^2}{\int_{-\infty}^{+\infty}|H(f)|^2 df} \tag{80}$$
达极大值.

由于 $H(0)=\int_{-T}^{T}h(t)dt, \int_{-\infty}^{+\infty}|H(f)|^2 df=\int_{-T}^{T}h^2(t)dt$，因此(80)可写为
$$Q_1(h)=\frac{\left(\int_{-T}^{T}h(t)dt\right)^2}{\int_{-T}^{T}h^2(t)dt}.$$

根据许瓦兹不等式，
$$Q_1(h)\leqslant \int_{-T}^{T}dt=2T.$$
等式只有在
$$h(t)=1, \quad |t|\leqslant T \tag{81}$$
时才成立. 这说明，在标准(80)之下，最佳时窗函数为截尾时窗函数(81).

2) 在条件(1),(2)之下，求 $h_0(t)$ 使
$$Q_2(h)=\frac{\left|\int_{-\tau}^{\tau}H(f)df\right|^2}{\int_{-\infty}^{+\infty}|H(f)|^2 df} \tag{82}$$
达极大.

由于
$$\left|\int_{-\tau}^{\tau}H(f)df\right|^2=\left|\int_{-\tau}^{\tau}df\int_{-T}^{T}dt\, h(t)e^{-i2\pi tf}\right|^2$$

$$= \left(\int_{-T}^{T} h(t) \frac{\sin 2\pi\tau t}{\pi t} dt\right)^2$$

$$\leq \int_{-T}^{T} \frac{\sin^2 2\pi\tau t}{\pi^2 t^2} dt \int_{-T}^{T} h^2(t) dt,$$

等式只有在

$$h(t) = \frac{\sin 2\pi\tau t}{\pi t}, \quad |t| \leq T \tag{83}$$

时才成立,这说明,在标准(82)之下,最佳时窗函数为(83).(83)式为截尾 Daniell 类型时窗[5].

3) 在条件(1),(2)和条件

$$\begin{cases} h(0)=1, \quad h(t) \text{为实偶连续函数}, \\ H(f) \geq 0 \end{cases} \tag{84}$$

之下,求 $h_0(t)$ 使

$$Q_3(h) = \int_{-\infty}^{+\infty} f^2 H(f) df \tag{85}$$

达极小.

这个问题的最佳时窗函数也可找出解析表达式[8].

上面我们讨论了在三个标准下的三个最佳时窗函数. 标准(80)只对 $H(f)$ 在一个特殊点 $f=0$ 的值提出要求,而没有考虑一个局部(区间)的性质,标准(80)实际上是面积平方与能量之比,这是两个不同量之间的比. 标准(85),没有考虑区间 $(-\tau,\tau)$,因此在应用中,不如标准(3)的目的性强,综上所述,对提高分辨率而言,标准(3)是一个较好的标准.

参考文献

[1] A. Eberhard, An optimal discrete window for the calculation of power spectra, IEEE. AU-21:1(1973),37—43.

[2] H. Babic and G. C. Temes, Optimum low-order window for discrete fourier trausform systems, 1976, IEEE, ASSP-24:6(1976),512—517.

[3] D. Slepian, H . O. Pollack, H. T. Landow, Prolate Spheroidal wave functions, Fourier analysis and Uncertainty Priciple I and II, *Bell, System Tech. J.*, 40:1(1961),43—84.

[4] С. Г. 米赫林著、陈传璋、卢鹤绂译,积分方程及其应用,商务印书馆,北京,1958 年.

[5] E. J. Haunan, Time series analysis. Methuen&Co Ltd. London,1960.

[6] A. Papoulis, The fourier integrel and Its Applications, McGraw-Hill Book Company,1962.

[7] Robere K. Otnes and Loren Enochson, Digital Time Series Analysis, *John Wiley and Sons*, 1972.

[8] A. Papoulis, Minimum-Bias Windows for High-Resolution Spectral Estimates, *IEEE*, IT-19:1(1973), 9—12.

[9] J. F. Kaiser, Digital Filters, in System Analysis by Digital Computer, F, F, Kuo and J. F. Kaiser, Eds., New York: Wiley, 1966.

原文载于《应用数学学报》,第2卷,第2期,1979年5月,119—131.

关于多维平稳序列奇异性和 WOLD 分解的谱表示

程 乾 生
北京大学数学系

对于多维平稳序列的正则性,文献[2,6,7]已做了讨论. 对于多维平稳序列的奇异性,文献[4]就某些特殊情况进行了讨论. 对于多维平稳序列 WOLD 分解的谱表示,文献[8,9]就几种特殊情况进行了讨论. 江泽培教授(见文献[1])对多维平稳序列的正交分解、奇异性和 WOLD 分解的谱表示,进行了系统的全面的研究,给出了奇异性的充要条件、WOLD 分解的谱表示和正交分解的结果. 正如江泽培教授所指出的(见文献[1] pp.271、278,或参看文献[4]),一般的平稳序列的预测问题,总可化为谱函数为绝对连续的平稳序列的预测问题. 因此,文献[1]所论述的是多维平稳过程的一般理论.

本文的目的,在于直接利用同构空间的方法,给出与文献[1]不同的多维平稳序列奇异性的充要条件和 WOLD 分解的谱表示.

设 $x_t=(x_1(t),\cdots,x_n(t)),t=0,\pm 1,\cdots$ 是一个 n 维平稳序列. x_t 的谱函数矩阵为 $F(\lambda)=[F_{ij}(\lambda)]_{n\times n}$,谱密度矩阵为 $f_\lambda=[f_\lambda^{ij}]=[f_{ij}(\lambda)]_{n\times n}$(参见文献[1]或[2]). 用记号 H_x 表示 $x_k(t)(k=1,\cdots,n;t=0,\pm 1,\cdots)$ 所组成的希尔伯特空间. 在 H_x 上可确定一个映射到自身的酉算子 U,使

$$U_{x_k}(t)=x_k(t+1) \quad (k=1,\cdots,n;t=0,\pm 1,\cdots)$$

(见文献[3]的定理 1).

正如文献[1] §2 所指出的,本文总假定平稳序列的谱函数是绝对连续的. 当 λ 固定时,可以把 f_λ 看成是 n 维酉空间 A 上的非负定的埃尔米特(Hermite)算子. 它在 $B_\lambda=f_\lambda A$ 上有逆算子 f_λ^{-1},这个逆算子仍然是非负定的埃尔米特算子.

设 $\mathfrak{U}$ 是由 n 维向量 $\boldsymbol{a}(\lambda)=(a_1(\lambda),\cdots,a_n(\lambda))$ 组成的空间,其中 $a_k(\lambda)(1\leqslant k\leqslant n)$ 皆是 $[-\pi,\pi]$ 上的可测函数,并且满足

$$\int_{-\pi}^{\pi}(f_\lambda \boldsymbol{a}(\lambda),\boldsymbol{a}(\lambda))d\lambda<+\infty.$$

在 $\mathfrak{U}$ 中定义内积

$$\langle a(\lambda), a'(h)\rangle = \int_{-\pi}^{\pi} (f_\lambda a(\lambda), a'(\lambda)) d\lambda,$$

其中 $a(\lambda), a'(\lambda)$ 皆属于 $\mathfrak{U}$. 于是 $\mathfrak{U}$ 成为希尔伯特空间.

在文献[2]中证明了: H_x 与 $\mathfrak{U}$ 同构, 并且可以有下面的对应关系:

$$x_k(t) \leftrightarrow e^{i\lambda t} e_k \quad (k=1,\cdots,n; t=0,\pm 1,\cdots), \tag{1}$$

其中

$$e_k = (\delta_k^1, \cdots, \delta_k^n), \quad \delta_k^j = \begin{cases} 1, & \text{当 } j=k \text{ 时}, \\ 0, & \text{当 } j\neq k \text{ 时}. \end{cases}$$

以后, 我们称满足关系(1)的 H_x 中元素与 $\mathfrak{U}$ 中元素的同构对应关系为对应关系(1). 相应于 H_x 上的酉算子 U, 在对应关系(1)下, 有 $\mathfrak{U}$ 上的酉算子 U' 使

$$U'a(\lambda) = e^{i\lambda} a(\lambda)$$

(见文献[2]第二章).

用记号 $H_x(t)$ 表示 $x_k(t)(k=1,\cdots,n)$ 所张成的子空间, 记号 $H_x^-(t)$ 表示 $x_k(s)(k=1,\cdots,n; s<t)$ 所张成的子空间. 记 $S_x = \bigcap_t H_x^-(t)$. 如果 $S_x = H_x$, 则称平稳序列 x_t 是奇异的. 如果 $S_x = \varnothing$, 则称平稳序列 x_t 是正则的.

下面我们要用到一些解析函数的知识. 关于 H_δ 类函数的概念和性质见文献[5](这里 $\delta > 0$). 两个 H_δ 类函数之比(要求分母不恒为 0)称为 N_δ 类函数. 因为 H_δ 类函数可表示为两个 B 类函数(即单位圆 $|z|<1$ 内有界解析函数; 有时为了方便, 也称它为 H_∞ 类函数)之比, 所以 N_δ 类函数也可表示为两个 H_∞ 类函数之比. 因而 N_δ 类 $= N_\infty$ 类. 这说明 N_δ 类和指示 δ 无关, 并且不难知, N_δ 类函数对于普通的加、减、乘、除运算是封闭的(即经过运算后仍然为 N_δ 类函数).

定理 1' 平稳序列 x_t 为非奇异的充要条件是存在一个 n 维向量

$$\boldsymbol{b}(\lambda) = (b_1(\lambda), \cdots, b_n(\lambda))$$

使

$$(f_\lambda^{-1} \boldsymbol{b}(\lambda), \boldsymbol{b}(\lambda)) \in L_1(-\pi, \pi), \tag{2}$$

其中,

$$\left.\begin{array}{l} \text{i) } \boldsymbol{b}(\lambda) = (b_1(\lambda), \cdots, b_n(\lambda)) \in B_\lambda; \\ \text{ii) } b_j(\lambda) (1 \leqslant j \leqslant h) \text{ 皆为 } H_1 \text{ 类函数之边值}; \\ \text{iii) } \boldsymbol{b}(\lambda) \text{ 不恒为 } 0. \end{array}\right\} \tag{3}$$

证明 必要性 因为平稳序列 x_t 为非奇异的, 所以 $H_x^-(0) \neq H_x$. 因而 $H_x \ominus H_x^-(0)$ 含有非 0 元素. 设 $\mathfrak{U}$ 的子空间 $\mathfrak{U} \ominus \mathfrak{U}^-(0)$ 在对应关系(1)下与 $H_x \ominus H_x^-(0)$ 相对应. 由于同构关系, 必有非 0 元素 $a(\lambda) \in \mathfrak{U} \ominus \mathfrak{U}^-(0)$, 也即 $a(\lambda)$ 满足

$$\int_{-\pi}^{\pi}(f_\lambda \boldsymbol{a}(h),e^{i\lambda t}\boldsymbol{e}_k)d\lambda = \int_{-\pi}^{\pi}e^{-i\lambda t}(f_\lambda \boldsymbol{a}(\lambda),\boldsymbol{e}_k)d\lambda = 0 \quad (1\leqslant k\leqslant n, t<0), \quad (4)$$

$$0 < \int_{-\pi}^{\pi}(f_\lambda \boldsymbol{a}(\lambda),\boldsymbol{a}(\lambda))d\lambda < +\infty. \tag{5}$$

令 $\boldsymbol{b}(\lambda)=f_\lambda \boldsymbol{a}(\lambda)=(b_1(\lambda),\cdots,b_n(\lambda))$，其中 $b_j(\lambda)=(f_\lambda \boldsymbol{a}(\lambda),\boldsymbol{e}_j)$. 由于(4)，按照文献[5]的第二章5.4定理1，我们知 $b_j(\lambda)(1\leqslant j\leqslant n)$ 皆为 H_1 类函数之边值. 再由(5)知, $\boldsymbol{b}(\lambda)$ 满足(2)且 $\boldsymbol{b}(\lambda)\not\equiv 0$. 于是必要性成立.

充分性 设 $\boldsymbol{b}(\lambda)$ 满足(2),(3). 令 $\boldsymbol{a}(\lambda)=f_\lambda^{-1}\boldsymbol{b}(\lambda)$. 由于 $b_j(\lambda)$ 皆为 H_1 类函数之边值，按照文献[5], $\boldsymbol{a}(\lambda)$ 满足(4). 由(2)得

$$\int_{-\pi}^{\pi}(f_\lambda \boldsymbol{a}(\lambda),\boldsymbol{a}(\lambda))d\lambda = \int_{-\pi}^{\pi}(f_\lambda^{-1}\boldsymbol{b}(\lambda),\boldsymbol{b}(\lambda))d\lambda < +\infty.$$

由(3)ⅲ)可得

$$\int_{-\pi}^{\pi}(f_\lambda \boldsymbol{a}(\lambda),\boldsymbol{a}(\lambda))d\lambda > 0.$$

事实上，假若不然，有

$$\int_{-\pi}^{\pi}(f_\lambda^{1/2}\boldsymbol{a}(\lambda),f_\lambda^{1/2}\boldsymbol{a}(\lambda))d\lambda = \int_{-\pi}^{\pi}(f_\lambda \boldsymbol{a}(\lambda),\boldsymbol{a}(\lambda))d\lambda = 0,$$

则必有 $f_\lambda^{1/2}\boldsymbol{a}(\lambda)=o(p,p)$. 于是 $\boldsymbol{b}(\lambda)=f_\lambda \boldsymbol{a}(\lambda)=f_\lambda^{1/2}(f_\lambda^{1/2}\boldsymbol{a}(\lambda))=o(p,p)$. 这与条件(3)ⅲ)相矛盾. 故 $\boldsymbol{a}(\lambda)$ 满足(5). 这表明有非 0 元素 $\boldsymbol{a}(\lambda)\in\mathfrak{U}\ominus\mathfrak{U}^-(0)$. 按照对应关系(1), $H_x\ominus H_x^-(0)$ 也必含有非 0 元素，这表明平稳序列 x_t 为非奇异的. 故充分性成立. 定理1′证毕.

由定理1′的证明知，空间 $H_x\ominus H_x(0), \mathfrak{U}\ominus\mathfrak{U}^-(0)$ 及 $\mathfrak{R}$ 是相互同构的, $\mathfrak{R}$ 是由 n 维向量 $\boldsymbol{b}(\lambda)=(b_1(\lambda),\cdots,b_n(\lambda))$ 组成的空间，其中 $\boldsymbol{b}(\lambda)\in B_\lambda, b_j(\lambda)$ 皆为 H_1 类函数之边值, $(f_\lambda^{-1}\boldsymbol{b}(\lambda),\boldsymbol{b}(\lambda))\in L_1(-\pi,\pi)$. 在 $\mathfrak{R}$ 中的内积为

$$\{\boldsymbol{b}(\lambda),\boldsymbol{b}'(\lambda)\} = \int_{-\pi}^{\pi}(\boldsymbol{b}(\lambda),f_\lambda^{-1}\boldsymbol{b}'(\lambda))d\lambda.$$

定理1 平稳序列 x_t 为非奇异的充要条件是存在一个 n 维向量

$$\boldsymbol{b}(\lambda)=(b_1(\lambda),\cdots,b_n(\lambda))$$

使

$$(f_\lambda^{-1}\boldsymbol{b}(\lambda),\boldsymbol{b}(\lambda))\in L_1(-\pi,\pi), \tag{6}$$

其中

$$\left.\begin{array}{l} \text{ⅰ)} \boldsymbol{b}(\lambda)\in B_\lambda, \\ \text{ⅱ)} b_j(\lambda)(1\leqslant j\leqslant n)\text{皆为 }N_\delta\text{类函数之边值}, \\ \text{ⅲ)} \boldsymbol{b}(\lambda)\text{不恒为 }0. \end{array}\right\} \tag{7}$$

证明 必要性 由定理 1′ 即知.

充分性 设 $b(\lambda)$ 满足 (6),(7). 因为 N_δ 类函数可表为两个有界解析函数之比,所以存在单位圆 $|z|<1$ 内的有界解析函数 $\varphi_j(z),\psi_j(z)(1\leqslant j\leqslant n)$ 使 $b_j(\lambda)=\dfrac{\varphi_j(e^{i\lambda})}{\psi_j(e^{i\lambda})}$,其中 $\psi_j(\lambda)(1\leqslant j\leqslant n)$ 皆不恒为 0. 令 $c(\lambda)=\prod_{j=1}^{n}\psi_j(\lambda)$. 显然,$c(\lambda)b(\lambda)$ 满足 (3). 因为 $c(\lambda)$ 为有界解析函数之边值,所以存在正数 K 使 $|c(\lambda)|\leqslant K(p,p)$. 故有

$$(f_\lambda^{-1}c(\lambda)b(\lambda),c(\lambda)b(\lambda))=|c(\lambda)|^2(f_\lambda^{-1}b(\lambda),b(\lambda))\leqslant K^2(f_\lambda^{-1}b(\lambda),b(\lambda)).$$

这表明 $(f_\lambda^{-1}c(\lambda)b(\lambda),c(\lambda)b(\lambda))\in L_1(-\pi,\pi)$. 所以 $c(\lambda)b(\lambda)$ 也满足 (2). 按照定理 1′,平稳序列 x_t 是非奇异的. 故充分性成立. 定理 1 证毕.

从定理 1 的充分性的证明知道,对定理 1 中的 $b_j(\lambda)$ 也可要求皆为单位圆内有界解析函数之边值.

定理 2 平稳序列 x_t 为奇异的充要条件是对任何 n 维向量 $b(\lambda)=(b_1(\lambda),\cdots,b_n(\lambda))$,只要它满足

ⅰ) $b(\lambda)\in B_\lambda$,

ⅱ) $b_j(\lambda)(1\leqslant j\leqslant n)$ 皆为 N_δ 类函数之边值,

ⅲ) $b(\lambda)$ 不恒为 0,

则恒有

$$(f_\lambda^{-1}b(\lambda),b(\lambda))\overline{\in} L_1(-\pi,\pi).$$

定理 2 是定理 1 的直接推论.

下面讨论多维平稳序列的 WOLD 分解谱表示.

对任何平稳序列 x_t 都有 WOLD 分解

$$x_t=x_t^{(r)}+x_t^{(s)},$$

其中 $S_x(r)=0,S_x(s)=H_x(t)=S_x$ (见文献[2]定理 9).

我们称 $x_t^{(r)}$ 为 x_t 正则分量,称 $x_t^{(s)}$ 为 x_t 的奇异分量. 设 $x_t^{(r)}$ 是秩为 ρ 的正则序列. 我们称 ρ 为平稳序列 x_t 的秩. 平稳序列 x_t 的非奇异性等价于 $\rho\geqslant 1$.

令

$$D_x(t)=H_x(t)\ominus H_x^-(t),$$
$$D_{x^{(r)}}(t)=H_{x^{(r)}}(t)\ominus H_{x^{(r)}}^-(t).$$

由于 $H_x(s)=S_x$,所以 $D_x(t)=D_{x^{(r)}}(t)$. 对于正则序列 $x_t^{(r)}$ 而言,$D_{x^{(r)}}(t)$ 的维数等于它的秩 ρ(见文献[2]第二章).

设

$$\left.\begin{array}{l} u_1(0),\cdots,u_\rho(0) \text{为子空间 } D_x(0) \text{的完备标准正交基;} \\ a_j(\lambda) \text{为 }\mathfrak{U}\text{ 中与 }u_j(0)\text{相对应的元素}(1\leqslant j\leqslant\rho). \end{array}\right\} \quad (8)$$

令
$$u_j(t) = U^t u_j(0) \quad (1 \leqslant j \leqslant \rho, t = 0, \pm 1, \cdots). \tag{9}$$

对平稳序列 $(u_1(t), \cdots, u_\rho(t))$，它的谱密度矩阵为
$$[(f_\lambda \boldsymbol{a}_j(\lambda), \boldsymbol{a}_k(\lambda))]_{1 \leqslant j,k \leqslant \rho} = \frac{1}{2\pi} I_\rho, \tag{10}$$

其中 $I_\rho = \begin{bmatrix} 1 & & 0 \\ & \ddots & \\ 0 & & 1 \end{bmatrix}$ (参见文献[2]).

由于 $H_x \ominus H_x^-(0) = \sum\limits_{t=0}^{+\infty} D_x(t)$，所以 $\{u_j(t), 1 \leqslant j \leqslant \rho, t \geqslant 0\}$ 是空间 $H_x \ominus H_x^-(0)$ 的完备标准正交基. 因而
$$\{U^t \boldsymbol{a}_j(\lambda) = e^{\lambda t} \boldsymbol{a}_j(\lambda), 1 \leqslant j \leqslant \rho, t \geqslant 0\} \text{ 是空间 } \mathfrak{U} \ominus \mathfrak{U}^-(0) \text{ 的完备标准正交基}. \tag{11}$$

定理 3 设平稳序列 x_t 的秩 $\rho \geqslant 1$，则存在矩阵
$$B = \begin{bmatrix} b_{11}(\lambda) & \cdots & b_{1n}(\lambda) \\ \vdots & & \vdots \\ b_{\rho 1}(\lambda) & \cdots & b_{\rho n}(\lambda) \end{bmatrix} = \begin{bmatrix} \boldsymbol{b}_1(\lambda) \\ \vdots \\ \boldsymbol{b}_\rho(\lambda) \end{bmatrix} \tag{12}$$

使
$$[(f_\lambda^{-1} \boldsymbol{b}_j(\lambda), \boldsymbol{b}_k(\lambda))]_{1 \leqslant j,k \leqslant \rho} = I_\rho, \tag{13}$$

其中
$$\left. \begin{array}{l} \text{i) } \boldsymbol{b}_j(\lambda) = (b_{j1}(\lambda), \cdots, b_{jn}(\lambda)) \in B_\lambda, \quad 1 \leqslant j \leqslant \rho, \\ \text{ii) } b_{ji}(\lambda) (1 \leqslant j \leqslant \rho, 1 \leqslant i \leqslant n) \text{ 皆为 } N_\delta \text{ 类函数之边值}. \end{array} \right\} \tag{14}$$

证明 令
$$\boldsymbol{d}_j(\lambda) = f_\lambda \boldsymbol{a}_j(\lambda) = (d_{j1}(\lambda), \cdots, d_{jn}(\lambda)) \quad (1 \leqslant j \leqslant \rho), \tag{15}$$

其中 $\boldsymbol{a}_j(\lambda)$ 的意义见(8). 因为 $\boldsymbol{a}_j(\lambda) \in \mathfrak{U} \ominus \mathfrak{U}^-(0) (1 \leqslant j \leqslant \rho)$，所以 $\boldsymbol{d}_j(\lambda)$ 的分量
$$d_{jk}(\lambda) \quad (1 \leqslant k \leqslant n)$$
皆是 H_1 类函数之边值(参看定理 $1'$ 的证明).

令
$$D = \begin{bmatrix} \boldsymbol{d}_1(\lambda) \\ \vdots \\ \boldsymbol{d}_\rho(\lambda) \end{bmatrix}. \tag{16}$$

按照(10)，有
$$[(f_\lambda^{-1} \boldsymbol{d}_j(\lambda), \boldsymbol{d}_k(\lambda))]_{1 \leqslant j,k \leqslant \rho} = \frac{1}{2\pi} I_\rho. \tag{17}$$

我们取 $B = \sqrt{2\pi} D$. 由上知，$\sqrt{2\pi} D$ 满足(13),(14)，即矩阵 B 就是我们所要求

的. 定理 3 证毕.

定理 4 设平稳序列 x_t 的秩 $\rho \geqslant 1$. 若形如(12)的矩阵 B 满足(13),(14),则

1) B 的秩几乎处处为 ρ;
2) 必存在 B 的子矩阵

$$B_{\rho \times \rho} = \begin{bmatrix} b_1 i_1(\lambda) & \cdots & b_1 i_\rho(\lambda) \\ \vdots & & \vdots \\ b_\rho i_1(\lambda) & \cdots & b_\rho i_\rho(\lambda) \end{bmatrix},$$

它的秩几乎处处为 ρ, 并且矩阵 B 的其他各列由 $B_{\rho \times \rho}$ 唯一确定, 即对任何满足 (13),(14) 的矩阵 $\widetilde{B}$, 有

$$B_{\rho \times \rho}^{-1} B = \widetilde{B}_{\rho \times \rho}^{-1} \widetilde{B}, \tag{18}$$

其中 $\widetilde{B}_{\rho \times \rho}$ 的意义与 $B_{\rho \times \rho}$ 相同;

3) 矩阵 B 与 x_t 的正则部分 $x_t^{(r)}$ 的谱特征 $\Phi_x^{x^r}$ 和谱密度矩阵 $[f_\lambda^{x^r x^r}]$ 有如下关系:

$$\Phi_x^{x^r} = \overline{B}'[f_\lambda^{-1} B], \quad [f_\lambda^{x^r x^r}] = \overline{B}' B, \tag{19}$$

其中

$$[f_\lambda^{-1} B] = \begin{bmatrix} f_\lambda^{-1} \boldsymbol{b}_1(\lambda) \\ \vdots \\ f_\lambda^{-1} \boldsymbol{b}_\rho(\lambda) \end{bmatrix}. \tag{20}$$

(关于谱特征 $\Phi_x^{x^r}$ 的意义, 参见文献[2]第二章).

证明 首先, 我们讨论矩阵 D(见(16))与 $\Phi_x^{x^r}$ 和 $[f_\lambda^{x^r x^r}]$ 的关系.

对正则部分 $x_t^{(r)}$ 有 $H_{x^{(r)}}^-(t+1) = \sum_{\tau=0}^{+\infty} D_{x^{(r)}}(t-\tau)$. 由于 $D_x(t) = D_{x^{(r)}}(t)$, 再根据(8)和(9), 我们得

$$x_k^{(r)}(t) = \sum_{j=1}^{\rho} \sum_{\tau=0}^{+\infty} \alpha_j^k(\tau) u_j(t-\tau) \quad (1 \leqslant k \leqslant n, \quad t = 0, \pm 1, \cdots), \tag{21}$$

其中

$$\alpha_j^k(\tau) = (x_k(0), u_j(-\tau)). \tag{22}$$

根据(22)和(15)有

$$\overline{\alpha_j^k(\tau)} = (u_j(-\tau), x_k(0)) = \int_{-\pi}^{\pi} e^{-i\lambda\tau} (f_\lambda \boldsymbol{a}_j(\lambda), \boldsymbol{e}_k) d\lambda$$

$$= \int_{-\pi}^{\pi} e^{-i\lambda\tau} d_{jk}(\lambda) d\lambda. \tag{23}$$

由(21)式

$$\sum_{j=1}^{\rho}\sum_{\tau=0}^{+\infty}|\alpha_j^k(\tau)|^2=(x_k^{(r)}(t),x_k^{(r)}(t))<+\infty,$$

所以 $\sum_{\tau=0}^{+\infty}\frac{1}{2\pi}\overline{\alpha_j^k(\tau)}e^{i\lambda\tau}$ 为 H_2 类函数之边值. 由(23)可知

$$d_{jk}(\lambda)=\sum_{\tau=0}^{+\infty}\frac{1}{2\pi}\overline{\alpha_j^k(\tau)}e^{i\lambda\tau} \qquad (1\leqslant j\leqslant\rho,1\leqslant k\leqslant n). \qquad (24)$$

设 $\mathfrak{U}$ 中的 $a_k^{(r)}(\lambda)$ 在对应关系(1)下与 $x_k^{(r)}(0)$ 相对应 $(1\leqslant k\leqslant n)$. 由(21)和(24)知

$$a_k^{(r)}(\lambda)=\sum_{j=1}^{\rho}\sum_{\tau=0}^{+\infty}\alpha_j^k(\tau)e^{-i\lambda\tau}a_j(\lambda)=\sum_{j=1}^{\rho}2\pi\overline{d_{jk}(k)}a_j(\lambda) \quad (1\leqslant k\leqslant n). \tag{25}$$

依照(20)的记法,(17)可写为

$$[f_\lambda^{-1}D]\overline{D}'=\frac{1}{2\pi}I_\rho. \tag{26}$$

所以根据(25)和(26)便得

$$\Phi_x^{r}=\begin{bmatrix}a_1^{(r)}(\lambda)\\ \vdots\\ a_\rho^{(r)}(\lambda)\end{bmatrix}=2\pi\overline{D}'\begin{bmatrix}a_1(\lambda)\\ \vdots\\ a_\rho(\lambda)\end{bmatrix}=2\pi\overline{D}'[f_\lambda^{-1}D]. \tag{27}$$

$$[f_\lambda^{r^rx^r}]=\Phi_x^{r}[f_\lambda^{xx}]\overline{\Phi_x^{r}}'=4\pi^2\overline{D}'[f_\lambda^{-1}D][f_\lambda^{xx}]\overline{[f_\lambda^{-1}D]}'D=4\pi^2\overline{D}'[f_\lambda^{-1}D]\overline{D}'D=2\pi\overline{D}'D \tag{28}$$

(参见文献[2]第二章).

其次,我们转而讨论矩阵 B 与矩阵 D 的关系. 取不恒为 0 的单位圆 $|z|<1$ 内有界解析函数之边值 $c_j(\lambda)$ 使 $c_j(\lambda)b_j(\lambda)$ 的每一个分量皆是 H_1 类函数边值 $(1\leqslant j\leqslant\rho)$ (参看定理 1 的证明). 于是 $f_\lambda^{-1}c_j(\lambda)b_j(\lambda)\in\mathfrak{U}\ominus\mathfrak{U}^-(0)(1\leqslant j\leqslant\rho)$ (见定理 1′ 的证明). 因为 $\{e^{i\lambda t}a_j(\lambda),1\leqslant j\leqslant\rho,t\geqslant 0\}$ 是 $\mathfrak{U}\ominus\mathfrak{U}^-(0)$ 的完备标准正交基(见(11)),所以

$$f_\lambda^{-1}c_k(\lambda)b_k(\lambda)=\sum_{j=1}^{\rho}\sum_{\tau=0}^{+\infty}\beta_j^k(\tau)e^{i\lambda\tau}a_j(\lambda)=\sum_{j=1}^{\rho}\varphi_{kj}(\lambda)a_j(\lambda) \quad (1\leqslant k\leqslant\rho), \tag{29}$$

其中

$$\beta_j^k(\tau)=\{f_\lambda^{-1}c_k(\lambda)b_k(\lambda),e^{i\lambda\tau}a_j(\lambda)\},$$
$$\varphi_{kj}(\lambda)=\sum_{\tau=0}^{+\infty}\beta_j^k(\tau)e^{i\lambda\tau}.$$

因为

$$\sum_{j=1}^{\rho}\sum_{\tau=0}^{+\infty}|\beta_j^k(\tau)|^2 = \{f_\lambda^{-1}c_k(\lambda)\boldsymbol{b}_k(\lambda), f_\lambda^{-1}c_k(\lambda)\boldsymbol{b}_k(\lambda)\} < +\infty,$$

所以 $\varphi_{kj}(\lambda)$ 是 H_2 类函数之边值.

由(29)得

$$c_k(\lambda)\boldsymbol{b}_k(\lambda) = f_\lambda(f_\lambda^{-1}c_k(\lambda)\boldsymbol{b}_k(\lambda)) = \sum_{j=1}^{\rho}\varphi_{kj}(\lambda)\boldsymbol{d}_j(\lambda),$$

即

$$\boldsymbol{b}_k(\lambda) = \sum_{j=1}^{\rho}\frac{\varphi_{kj}(\lambda)}{c_k(\lambda)}\boldsymbol{d}_j(\lambda) \qquad (1 \leqslant k \leqslant \rho). \tag{30}$$

记

$$\Phi = \begin{bmatrix} \dfrac{\varphi_{11}(\lambda)}{c_1(\lambda)} & \cdots & \dfrac{\varphi_{1\rho}(\lambda)}{c_1(\lambda)} \\ \vdots & & \vdots \\ \dfrac{\varphi_{\rho 1}(\lambda)}{c_\rho(\lambda)} & \cdots & \dfrac{\varphi_{\rho\rho}(\lambda)}{c_\rho(\lambda)} \end{bmatrix}. \tag{31}$$

于是(30)便成为

$$B = \Phi D. \tag{32}$$

最后,我们来证明定理 4 本身. 我们要注意这样的事实: N_δ 类函数之边值对普通的乘法及加减法运算是封闭的; 另外, 由于单位圆内有界解析函数之边值有唯一性定理(见文献[5]), 因而 N_δ 类函数之边值也有唯一性定理, 即当某个 N_δ 类函数边值在一个测度大于 0 的勒贝格可测集上取 0 值时, 则此 N_δ 类函数边值几乎处处为 0.

若矩阵 B 有某个 ρ 阶子矩阵在某个测度大于 0 的勒贝格可测集 E 上秩为 ρ, 即这个子矩阵的行列式的值在 E 上不取 0 值, 由唯一性定理知, 这个子矩阵的秩几乎处处为 ρ, 所以矩阵 B 的秩也几乎处处为 ρ, 若矩阵 B 所有的 ρ 阶子矩阵的秩皆几乎处处小于 ρ, 于是必有某个 $j_0(1 \leqslant j_0 \leqslant \rho)$ 使

$$\boldsymbol{b}_{j_0}(\lambda) = \sum_{\substack{j=1 \\ j \neq j_0}}^{\rho} r_j(\lambda)\boldsymbol{b}_j(\lambda).$$

根据(13)得 $(f_\lambda^{-1}\boldsymbol{b}_{j_0}(\lambda), \boldsymbol{b}_{j_0}(\lambda)) = 1(p, p)$ 和 $r_j(\lambda) = 0(p, p)(1 \leqslant j \leqslant \rho, j \neq j_0)$, 也即 $\boldsymbol{b}_{j_0}(\lambda) = 0(p, p)$, 这和 $(f_\lambda^{-1}\boldsymbol{b}_{j_0}(\lambda), \boldsymbol{b}_{j_0}(\lambda)) = 1(p, p)$ 相矛盾. 这说明上述第二种情况不可能出现. 因此, 矩阵 B 的秩几乎处处为 ρ.

由上知, 必有 B 的某个子矩阵

$$B_{\rho \times \rho} = \begin{bmatrix} b_{1i_1}(\lambda) & \cdots & b_{1i_\rho}(\lambda) \\ \vdots & & \vdots \\ b_{\rho i_1}(\lambda) & \cdots & b_{\rho i_\rho}(\lambda) \end{bmatrix},$$

它的秩几乎处处为 ρ.

取

$$D_{\rho\times\rho}=\begin{bmatrix} d_{1i1}(\lambda) & \cdots & d_{1i\rho}(\lambda) \\ \vdots & & \vdots \\ d_{\rho i1}(\lambda) & \cdots & d_{\rho i\rho}(\lambda) \end{bmatrix}.$$

由(32)知

$$B_{\rho\times\rho}=\Phi D_{\rho\times\rho},$$

所以

$$\Phi=B_{\rho\times\rho}D_{\rho\times\rho}^{-1}, \tag{33}$$

因此,根据(32)和(33)有

$$B_{\rho\times\rho}^{-1}B=B_{\rho\times\rho}^{-1}\Phi D=B_{\rho\times\rho}^{-1}(B_{\rho\times\rho}D_{\rho\times\rho}^{-1})D=D_{\rho\times\rho}^{-1}D. \tag{34}$$

由于 D 是事先选择好的矩阵,所以(34)与(18)等价.

由(32)得

$$D=\Phi^{-1}B. \tag{35}$$

再由(13)和(26)便得

$$\frac{1}{2\pi}I_\rho=[f_\lambda^{-1}D]\overline{D}'=\Phi^{-1}[f_\lambda^{-1}B]\overline{B}'\overline{\Phi^{-1}}'=\Phi^{-1'}\overline{\Phi^{-1}}'. \tag{36}$$

综合(27),(28),(35),(36)便得

$$\Phi_x^{x^r}=\overline{B}'[f_\lambda^{-1}B], \quad [f_\lambda^{x^r x^r}]=\overline{B}'B.$$

至此,定理 4 证毕.

关于 x_t 的奇异部分 $x_t^{(s)}$ 的谱特征 $\Phi_x^{x^s}$ 和谱密度矩阵 $[f_\lambda^{x^s x^s}]$,有

$$\Phi_x^{x^s}=I_n-\Phi_x^{x^r}, \quad [f_\lambda^{x^s x^s}]=[f_\lambda^{xx}]-[f_\lambda^{x^r x^r}].$$

因此,定理 4 在理论上给出了 WOLD 分解的谱表示.

为了说明上述定理的意义,我们给出两个特殊情况下的命题.

命题 1 设一维平稳序列 $x(t)$ 具有绝对连续谱函数,谱密度为 $f(\lambda)$. 则下面三个命题是等价的:

1) 平稳序列 $x(t)$ 是非奇异的;
2) $\log f(\lambda)\in L_1(-\pi,\pi)$;
3) 存在不恒为 0 的 N_δ 类函数之边值 $\varphi(\lambda)$,使

$$\frac{|\varphi(\lambda)|^2}{f(\lambda)}\in L_1(-\pi,\pi).$$

证明 由文献[3]知,1)与 2)是等价的. 根据定理 1,1)与 3)是等价的. 因此,1),2),3)彼此等价. 证毕.

现在我们给出命题 1 的一个分析证明.

命题 1′ 设 $f(\lambda) \geqslant 0, f(\lambda) \in L_1(-\pi, \pi)$. 则下面三个命题等价:

1) 存在不恒为 0 的 N_δ 类函数之边值 $\varphi(\lambda)$, 使
$$\frac{|\varphi(\lambda)|^2}{f(\lambda)} \in L_1(-\pi, \pi);$$

2) 存在 $g(\lambda) \geqslant 0$, 它满足 $\log g(\lambda) \in L_1(-\pi, \pi)$, 使
$$\frac{g(\lambda)}{f(\lambda)} \in L_1(-\pi, \pi);$$

3) $\log f(\lambda) \in L_1(-\pi, \pi)$.

证明 由 1) 可以导出 2). 只要令 $g(\lambda) = |\varphi(\lambda)|^2$ 就行了.

由 2) 可以导出 3).

令
$$E = \left\{\lambda : \frac{g(\lambda)}{f(\lambda)} \leqslant 1\right\}, \quad F = \left\{\lambda : \frac{g(\lambda)}{f(\lambda)} > 1\right\}.$$

在 E 上有 $0 \leqslant \log f(\lambda) - \log g(\lambda) \leqslant f(\lambda) - \log g(\lambda)$. 由于 $\log g(\lambda)$ 和 $f(\lambda) - \log g(\lambda)$ 在 E 上可积, 所以 $\log f(\lambda)$ 也在 E 上可积.

在 F 上有 $0 \leqslant \log g(\lambda) - \log f(\lambda) \leqslant \frac{g(\lambda)}{f(\lambda)}$. 由于 $\log g(\lambda)$ 和 $\frac{g(\lambda)}{f(\lambda)}$ 在 F 上可积, 所以 $\log f(\lambda)$ 也在 F 上可积. 因此, $\log f(\lambda) \in L_1(-\pi, \pi)$.

由 3) 可以导出 2). 只要令 $g(\lambda) = f(\lambda)$ 就行了.

由 2) 可以导出 1). 做 A 类函数
$$\Gamma(z) = \exp\left\{\frac{1}{2\pi}\int_{-\pi}^{\pi} \log \sqrt{g(\theta)}\, \frac{e^{i\theta} + z}{e^{i\theta} - z} d\theta\right\}$$

(参见文献[5]). 因为 $|\Gamma(e^{i\lambda})|^2 = g(\lambda)$, 所以
$$\frac{|\Gamma(e^{i\lambda})|^2}{f(\lambda)} = \frac{g(\lambda)}{f(\lambda)} \in L_1(-\pi, \pi).$$

至此, 命题 1′ 证毕.

命题 2 设平稳序列 $x_t = (x_1(t), \cdots, x_n(t))$ 的子序列 $x'_t = (x_1(t), \cdots, x_{n-1}(t))$ 是奇异的, 且 $d_{n-1}(\lambda) > 0(p, p)$. 则 x_t 为非奇异的充要条件是

$$\log \frac{d_n(\lambda)}{d_{n-1}(\lambda)} \in L_1(-\pi, \pi)^{①}, \tag{37}$$

其中 $d_k(\lambda)(k = 1, \cdots, n)$ 是矩阵 $[f_\lambda^{xx}]_{n \times n}$ 的 k 阶主子式.

---

① 令 $\log 0 = -\infty$.

当条件(37)满足时,有

$$[f_\lambda^{x^r x^r}] = \begin{bmatrix} 0 & & & \\ & \ddots & & \\ & & 0 & \\ & & & \dfrac{d_n(\lambda)}{d_{n-1}(\lambda)} \end{bmatrix}, \qquad (38)$$

$$\Phi_x^{x^r} = \dfrac{d_n(\lambda)}{d_{n-1}(\lambda)} \begin{bmatrix} 0 & & & \\ & \ddots & & \\ & & 0 & \\ & & & 1 \end{bmatrix} [f_\lambda^{xx}]^{-1}. \qquad (39)$$

证明 对任何 $\boldsymbol{b}(\lambda)=(b_1(\lambda),\cdots,b_n(\lambda))$,它满足:$\boldsymbol{b}(\lambda)\in B_\lambda, b_j(\lambda)(j=1,\cdots,n)$ 皆为 N_δ 类函数之边值,且使 $(f_\lambda^{-1}\boldsymbol{b}(\lambda),\boldsymbol{b}(\lambda))\in L_1(-\pi,\pi)$,则必有

$$b_j(\lambda)\equiv 0 \qquad (j=1,\cdots,n-1).$$

我们来证明这一点. 不妨假定 $b_j(\lambda)(j=1,\cdots,n)$ 皆是 H_1 类函数之边值[①]. 令 $\boldsymbol{a}(\lambda)=f_\lambda^{-1}\boldsymbol{b}(\lambda)$. 显然 $\boldsymbol{a}(\lambda)\in \mathfrak{U}\ominus\mathfrak{U}'(0)$. 因为 x_t' 是奇异的,故 $H_{x'}=H_{x'}^-(0)$,所以空间 $H_x\ominus H_x^-(0)$ 与空间 $H_{x'}$ 垂直. 这个性质反映到同构空间 $\mathfrak{U}$ 中就得到

$$\int_{-\pi}^{\pi} e^{i\lambda t} b_j(\lambda) d\lambda = \int_{-\pi}^{\pi} e^{i\lambda t}(f_\lambda \boldsymbol{a}(\lambda),e_j) d\lambda$$
$$= \int_{-\pi}^{\pi}(f_\lambda \boldsymbol{a}(\lambda),e^{-i\lambda t}e_j) d\lambda = 0$$
$$(j=1,\cdots,n-1; t=0,\pm 1,\cdots).$$

所以 $b_j(\lambda)\equiv 0(j=1,\cdots,n-1)$.

因此,必然有 x_t 的秩 $\rho\leqslant 1$. 实际上,根据上面的讨论知,任何满足条件 (13),(14) 的矩阵 B,其秩只能几乎处处 $\leqslant 1$. 所以必有 $\rho\leqslant 1$.

如果 x_t 为非奇异的,则 x_t 的秩 $\rho=1$. 根据定理3,定理4,有不恒为 0 的 n 维向量 $\boldsymbol{b}(\lambda)=(0,\cdots,0,b_n(\lambda))$ 满足(13),(14)和(19). 因为有不恒为 0 的向量

$$\boldsymbol{b}(\lambda)=(0,\cdots,0,b_n(\lambda))\in B_\lambda,$$

又由于 $d_{n-1}(\lambda)>0(p,p)$,所以 $d_n(\lambda)>0(p,p)$. 在现在的具体情况下,(13)变为

$$[(f_\lambda^{-1}\boldsymbol{b}(\lambda),\boldsymbol{b}(\lambda))]=|b_n(\lambda)|^2 \dfrac{d_{n-1}(\lambda)}{d_n(\lambda)}=1. \qquad (40)$$

由(40)和(19)知,(37),(38),(39)皆成立.

如果(37)成立,根据命题 1′,存在不恒为 0 的 N_δ 类函数之边值 $\varphi(\lambda)$ 使

[①] 参见定理1的证明.

$$\frac{|\varphi(\lambda)|^2}{d_n(\lambda)/d_{n-1}(\lambda)} \in L_1(-\pi,\pi).$$

令 $b(\lambda)=(0,\cdots,0,\varphi(\lambda))$. 显然有

$$(f_\lambda^{-1}b(\lambda),b(\lambda)) = \frac{|\varphi(\lambda)|^2}{d_n(\lambda)/d_{n-1}(\lambda)} \in L_1(-\pi,\pi).$$

根据定理 1, x_t 为非奇异的. 命题 2 证毕.

以上结果是作者在学生时期(1963年年初)得到的,现在整理成此文. 江泽培教授看阅了此文,并提出了宝贵意见. 作者对江泽培教授的热情支持和帮助,表示衷心的感谢.

参考文献

[1] 江泽培,多维平稳过程的预测理论,Ⅰ和Ⅱ,数学学报,13(1963),269—298;14(1964),438—450.

[2] Розанов Ю. А., Спектраłьная теория многомерных стационарных случайных процессов с днскретных временем, успехц Матем. Наук,13:2(80)(1958),93—142.

[3] Колмогоров А. Н., Стационарные последовательности в гильбертовском пространстве, Бюлл. М. Т. у,2,вып. 6. (1941). 1—40.

[4] Матвеев Р. Ф., О сигулярных многмерных стационарных процессах, Тер. Вероят. ц ее прцмен.,V:1(1960),38—44.

[5] Привалов,И. И.,解析函数的边界性质(中译本),科学出版社,1956 年.

[6] Матвев Р. Ф.,О регулярности многомерных стационарных спучайных процессов с дискрстным временем. ДАН,126:4(1959). 713—715.

[7] Розанов, Ю. А., Слектралвные свойства многомерных стационарных процессов и граничные свойства аналитических матриц, Теор. Вероят. и ее притен.,V:4(1960),399—413.

[8] Wiener N. & Masani P.,The prediction theory of multivariates stochastic processes I, Acta Math. 98,1—2(1957),111—150.

[9] Masani,P.,Cramer's theorem of monotone matrix-valued functions and the wold decomposition,probability and statistics(The Herold Cramér Volume),New York,175—189,1959.

原文载于《数学学报》,第 3 卷,第 5 期,1980 年 9 月,684—694.

Z-Transform Models and Data Extrapolation Formulas in the Maximum Entropy Methods of Power Spectral Analysis

Cheng Qiansheng

Department of Mathematics, Peking University

I. Introduction and Problems

The maximum entropy method (MEM) of spectral analysis is a nonlinear method. Let $S(\omega)$ be a power spectrum. Now there are two different definitions of entropy:

$$E_1 = \frac{1}{2\pi}\int_{-\pi}^{\pi} \ln S(\omega)d\omega$$

(see Ref. [1]) and

$$E_2 = \frac{1}{2\pi}\int_{-\pi}^{\pi} S(\omega)\ln S(\omega)d\omega$$

(see Refs. [2,3]).

We can deduce an all-pole model from MEM with E_1. But, up to now theoretical work has hardly been done for MEM with E_2, even its corresponding Z-transform model has not been given yet. In Ref. [4] the original signal is assumed to be a rational or an all-pole Z-transform model. Unfortunately it is false. Hence, it is necessary to discuss the rational model and the exponential model and their data extra-polation in the maximum entropy methods.

We first give some known fromulas which will be used later.

Let γ_k be the autocorrelation function whose F T is $S(\omega)$, x_k be the minimum phase signal which satisfies that the amplitude spectrum of x_k is $\sqrt{S(\omega)}$ and $x_0 > 0$. It follows from Ref. [6] that

$$X(z) = \sum_{k=0}^{\infty} x_k z^k = \exp\{c_0 + c_1 z + c_2 z^2 + \cdots\}, \qquad (1)$$

where

$$\begin{cases} c_0 = \dfrac{1}{4\pi}\int_{-\pi}^{\pi} \ln S(\omega)d\omega, \\ c_k = \dfrac{1}{2\pi}\int_{-\pi}^{\pi} (\ln S(\omega))e^{ik\omega}d\omega, \quad k \geq 1. \end{cases} \quad (2)$$

The relations between x_k and c_k are as follows:

$$\begin{cases} x_0 = e^{c_0}, \\ x_k = \dfrac{1}{k}\sum_{l=0}^{k-1}(k-l)c_{k-l}x_l, \quad k \geq 1, \end{cases} \quad (3)$$

$$\begin{cases} c_0 = \ln x_0, \\ c_k = \dfrac{1}{x_0}\left(x_k - \dfrac{1}{k}\sum_{i=0}^{k-1}(k-l)c_{k-l}x_l\right), \quad k \geq 1 \end{cases} \quad (4)$$

(see pp. 413—146, Ref. [6]). By (3) and (4), we can obtain $x_0, x_1, \cdots, x_k$ from $c_0, c_1, \cdots, c_k$ and *vice versa*.

The two problems we shall solve are as follows.

Problem 1. Given $r_k(|k|\leq n)$ and $c_k(1\leq k\leq m)$, under the conditions

$$\dfrac{1}{2\pi}\int_{-\pi}^{\pi} S(\omega)e^{ik\omega}d\omega = r_k, \quad |k|\leq n, \quad (5)$$

and

$$\dfrac{1}{2\pi}\int_{-\pi}^{\pi} \ln S(\omega)e^{ik\omega}d\omega = c_k, \quad 0<|k|<m, \quad (6)$$

find $S(\omega)$ maximizing E_1 (when $m=0$, Condition (6) is removed).

Problem 2. Given $r_k(|k|\leq n)$, under the condition

$$\dfrac{1}{2\pi}\int_{-\pi}^{\pi} S(\omega)e^{ik\omega}d\omega = r_k, \quad |k|\leq n,$$

find $S(\omega)$ maximizing E_2.

II. The Rational Model for Problem 1

By using the variational method, the solution $S(\omega)$ of Problem 1 satisfies

$$\dfrac{\partial}{\partial S}\left(\ln S + \sum_{\substack{k=-m \\ k\neq 0}}^{m}\mu_k \ln Se^{ik\omega} + \sum_{k=-n}^{n}\lambda_k Se^{ik\omega}\right) = 0,$$

where μ_k and λ_k are constants. Then we can obtain

$$S(\omega) = \dfrac{1+\sum\limits_{0<|k|\leq m}\mu_k e^{ik\omega}}{-\sum\limits_{k=-n}^{n}\lambda_k e^{ik\omega}}. \quad (7)$$

Since $S(\omega) \geqslant 0$, the Z-transform $X(Z)$ of the minimum-phase signal of $S(\omega)$ is

$$X(z) = \frac{\sum_{l=0}^{m} a_l z^l}{\sum_{l=0}^{n} b_l z^l}, \quad (b_0 = 1). \tag{8}$$

We see that Problem 1 produces a rational or a zero-pole model. Thus Problem 1 is turned into finding a_l and b_l subjected to (5) and (6).

Set

$$a_0 = 1, \quad a_k = \frac{1}{k} \sum_{l=0}^{k-1} (k-l) c_{k-l} a_l, \quad 1 \leqslant k \leqslant m. \tag{9}$$

By (9), $a_k (1 \leqslant k \leqslant m)$ is determined by $c_k (1 \leqslant k \leqslant m)$ and *vice versa*.

According to (1), (9) and (3), we know that

$$x_k = x_0 a_k, \quad 1 \leqslant k \leqslant m. \tag{10}$$

Thus Problem 1 becomes the following

Problem 1'. Given $r_k (0 \leqslant k \leqslant n)$ and $a_k (1 \leqslant k \leqslant m)$, under Condition (10) and

$$\sum_{l=0}^{\infty} x_{k+l} x_l = r_k, \quad 0 \leqslant k \leqslant n, \tag{11}$$

find x_k such that $x_0 = \exp\left\{\frac{1}{4\pi} \int_{-\pi}^{\pi} \ln S(\omega) d\omega\right\}$ (see(3)) arrives at the maximum value.

It follows from (8) that

$$a_k = b_k * x_k = \sum_{l=0}^{n} b_l x_{k-l}, \tag{12}$$

$$a_k * x_{-k} = b_k * x_k * x_{-k} = b_k * r_k. \tag{13}$$

Substituting (12) into (13) and taking k from 0 to n, we get

$$\sum_{l=0}^{n} b_l \left(\sum_{s=0}^{m} x_{s-l} x_{s-k} \right) = \sum_{l=0}^{n} b_l r_{k-l}, \quad 0 \leqslant k \leqslant n. \tag{14}$$

It is known that the $(n+1) \times (n+1)$ matrix $[r_{k-l}]_{0 \leqslant l, k \leqslant n}$ is positive definite and the $(n+1) \times (n+1)$ matrix

$$\left[\sum_{s=0}^{m} x_{s-l} x_{s-k} \right]_{0 \leqslant l, k \leqslant n} = x_0^2 \left[\sum_{s=0}^{m} a_{s-l} a_{s-k} \right]_{0 \leqslant l, k \leqslant n}$$

is positive semi-definite (here $a_k = 0$ if $k \leqslant 0$). We take x_0 such that

$$Q(x_0) = [r_{k-l}]_{0 \leqslant l, k \leqslant n} - x_0^2 \left[\sum_{s=0}^{m} a_{s-l} a_{s-k} \right]_{0 \leqslant l, k \leqslant n}$$

is positive semi-definite and $\det Q(x_0) = 0$. After x_0 is determined, b_k is obtained from (14) (refer to Ref. [7] for the algorithm). Then, we can compute a_k ($0 \leqslant k \leqslant m$) from (12) and (10). After that, we can get the recursion formula of x_k from (12) and the recursion formula of r_k from (13).

We see that Problem 1 gives the maximum entropy interpretation of a rational or zero-pole model. Burg's maximum entropy problem is only its special case ($m=0$).

III. The Exponential Model for Problem 2

For the functional of $S(\omega)$

$$\int_{-\pi}^{\pi} \left(S(\omega) \ln S(\omega) + \sum_{k=-n}^{n} \lambda_k S(\omega) e^{ik\omega} \right) d\omega$$

taking the variation and letting it be zero, we get

$$S(\omega) = \exp\left\{-1 + \sum_{l=-n}^{n} \lambda_l e^{il\omega}\right\}. \tag{15}$$

Since $S(\omega) \geqslant 0$, then $\bar{\lambda}_l = \lambda_{-l}$, λ_0 is real. It follows from (2) that

$$\begin{cases} c_0 = \dfrac{\lambda_0 - 1}{2}, \\ c_k = \lambda_{-k} = \bar{\lambda}_k, & 1 \leqslant k \leqslant n, \\ c_k = 0, & k > n. \end{cases} \tag{16}$$

According to (1), the Z-transform of the minimum phase signal of $S(\omega)$ is

$$X(z) = \exp\left\{\dfrac{\lambda_0 - 1}{2} + \sum_{k=1}^{n} \bar{\lambda}_k z^k\right\} \tag{17}$$

$$= \exp\{c_0 + c_1 z + \cdots + c_n z^n\}. \tag{17}'$$

This is an exponential Z-transform model. We call it an n-order exponential model. So Problem 2 produces an exponential model.

From (17) we know that $X(z)$ and $1/X(z)$ are analytical on the Z-plane. Therefore $X(z)$ cannot be a rational model. So the assumption about $X(z)$ in Ref. [4] is false.

By using (16), (3) and (4), $x_0, x_1, \cdots, x_n$ can be determined by $\lambda_0, \lambda_1, \cdots, \lambda_n$ and *vice versa*.

Now we present an extrapolation formula of r_k when λ_k and r_k ($|k| \leqslant n$) are given.

Since
$$\sum_{k=-\infty}^{\infty} r_k e^{-ik\omega} = S(\omega), \tag{18}$$
we take the derivatives of both sides of Eqs. (18) and (15) to obtain
$$\sum_{k=-\infty}^{\infty} kr_k e^{-ik\omega} = \left(\sum_{k=-\infty}^{\infty} r_k e^{-ik\omega}\right)\left(\sum_{l=-n}^{n} l\bar{\lambda}_l e^{-il\omega}\right). \tag{19}$$
By comparison of coefficients, we get
$$kr_k = \sum_{l=-n}^{n} l\bar{\lambda}_l r_{k-l}. \tag{20}$$
Suppose $\lambda_n \neq 0$, then we obtain the following recursion formula:
$$r_{k+n} = \frac{-1}{n\bar{\lambda}_n}\left(kr_k - \sum_{l=-n+1}^{n} l\bar{\lambda}_l r_{k-l}\right), \quad k \geq 1. \tag{21}$$

We shall simply discuss three problems below.

(i) Find $\lambda_k(|k| \geq n)$ from autocorrelation values of $r_k(|k| \geq n)$.

Substitution of (15) into the constraint conditions of Problem 2 yields a set of equations about λ_k. It is difficult to find the explicit solution of the set of equations.

(ii) The relationship between power spectra and n-order exponential models. From (1), any power spectrum $S(\omega)$, satisfying $\ln S(\omega) \in L_1(-\pi, \pi)$, corresponds to an infinite-order exponential model. We can prove that given $S(\omega)$ and $\varepsilon > 0$, there exists an n-order (finite order) exponential model $X(z)$ (see (17)′) satisfying
$$\int_{-\pi}^{\pi} \left| \sqrt{S(\omega)} - |X(e^{-i\omega})| \right| d\omega < \varepsilon,$$
or
$$\int_{-\pi}^{\pi} \left| S(\omega) - |X(e^{-i\omega})|^2 \right| d\omega < \varepsilon.$$

(iii) The measured data y_k. If y_k is random, or determined but not of minimum phase, then the formulas (16), (3) and (4) cannot be used. By the way, it is not necessary to turn y_k into $a^k y_k$ (see (19) in Ref. [4]), because it can be proved that the result of this extrapolation method in Ref. [4] is the same as the directly extrapolated y_k.

Finally we point out that the above discussion suits to complex signals, too. Multi-dimensional maximum entropy problem similar to Problem 2 will be discussed in another paper.

References

[1] Haykin, S. Ed. , *Nonlinear Methods of Spectral Analysis*, Springer-Verlag, Berlin, Heidelberg, New York, 1979.
[2] Friedem, B. R. , *J. Ope. Soc. Amer.* , 62 (1972), 511—518.
[3] Gull, S. F. & Daniell, G. J. , *Nature*, 272(1978), 686—690.
[4] Wu, N-L. , *IEEE*, ASSP-31(1988), 486—491.
[5] 蔡长年,汪润生,信息论,人民邮电出版社,北京,1962.
[6] 程乾生,信号数字处理的数学原理,石油工业出版社,北京,1979.
[7] Muillis, C. T. & Roberts, R. A. , *IEEE*, ASSP-24(1976),226—238.

原文载于《科学通报》,Vol. 30,No. 4,April 1985,436—440.

Rank of a Class of Autocorrelation Matrixes in Spectral Estimation

Cheng Qiansheng

Institute of Mathematics and Department of Mathematics, Peking University

Keyword: stationary random signal; rank of autocorrelation matrix; exponential model

Let x_t be a real stationary random signal with mean zero and autocorrelation function

$$r_x(t) = Ex_{t+s}x_s = \frac{1}{2\pi}\int_{\pi}^{-\pi} e^{it\omega} dF(\omega), \tag{1}$$

where $F(\omega)$ is the spectral function of x_t[1]. Consider a class of autocorrelation matrixes

$$Q_{L\times M} = (r_x(t-l) - r_x(t+l))_{1\leqslant t\leqslant L, 1\leqslant l\leqslant M}$$

$$= \begin{bmatrix} r_x(0)-r_x(2) & r_x(-1)-r_x(3) & \cdots & r_x(1-m)-r_x(1+m) \\ \vdots & \vdots & \cdots & \\ r_x(L-1)-r_x(L+1) & r_x(-2)-r_x(L+2) & \cdots & r_x(L-M)-r_x(L+M) \end{bmatrix}. \tag{2}$$

In order to study the rank of $Q_{L\times M}$, we assume that $F(\omega)$ satisfies the following condition

$$\int_G dF(\omega) > 0, \tag{3}$$

where $G = [-\pi, \pi] - \{\omega_1, \omega_2, \cdots, \omega_n\}$, $\omega_k \in [-\pi, \pi]$, $1\leqslant k\leqslant n$, n is an arbitrary positive integer.

Theorem 1. *Under condition (3),*

$$\text{rank}(Q_{M\times M}) = M. \tag{4}$$

Proof. Let a be a non-zero M-dimensional vector:

$$a = (a_1, a_2, \cdots, a_M)^T,$$

$$A(\omega) = \sum_{t=1}^{M} a_t e^{it\omega}.$$

Set
$$q = a^T Q_{M \times M} a. \tag{5}$$

Then
$$\begin{aligned} q &= \sum_{t,l=1}^{M} a_t a_l (r_x(t-l) - r_x(t+l)) \\ &= \sum_{t,l=1}^{M} a_t a_l \frac{1}{2\pi} \int_{-\pi}^{\pi} (e^{i(t-l)\omega} - e^{i(t+l)\omega}) dF(\omega) \\ &= \frac{1}{2\pi} \int_{-\pi}^{\pi} (|A(\omega)|^2 - A^2(\omega)) dF(\omega). \end{aligned}$$

We note that q is real. Therefore,
$$\begin{aligned} q &= \mathrm{Re}(q) \\ &= \frac{1}{2\pi} \int_{-\pi}^{\pi} (|A(\omega)|^2 - \mathrm{Re}(A^2(\omega))) dF(\omega), \end{aligned} \tag{6}$$

where Re denotes the real part. It is evident that
$$|A(\omega)|^2 - \mathrm{Re}(A^2(\omega)) \geqslant 0. \tag{7}$$

We consider the equation
$$|A(\omega)|^2 - \mathrm{Re}(A^2(\omega)) = 0. \tag{8}$$

Eq. (8) is equivalent to
$$|A(\omega)|^2 = A^2(\omega). \tag{9}$$

From Eq. (9) it follows that
$$\begin{cases} A(\omega) = 0, & (10) \\ \overline{A(\omega)} = A(\omega). & (11) \end{cases}$$

It is well known that there exist at most L zeros for an L-order trigonometric polynomial. Hence, formula (10) has at most M zeros and formula (11) has at most $2M$ zeros. We denote the zeros of (10) and (11) by $\{\omega_1, \omega_2, \cdots, \omega_n\}$, $n \leqslant 3M$.

Set
$$G = [-\pi, \pi] - \{\omega_1, \omega_2, \cdots, \omega_n\}.$$

We have
$$|A(\omega)|^2 - \mathrm{Re}(A^2(\omega)) > 0, \quad \omega \in G. \tag{12}$$

By (6), (7), (12) and (3), we get
$$q \geqslant \frac{1}{2\pi} \int_{G} (|A(\omega)|^2 - \mathrm{Re}(A^2(\omega))) dF(\omega) > 0,$$

that is,

$$a^T Q_{M\times M} a > 0.$$
Then (4) holds. Theorem 1 is proved.

Theorem 2. *Under condition* (3),
$$\text{rank}(Q_{L\times M}) = \min(L,M). \tag{13}$$
Proof. Let $K=\min(L,M)$. It is clear that
$$\text{rank}(Q_{K\times K}) \leqslant \text{rank}(Q_{L\times M}) \leqslant K.$$
It follows from Theorem 1 that
$$\text{rank}(Q_{K\times K}) = K.$$
Thus Theorem 2 is proved.

We make the following remarks.

(ⅰ) When the spectral function $F(\omega)$ is absolutely continuous, formula (3) holds. It shows that condition (3) is very weak.

(ⅱ) Application of $Q_{L\times M}$ to spectral estimation. We know that there are two important parameter models in time series: autoregressive moving average model (ARMA model) and exponential model (EX model). Both of them are the maximum entropy spectral models[2]. A stationary time series is said to be an EX model if its spectral function is absolutely continuous and its spectral density has the following representation
$$F'(\omega) = S(\omega) = \sigma^2 \exp\left\{\sum_{l=1}^{p} 2\lambda_p \cos l\omega\right\}.$$
Ref. [3] studies the frequency domain method for estimating parameters. Ref. [4] discusses the time domain method. Refs. [2] and [4] give the parameter equation.
$$\sum_{l=1}^{p} \lambda_l(r_x(t-l) - r_x(t+l)) = \text{tr}_x(t), \quad t \geqslant 1. \tag{14}$$
The matrix $Q_{L\times M}$ is the coefficient matrix of (14). Hence the rank of the matrix $Q_{L\times M}$ is very important for solving Eq. (14).

References

[1] Priestley, M. B., *Spectral Analysis and Time Series* (Vol. 1), Academic Press, London, New York, 1981.
[2] 程乾生,科学通报,**29**(1984),19:1210—1213.
[3] Bloomfield, P., *Biometrika*, **60**(1973), 217—238.
[4] 程乾生,信号处理,**5**(1989),2:65—67.

原文载于 Chinese Science Bulletin, Vol. 36, No. 1, January 1991, 72—74.

多谱估计的参数方法

程 乾 生

北京大学数学系

摘要：本文讨论多谱估计的参数方法. 对非最小相位 AR, MA, ARMA 模型, 利用高阶累量或高阶矩, 给出了用于估计参数的参数方程.

一、引言

设信号 x_t 为
$$x_t = w_t * u_t, \tag{1}$$
其中 u_t 为驱动白噪, 是一个均值为零的相互独立、相同分布的随机序列, w_t 为系统响应序列, 它是能量有限的, 即
$$0 < \sum_t w_t^2 < \infty. \tag{2}$$

许多实际信号都可以表示为上述模型, 如语音信号[1]和地震信号[2]. 在地震信号中, x_t 表示地震信号, w_t 为地震子波, u_t 为反射系数.

由 x_t 估计 w_t 称为系统识别问题, 由 x_t 估计 u_t 称为反褶积问题. 这两个问题在实际中是非常重要的问题. 这两个问题实质上是一个问题, 因为由 x_t 和 w_t 很容易估计出 u_t, 而由 x_t 和 u_t 也很容易估计出 w_t.

我们知道, 当 w_t 为最小相位信号时, 可以利用 x_t 的自相关函数由 x_t 估计出 u_t 或 w_t[1—3]. x_t 的自相关函数仅包含 w_t 的振幅谱信息而不包含 w_t 的相位信息 (当 w_t 对最小相位时, w_t 的相位信息由 x_t 的振幅谱唯一确定[3]). 因此, 当 w_t 为非最小相位时, 由 x_t 的自相关函数不能估计出 w_t. 从随机过程角度来看, 一个零均值的高斯过程由它的自相关函数唯一确定它的概率分布. 因此, 由高斯随机信号 x_t 是不能估计出系统响应序列 w_t 的相位信息的. 所以, 问题的确切提法是: 如何由非高斯随机信号 x_t 估计非最小相位序列 w_t 或驱动噪声 u_t.

为了从 x_t 提取 w_t 和 u_t, 目前有三种方法. 第一, 最小熵方法[4,5]. 基本思想是考虑一个目标函数, 然后求反褶积因子使得最小化目标函数. 这是一个非线性最优化问题. 第二, 高阶谱方法[6—9]. 基本思想是先求高阶谱, 然后由此求 w_t

的相位谱,最后恢复 w_t. 这种方法要用到 FFT. 第三,参数方法[10]. 这是一种时域方法. 基本思想是,在系统响应为有理函数的假设下,由高阶累量给出参数方程(线性方程),然后解参数方程求出参数.

在文献[10]中,仅给出了用三阶或四阶累量表示的参数方程,而且没有给出证明. 在本文中,将用任意高阶累量或高阶矩给出参数方程,并给出严格的证明. 在第二节,给出基本定理,在第三节,利用基本定理给出 AR, MA 和 ARMA 模型的参数方程.

二、基本定理

在这一节,我们先给出关于系统响应序列 w_t 的基本引理,然后给出关于高阶累量的基本定理,最后给出关于高阶矩的基本定理.

1. 基本引理

设系统响应序列 w_t 的 z 变换可表示为

$$W(z) = \sum_{t=-\infty}^{\infty} w_t z_t = \frac{B(z)}{A(z)}, \tag{3}$$

其中 $A(z)=1+a_1z+\cdots+a_pz^p$, $B(z)=b_0+b_1z+\cdots+b_qz^q$.

我们知道[3]: 当 $A(z)\neq 0 (|z|\leqslant 1)$ 时,

$$w_t=0, \quad t<0; \tag{4}$$

当 $A(z)\neq 0 (|z|\geqslant 1)$ 时,

$$w_t=0, \quad t>q-p. \tag{5}$$

令

$$r_w(\tau_1,\tau_2,\cdots,\tau_k) = \sum_t w_t w_{t+\tau_1}\cdots w_{t+\tau_k}. \tag{6}$$

由式(3)知

$$w_t * a_t = b_t,$$

即

$$w_t + a_1 w_{t-1} + \cdots + a_p w_{t-p} = b_t. \tag{7}$$

用 $w_{t+\tau_1} w_{t+\tau_2}\cdots w_{t+\tau_k}$ 乘(7)式两边,再对 t 求和,得

$$\sum_{j=0}^{p} a_j r_w(\tau_1+j,\tau_2+j,\cdots,\tau_k+j) = \sum_{t=0}^{q} b_t w_{t+\tau_1} w_{t+\tau_2}\cdots w_{t+\tau_k}. \tag{8}$$

在条件(4)之下,当

$$\tau_{k_0} = \min(\tau_1,\tau_2,\cdots,\tau_k) < -q \tag{9}$$

时, 有 $t+\tau_{k_0}<0$(其中 t 满足 $0\leq t\leq q$), 再由(4)得

$$\sum_{t=0}^{q} b_t w_{t+\tau_1} w_{t+\tau_2} \cdots w_{t+\tau_k} = 0. \tag{10}$$

在条件(5)之下, 当

$$\tau_{k_0} = \max(\tau_1, \tau_2, \cdots, \tau_k) > q-p \tag{11}$$

时, 有 $t+\tau_{k_0}>q-p$(其中 t 满足 $t\geq 0$), 再由(5)得

$$\sum_{t=0}^{q} b_t w_{t+\tau_1} w_{t+\tau_2} \cdots w_{t+\tau_k} = 0. \tag{12}$$

由式(8)—(12)我们得到基本引理.

基本引理 设 w_t 的 z 变换为有理函数(3)式. 当 $A(z)\neq 0(|z|\leq 1)$ 时,

$$\sum_{j=0}^{p} a_j r_w(\tau_1+j, \cdots, \tau_k+j) = 0, \quad \min(\tau_1, \cdots, \tau_k) < -q; \tag{13}$$

当 $A(z)\neq 0(|z|\geq 1)$ 时,

$$\sum_{j=0}^{p} a_j r_w(\tau_1+j, \cdots, \tau_k+j) = 0, \quad \max(\tau_1, \cdots, \tau_k) > q-p. \tag{14}$$

2. 基本定理 1

平稳信号 x_t 的 $k+1$ 阶累量定义为

$$c_x(\tau_1, \cdots, \tau_k) = (-1)^{k+1} \frac{\partial^{k+1}}{\partial \lambda_1 \cdots \partial \lambda_{k+1}} \log\big(E\exp\big(i\sum_{j=1}^{k} x_{\tau_k}\lambda_j + i x_{\tau_0}\lambda_{k+1}\big)\big)$$
$$(\lambda_1 = \lambda_2 = \cdots = \lambda_{k+1} = 0). \tag{15}$$

由高阶谱理论知[6]

$$c_x(\tau_1, \cdots, \tau_k) = c_u(k) r_w(\tau_1, \cdots, \tau_k), \tag{16}$$

其中 $c_u(k) = c_u(0, \cdots, 0)$, $r_w(\tau_1, \cdots, \tau_k)$ 由(6)确定.

由式(16)和基本引理, 我们可以得到下面的关于 $k+1$ 阶累量表示的参数方程的基本定理 1.

基本定理 1 设随机信号 x_t 由式(1)表示, w_t 的 z 变换由式(3)表示. 当 $A(z)\neq 0(|z|\leq 1)$ 时, 有

$$\sum_{j=0}^{p} a_j c_x(\tau_1+j, \cdots, \tau_k+j) = 0, \quad \min(\tau_1, \cdots, \tau_k) < -q; \tag{17}$$

当 $A(z)\neq 0(|z|\geq 1)$ 时,

$$\sum_{j=0}^{p} a_j c_x(\tau_1+j, \cdots, \tau_k+j) = 0, \quad \max(\tau_1, \cdots, \tau_k) > q-p. \tag{18}$$

3. 基本定理 2

平稳随机信号 x_t 的 $k+1$ 阶矩定义为

$$r_x(\tau_1,\cdots,\tau_k)=Ex_t x_{t+\tau_1}\cdots x_{t+\tau_k}. \tag{19}$$

由式(1)和(7)知

$$x_t * a_t = b_t * u_t, \tag{20}$$

用 $x_{t+\tau_1}\cdots x_{t+\tau_k}$ 乘(19)式的两边,再取数学期望,得

$$E\sum_{j=0}^{p}a_j x_{t-j}x_{t+\tau_1}\cdots x_{t+\tau_k}=E\sum_{j=0}^{q}b_j v_{t-j}x_{t+\tau_1}\cdots x_{t+\tau_k}. \tag{21}$$

由于

$$x_{t+\tau_k}=\sum_{l_k=-\infty}^{\infty}w_{l_k}u_{t+\tau_k-l_k},$$

所以

$$E\sum_{j=0}^{q}b_j u_{t-j}x_{t+\tau_1}\cdots x_{t+\tau_k}$$
$$=\sum_{j=0}^{q}\sum_{l_1=-\infty}^{\infty}\cdots\sum_{l_k=-\infty}^{\infty}b_j w_{l_1}\cdots w_{l_k}Eu_{t-j}u_{t+\tau_1-l_1}\cdots u_{t+\tau_k-l_k}. \tag{22}$$

当 $A(z)\neq 0(|z|\leq 1)$ 时,在 $l_i<0$ 时有 $w_{l_i}=0$,而在 $l_i\geq 0(1\leq i\leq k)$ 并且

$$\max(\tau_1,\cdots,\tau_k)<-q \tag{23}$$

时,有 $t-j>t+\tau_i-l_i(0\leq j\leq q)$,于是

$$Eu_{t-j}u_{t+\tau_i-l_1}\cdots u_{t+\tau_k}-l_k=Eu_{t-j}Eu_{t+\tau_1-l_1}\cdots u_{t+\tau_k}-l_k=0. \tag{24}$$

由式(22)得

$$E\sum_{j=0}^{q}b_j u_{t-j}x_{t+\tau_1}\cdots x_{t+\tau_k}=0. \tag{25}$$

当 $A(z)\neq 0(|z|\geq 1)$ 时,在 $l_i>q-p$ 时有 $w_{l_i}=0$,而在 $l_i\leq q-p(1\leq i\leq k)$ 并且

$$\min(\tau_1,\cdots,\tau_k)>q-p \tag{26}$$

时,有 $t-j<t+\tau_i-l_i(0\leq j\leq q-p)$,于是在这种情况下式(24)成立,进而知(25)也成立.

由此上分析和式(21),可得到高阶矩表示的参数方程.

基本定理 2 设随机信号 x_t 由(1)表示,w_t 的 z 变换为(3)表示. 当 $A(z)\neq 0(|z|\leq 1)$ 时,有

$$\sum_{j=0}^{p}a_j r_x(\tau_1+j,\cdots,\tau_k+j)=0, \quad \max(\tau_1,\cdots,\tau_k)<-q; \tag{27}$$

当 $A(z)\neq 0(|z|\geqslant 1)$ 时,有
$$\sum_{j=0}^{p} a_j r_x(\tau_1+j,\cdots,\tau_k+j) = 0, \quad \min(\tau_1,\cdots,\tau_k) > q-p. \tag{28}$$

三、多谱估计的参数方法

我们分 AR,MA 和 ARMA 三种模型讨论参数估计方法.

1. AR 模型

设系统响应序列 w_t 的 z 变换为
$$w(z)=1/A(z), \tag{29}$$
其中
$$\begin{cases} A(z)=1+a_1 z+\cdots+a_p z^p=C(z)D(z), \\ C(z)=(1-\alpha_1 z)\cdots(1-\alpha_{p_1} z), \quad |\alpha_1|>1,\cdots, \quad |\alpha_{p_1}|>1, \\ D(z)=(1-\beta_1 z)\cdots(1-\beta_{p-p_1} z), \quad |\beta_1|<1,\cdots, \quad |\beta_{p-p_1}|<1, \end{cases} \tag{30}$$
p_1 满足 $0\leqslant p_1\leqslant p$. 由上知,$C(z),D(z)$ 还可表示为
$$\begin{cases} C(z)=1+c_1 z+\cdots+c_{p_1} z^{p_1}, \\ D(z)=1+d_1 z+\cdots+d_{p-p_1} z^{p-p_1}. \end{cases} \tag{31}$$

参数估计的目的是由 x_t 估计出 $A(z)$ 的系数 $a_1,\cdots,a_p$,或者 $C(z)$ 和 $D(z)$ 的系数 $c_1,\cdots,c_{p_1},d_1,\cdots,d_{p-p_1}$.

首先,我们利用 x_t 的二阶矩即自相关函数按照通常的 Yule-Walker 方程求出最小相位 AR 参数 $\tilde{a}_t=(1,\tilde{a}_1,\cdots,\tilde{a}_p)$. 它的 z 变换 $\tilde{A}(z)=\sum_{j=0}^{p}\tilde{a}_j z^j$ 的根在单位圆外,而且它的振幅谱与 $a_t=(1,a_1,\cdots,a_p)$ 的振幅谱只相差一个常数. 因此,我们可得
$$\tilde{A}(z)=c_{p_1}^{-1} z^{p_1} C(z^{-1})D(z). \tag{32}$$
用 $\tilde{a}_t$ 对 x_t 进行滤波得 y_t,
$$y_t=\tilde{a}_t * x_t=\tilde{w}_t * u_t,$$
其中 $\tilde{w}_t$ 的 z 变换为
$$\tilde{w}(z)=\tilde{A}(z)w(z)=c_{p_1}^{-1} z^{p_1} C(z^{-1})/C(z). \tag{33}$$
上式分子、分母皆为 p_1 阶多项式,且 $C(z)\neq 0(|z|\geqslant 1)$. 由基本定理 1 和 2 可得
$$\sum_{j=0}^{p_1} c_j c_y(\tau_1+j,\cdots,\tau_k+j) = 0, \quad \max(\tau_1,\cdots,\tau_k) > 0, \tag{34}$$

$$\sum_{j=0}^{p_1} c_j r_y(\tau_1+j,\cdots,\tau_k+j) = 0, \quad \min(\tau_1,\cdots,\tau_k) > 0. \tag{35}$$

用最小平方法或 SVD 法解上述超定方程可得 c_j. 用 SVD 法的奇异值还可判断阶数 p_1.

在求出 c_j 后,可根据关系式(32)用最小平方法解出 d_k, $1 \leqslant k \leqslant p-p_1$.

由式(33)和(29)可得

$$c_{p_1}^{-1} z^{p_1} C(z^{-1}) A(z) = C(z) \widetilde{A}(z), \tag{36}$$

由上式可用最小平方法解出 a_k, $1 \leqslant k \leqslant p$.

2. MA 模型

设系统响应序列 w_t 的 z 变换为

$$w(z) = B(z) = 1 + b_1 z + \cdots + b_q z^q = E(z) F(z), \tag{37}$$

其中

$$E(z) = 1 + e_1 z + \cdots + e_{q_1} z^{q_1} = (1-\alpha_1 z)\cdots(1-\alpha_{q_1} z),$$
$$F(z) = 1 + f_1 z + \cdots + f_{q-q_1} z^{q-q_1} = (1-\beta_1 z)\cdots(1-\beta_{q-q_1} z),$$
$$|\alpha_1|>1,\cdots,|\alpha_{q_1}|>1, |\beta_1|<1,\cdots,|\beta_{q-q_1}|<1, \quad 0 \leqslant q_1 \leqslant q.$$

问题是由 x_t 求 $B(z)$ 的系数或者 $E(z)$ 和 $F(z)$ 的系数.

首先由 x_t 的自相关函数求最小相位 MA 参数 $\tilde{b}_t = (1, \tilde{b}_1, \cdots, \tilde{b}_q)$. 它的 z 变换 $\widetilde{B}(z)$ 的根在单位圆外,且 $\tilde{b}_t$ 和 b_t 的振幅谱差一常数. 因此可得

$$\widetilde{B}(z) = e_{q_1}^{-1} Z^{q_1} E(z^{-1}) F(z). \tag{38}$$

上式类似于式(32).

求 $\tilde{b}_t$ 的反信号 $\tilde{b}_t^{(-1)}$,即

$$\sum_{t=0}^{\infty} \tilde{b}_t^{(-1)} z^t = \frac{1}{\widetilde{B}z}. \tag{39}$$

用 $\tilde{b}_t^{(-1)}$ 对 x_t 进行滤波得

$$y_1 = \tilde{b}_t^{(-1)} * x_t = \widetilde{w}_t * u_t, \tag{40}$$

其中 $\widetilde{w}_t$ 的 z 变换为

$$\widetilde{w}(z) = \frac{B(z)}{\widetilde{B}(z)} = \frac{e_{q_1} E(z)}{z_{q_1} E(z^{-1})}. \tag{41}$$

上式分子、分母皆为 q_1 阶多项式,且分母多项式在 $|z| \leqslant 1$ 内无零点. 由基本定理中的公式(17)和(27)得

$$\sum_{j=0}^{q_t} e_{q_1-j} c_y(\tau_1+j,\cdots,\tau_k+j) = 0, \quad \min(\tau_1,\cdots,\tau_k) < -q_1, \tag{42}$$

$$\sum_{j=0}^{q_1} e_{q_1-j}\gamma_y(\tau_1+j,\cdots,\tau_k+j)=0, \quad \max(\tau_1,\cdots,\tau_k)<-q_1 \qquad (43)$$

由上述方程可求 $e_j, 1\leqslant j\leqslant q_1$.

由关系式(38)可求得 $f_j, 1\leqslant j\leqslant q-q_1$.

由式(41)可得
$$B(z)z^{q_1}E(z^{-1})=e_{q_1}E(z)\widetilde{B}(z) \qquad (44)$$

或
$$B(z)z^{q_1}E(z^{-1})\frac{1}{\widetilde{B}(z)}=e_{q_1}E(z) \qquad (45)$$

由式(44)或(45)可求出 $b_j, 1\leqslant j\leqslant q$.

在实践中,反信号 $\widetilde{b}_t^{(-1)}$ 可通过解 Yule-Walker 方程得到.

正如文献[10]所指出的,MA 模型的参数 b_j 可以通过高阶累量直接得到,例如用三阶累量 $c_x(\tau_1,\tau_1)$ 可得
$$b_j=c_x(q,j)/c_x(q,0), \quad 1\leqslant j\leqslant q. \qquad (46)$$

这种方法要求确切地知道 q,否则误差比较大.

3. ARMA 模型

设系统响应序列 w_t 的 z 变换为
$$w(z)=B(z)/A(z), \qquad (47)$$

其中
$$A(z)=C(z)D(z), \quad B(z)=E(z)F(z), \qquad (48)$$

这里 $C(z)$ 为 p_1 阶多项式,根在单位圆内;$D(z)$ 为 $p-p_1$ 阶多项式,根在单位圆外;$F(z)$ 为 q_1 阶多项式,根在单位圆内;$F(z)$ 为 $q-q_1$ 阶多项式,根在单位圆外.

如果已知 $p_1=0$ 或者 $p_1=p$,则由基本定理 1 或 2 可求出 $A(z)$ 的系数. 如果不知 p_1 是否等于 0 或 p,则我们要求 $w(z)$ 中不含全通因子,即不存在 $A(z)$ 中的因子 $G(z)$ 和 $B(z)$ 中的因子 $H(z)$,使
$$\left|\frac{G(e^{-iw})}{H(e^{-iw})}\right|\equiv 1. \qquad (49)$$

现在我们讨论 $A(z)$ 在单位圆内外皆可能有根时 $a_1,\cdots,a_p$ 的求法.

首先用 x_t 的自相关函数按照 Yule-Walker 方程求出最小相位 AR 参数 $\widetilde{a}_t=(1,\widetilde{a}_t,\cdots,\widetilde{a}_p)$.

然后用 $\widetilde{a}_t$ 对 x_t 进行滤波得 y_t

$$y_t = \tilde{a}_t * x_t.$$

再根据基本定理由下面方程求 $c_1, \cdots, c_{p_1}$：

$$\sum_{j=0}^{p_1} c_j c_x(\tau_1 + j, \cdots, \tau_k + j) = 0, \quad \max(\tau_1, \cdots, \tau_k) > q,$$

$$\sum_{j=0}^{p_1} c_j r_x(\tau_1 + j, \cdots, \tau_k + j) = 0, \quad \min(\tau_1, \cdots, \tau_k) > q.$$

由关系式

$$z^{p_1} C(z^{-1}) D(z) = \tilde{A}(z),$$

求出 $D(z)$ 的参数 $d_0, d_1, \cdots, d_{p-p_1}$. 由此可得 $A(z)$

$$A(z) = C(z) D(z) / d_0.$$

接着用 a_t（$A(z)$ 的系数）对 x_t 进行滤波得

$$z_t = a_t * x_t.$$

z_t 为 MA 模型. 用 MA 模型参数估计法可得 $B(z)$.

四、讨论和结论

1. 多谱估计的参数方法，其核心是高阶累量和高阶矩参数方程. 本文在基本定理中给出了任意阶累量和任意阶矩的参数方程，这对综合运用或有选择地运用不同阶数的累量或矩的参数方程提供了理论根据. 基本定理中的参数方程，可以看成是关于累量或矩的最一般的 Yule-Walker 方程.

2. 阶数 p 和 q 的判别. 对 AR 模型，MA 模型和不含全通因子（见式(49)）的 ARMA 模型，阶数 p 和 q 可以通过 x_t 的自相关函数判别. 利用高阶累量或高阶矩参数方程可判别在单位圆内多项式根的个数 q_1 和 p_1.

3. 本文的目的是研究多谱参数估计的理论和方法. 关于具体的计算和模拟试验，将另行讨论分析.

参考文献

[1] A. V. Oppenheim and R. W. Schafer: Digital Signal Processing. Prentice-Hall. Inc., New Jersey, 1975.

[2] M. T. Silvia and E. A. Robinson: Deconvolution of Geophysical Time Series in The Exploration for Oil and Natural gas, Elsevier Scientific Publishing Company. New York, 1979.

[3] 程乾生：信号数字处理的数学原理，石油工业出版社，北京，1979.

[4] R. A. Wiggins: Geoexploration, Vol. 16, pp. 21−35, 1978.

[5] 程乾生:北京大学数学所研究报告,No. 11,1989.
[6] M. Rosenblatt:Stationary Sequences and Random Fields,Birkhäuser Boston Inc. ,1985.
[7] D. R. Brillinger:Biometrika,Vol. 64,pp. 509—515,1977.
[8] K. S. Lii and M. Rosenblatt:Ann. Statist,Vol. 10,pp. 1195—1208,1982.
[9] T. Matsuoka and T. J. Ulrych:Proceeding of The IEEE. Vol. 72. No. 10,pp. 1408—1411,1984.
[10] 程乾生:多谱估计的 Cepstrum 分析和参数方法,中国信号处理学术年会大会报告,1986 年 10 月,中国,南京.

原文载于《电子学报》,第 1 期,1991 年 1 月,98—104.

时间序列分析

The Convolution-Type Matrix and the Property of the Complex Space l_2

Cheng Qiansheng
Department of Mathematics, Peking University

Abstract: In this paper, the property of convolution-type matrix is discussed. Using convolution-type matrix as an instrument, we investigate the property of the pure phase sequence and solve the consistence problem in the complex space l_2.

0. Symbol

We denote by l_2 the Hilbert space which consists of all summable complex sequence $x=(x(0),x(1),\cdots,x(t),\cdots)$. The inner product in l_2 is defined as $(x, y) = \sum_{t=0}^{\infty} x(t)\overline{y(t)}$, the norm of x as $\|x\| = \sqrt{(x, x)}$, the z-transformation of x as

$$X(z) = \sum_{t=0}^{\infty} x(t) z^t, \qquad (0.1)$$

and the frequence spectrum of x as $X(e^{-i\omega})$.

$$\gamma_{xy}(t) = \sum_{s=\max(0,-t)}^{\infty} x(t+s)\overline{y(s)} \qquad (0.2)$$

is called the crosscorrelation function of x and y. We have

$$\gamma_{xy}(t) = \frac{1}{2\pi} \int_{-\pi}^{\pi} X(e^{-i\omega})\overline{Y(e^{-i\omega})} e^{it\omega} d\omega. \qquad (0.3)$$

We call g a pure phase sequence or an all pass sequence if $g \in l_2$ and $|G(e^{-i\omega})|=1$ (a. e.) or $\gamma_{gg}(t)=\delta(t)$ where $\delta(t)$ is a kronecher function.

If $x, y \in l_2$, then $u = x * y$ denotes their convolution, and it is defined as

$$u(t) = x(t) * y(t) = \sum_{s=0}^{t} x(t-s)y(s), \quad t \geq 0. \qquad (0.4)$$

Let

$$\Delta_n(x) = \begin{pmatrix} x(0) & & & 0 \\ x(1) & x(0) & & \\ \vdots & \vdots & \ddots & \\ x(n) & x(n-1) & \cdots & x(0) \end{pmatrix}, \qquad (0.5)$$

$$y_n = (y(0), \cdots, y(n))', \quad u_n = (u(0), \cdots, u(n))', \qquad (0.6)$$

where prime denotes transport.

We can rewrite (0.4) as

$$\Delta_n(x) y_n = u_n, \qquad (0.7)$$

or

$$\Delta_n(y) x_n = u_n.$$

We call $\Delta_n(x)$ a convolution-type matrix.

Let Z^{n+1} be $n+1$ dimensional space. Then $\Delta_n(x)$ is a linear operator on Z^{n+1}; the norm of $\Delta_n(x)$ is as follows:

$$\|\Delta_n(x)\| = \sup_{\|y_n\|=1} \|\Delta_n(x) y_n\|. \qquad (0.8)$$

I. The Property of Convolution-Type Matrix

Let $x(0), x(1), \cdots, x(n), \cdots$ be a complex sequence.

Property 1. $\|\Delta_n(x)\| \geq \left[\sum_{t=0}^{n} |x(t)|^2\right]^{1/2}$.

Proof. Set

$$\delta_n = (1, 0, \cdots, 0)'_{n+1}, \qquad (1.1)$$

then

$$\|\Delta_n(x)\| \geq \|\Delta_n(x) \delta_n\| = \left[\sum_{t=0}^{n} |x(t)|^2\right]^{1/2}.$$

Property 2. $\|\Delta_n(x)\| \leq \|\Delta_{n+1}(x)\|$.

Proof. Take $y_n \in Z^{n+1}$ such that $\|y_n\| = 1$ and $\|\Delta_n(x) y_n\| = \|\Delta_n(x)\|$. Set $y'_{n+1} = (y'_n, 0)$, then

$$\|\Delta_{n+1}(x)\| \geq \|\Delta_n+1(x) y_{n+1}\| = \left[\|\Delta_n(x) y_n\|^2 + \left|\sum_{s=0}^{n+1} y(s) x(n+1-s)\right|^2\right]^{1/2}$$

$$\geq \|\Delta_n(x) y_n\| = \|\Delta_n(x)\|.$$

Property 2 is proved.

If $b_n \in Z^{n+1}$ and $B(z) = \sum_{t=0}^{n} b(t)z^t$ is not equal to zero in $|z|<1$, then b_n is called a minimum phase sequence.

Property 3. *There must exist a minimum phase sequence* $b_n \in Z^{n+1}$ ($b_n \neq 0$) *such that* $\|\Delta_n(x)b_n\| = \|\Delta_n(x)\| \cdot \|b_n\|$.

Proof. Take $d_n \in Z^{n+1}$ ($d_n \neq 0$) such that $\|\Delta_n(x)d_n\| = \|\Delta_n(x)\| \cdot \|d_n\|$. There exists a minimum phase sequence $b_n \in Z^{n+1}$ such that $D(z) = G(z)B(z)$, where $G(z)$ is the z-transformation of a pure phase sequence[2]. From [1], we have
$$\|\Delta_n(x)d_n\| \leqslant \|\Delta_n(x)b_n\|.$$
However $\|b_n\| = \|d_n\|$, $\|\Delta_n(x)b_n\| \leqslant \|\Delta_n(x)\| \cdot \|b_n\| = \|\Delta_n(x)\| \cdot \|d_n\|$. Therefore $\|\Delta_n(x)b_n\| = \|\Delta_n(x)\| \cdot \|b_n\|$. Property 3 is proved.

Property 4. *Let* $\|\Delta_n(x)\| = \|\Delta_{n+1}(x)\| = \cdots = \|\Delta_{n+k}(x)\|$ *and let* $b_n = (1, b(1), \cdots, b(n))'$ *be a minimum phase sequence such that* $\|\Delta_n(x)b_n\| = \|\Delta_n(x)\| \cdot \|b_n\|$, *then*
$$x(n+l) = -\sum_{t=1}^{n} b(t)x(n+l-t), \quad 1 \leqslant l \leqslant k. \tag{1.2}$$

Proof. Set $b_{n+k} = (b_n', 0, \cdots, 0)'$. Thus
$$\|\Delta_{n+k}(x)b_{n+k}\|^2 = \|\Delta_n(x)b_n\|^2 + \sum_{l=1}^{k}\left|\sum_{s=0}^{n} b(s)x(n+l-s)\right|^2$$
$$\leqslant \|\Delta_{n+k}(x)\|^2 \cdot \|b_{n+k}\|^2 = \|\Delta_n(x)\|^2 \cdot \|b_n\|^2,$$
from which we know that (1.2) holds. Property 4 is proved.

Setting
$$\nabla_n(x) = \begin{pmatrix} x(n) & x(n-1) & \cdots & x(0) \\ \vdots & \vdots & \ddots & \\ \vdots & \vdots & \ddots & \\ x(1) & x(0) & & 0 \\ x(0) & & & \end{pmatrix}, \tag{1.3}$$

we consider the equation
$$\nabla_n(x)\bar{b}_n = \lambda b_n, \tag{1.4}$$
where λ is a real number, $b_n \in Z^{n+1}$ and $\bar{b}_n$ denotes the conjugate vector of b_n.

Denote the real part and the imaginary part by Re and Im, then Eq. (1.4) is equivelant to

$$\begin{pmatrix} \mathrm{Re}\,\nabla_n(x) & \mathrm{Im}\,\nabla_n(x) \\ \mathrm{Im}\,\nabla_n(x) & -\mathrm{Re}\,\nabla_n(x) \end{pmatrix} \begin{pmatrix} \mathrm{Re}\,b_n \\ \mathrm{Im}\,b_n \end{pmatrix} = \lambda \begin{pmatrix} \mathrm{Re}\,b_n \\ \mathrm{Im}\,b_n \end{pmatrix}. \tag{1.5}$$

Since the matrix on the right side in (1.5) is a real symmetrical matrix, eigenvalues of (1.5) are real. Let $\lambda_1, \lambda_2, \cdots, \lambda_{2n+2}$ be the eigenvalues of (1.5) such that $|\lambda_1| \geqslant \|\lambda_2\| \geqslant \cdots \geqslant |\lambda_{2n+2}|$,

$$e_j = \begin{pmatrix} \mathrm{Re}\,d_n^{(j)} \\ \mathrm{Im}\,d_n^{(j)} \end{pmatrix}, \quad 1 \leqslant j \leqslant 2n+2,$$

is the eigenvector corresponding to λ_j; these $2n+2$ eigenvectors are perpendicular to each other.

Property 5. $\|\Delta_n(x)\| = |\lambda_1|$, $\|\Delta_n(x)\overline{d_n^{(1)}}\| = \|\Delta_n(x)\|$.

Proof. For any $\bar{b}_n \in Z^{n+1}$, we have

$$\|\Delta_n(x)b_n\| = \|\nabla_n(x)b_n\| = \left\| \begin{pmatrix} \mathrm{Re}\,\nabla_n(x) & \mathrm{Im}\,\nabla_n(x) \\ \mathrm{Im}\,\nabla_n(x) & -\mathrm{Re}\,\nabla_n(x) \end{pmatrix} \begin{pmatrix} \mathrm{Re}\,b_n \\ \mathrm{Im}\,b_n \end{pmatrix} \right\|. \tag{1.6}$$

Moreover, we know

$$\begin{pmatrix} \mathrm{Re}\,b_n \\ \mathrm{Im}\,b_n \end{pmatrix} = a_1 e_1 + a_2 e_2 + \cdots + d_{2n+2} e_{2n+2}. \tag{1.7}$$

From (1.5)—(1.7), we have

$$\|\Delta_n(x)b_n\|^2 = \|a_1 \lambda_1 e_1 + \cdots + a_{2n+2} \lambda_{2n+2} e_{2n+2}\|^2$$
$$= \sum_{j=1}^{2n+2} \lambda_j^2 a_j^2 \leqslant \lambda_1^2 \sum_{j=1}^{2n+2} a_j^2 = \lambda_1^2 \|b_n\|^2.$$

When $b_n = \overline{d_n^{(1)}}$, from (1.5) or (1.4) we know

$$\|\Delta_n(x)\overline{d_n^{(1)}}\| = |\lambda_1| \cdot \|d_n^{(1)}\| = |\lambda_1|.$$

Hence $\|\Delta_n(x)\| = |\lambda_1|$. Property 5 is proved.

Property 6. *Given $x(0), x(1), \cdots, x(n)$, there exists $x(n+1)$ such that*

$$\|\Delta_{n+1}(x)\| = \|\Delta_n(x)\|.$$

Proof. If $x(j) = 0$ $(0 \leqslant j \leqslant n)$, then setting $x(n+1) = 0$ we obtain

$$\|\Delta_{n+1}(x)\| = \|\Delta_n(x)\|.$$

If $x(j)$ $(0 \leqslant j \leqslant n)$ do not all vanish, then $\lambda_1 \neq 0$ and $d_n^{(1)}$ satisfies

$$\nabla_n(x)\overline{d_n^{(1)}} = \lambda_1 d_n^{(1)}. \tag{1.8}$$

Let $D(z) = \sum_{t=0}^{n} \overline{d_n^{(1)}(t)} Z^t$ and $B(z) = \sum_{t=0}^{n} b(t) Z^t$ being a minimum phase sequence and satisfying $|B(e^{-i\omega})| = |D(e^{-i\omega})|$. Consequently, $D(z) = G(z)B(z)$,

where $G(z)$ is a z-transformation of a pure phase sequence. From Property 5 and the proof of Property 3, it follows that
$$\|\Delta_n(x)b_n\| = |\lambda_1|. \tag{1.9}$$
Let
$$a_n = \lambda_1^{-1}\Delta_n(x)b_n, \tag{1.10}$$
$$A(z) = \sum_{t=0}^{n} a(t)Z^t,$$
$$Y(z) = \sum_{t=0}^{\infty} y(t)Z^t = \frac{A(z)}{B(z)}, \quad |z|<1. \tag{1.11}$$
By (1.10) and (1.11), we have
$$y(t) = \lambda_1^{-1} x(t), \quad 0 \leqslant t \leqslant n. \tag{1.12}$$
We note that
$$D(z)Y(z) = G(z)B(z)\frac{A(z)}{B(z)} = G(z)A(z). \tag{1.13}$$
From (1.9) and (1.10), it follows that
$$\|a_n\| = 1. \tag{1.14}$$
From (1.12) and Property 5, we have
$$\|\Delta_n(y)\overline{d_n^{(1)}}\| = |\lambda_1^{-1}| \cdot \|\Delta_n(x)\overline{d_n^{(1)}}\| = 1. \tag{1.15}$$
According to Theorem 2.4 of [1], from (1.13)—(1.15) and (1.8) it follows that
$$G(z)A(z) = \sum_{t=0}^{n} d_n^{(1)}(t) Z^{n-t}.$$
Hence $|A(e^{-i\omega})| = |D(e^{-i\omega})|$. Therefore, it follows from (1.11) that $y(t)$ is a pure phase sequence. From [1], we know $\|\Delta_{n+1}(y)h_{n+1}\| \leqslant \|h_{n+1}\|$ for any $h_{n+1} \in Z^{n+2}$, that is, $\|\Delta_{n+1}(y)\| \leqslant 1$. Moreover, from (1.5) and Property 2, we have $\|\Delta_{n+1}(y)\| = \|\Delta_n(y)\|$. Setting $x(n+1) = \lambda_1 y(n+1)$, we obtain $\|\Delta_{n+1}(x)\| = \|\Delta_n(x)\|$. Property 6 is proved.

Property 7. $\big| \|\Delta_n(x)\| - \|\Delta_n(\tilde{x})\| \big| \leqslant \sqrt{n+1} \sum_{j=0}^{n} |x(j) - \tilde{x}(j)|.$

Proof. When $y_n \in Z^{n+1}$ and $\|y_n\| = 1$, we have
$$\big| \|\Delta_n(x)y_n\| - \|\Delta_n(\tilde{x})y_n\| \big|$$
$$\leqslant \|\Delta_n(x-\tilde{x})y_n\|$$
$$\leqslant \Big[|x(0)-\tilde{x}(0)|^2 + \cdots + \Big(\sum_{j=0}^{n} |x(j)-\tilde{x}(j)|\Big)^2 \Big]^{1/2}$$

$$\leqslant \sqrt{n+1} \sum_{j=0}^{n} |x(j) - \tilde{x}(j)|.$$

Hence

$$\|\Delta_n(x)y_n\| \leqslant \|\Delta_n(\tilde{x})y_n\| + \sqrt{n+1} \sum_{j=0}^{n} |x(j) - \tilde{x}(j)|$$

$$\leqslant \|\Delta_n(\tilde{x})\| + \sqrt{n+1} \sum_{j=0}^{n} |x(j) - \tilde{x}(j)|.$$

From this we have

$$\|\Delta_n(x)\| \leqslant \|\Delta_n(\tilde{x})\| + \sqrt{n+1} \sum_{j=0}^{n} |x(j) - \tilde{x}(j)|.$$

In the same way, we can obtain

$$\|\Delta_n(\tilde{x})\| \leqslant \|\Delta_n(x)\| + \sqrt{n+1} \sum_{j=0}^{n} |x(j) - \tilde{x}(j)|.$$

Therefore, Property 7 holds.

Theorem 1 (Property 8). *Let $x(0), x(1), \cdots, x(n)$ be $n+1$ complex numbers, $\beta \geqslant \|\Delta_n(x)\|$, then there exists $x(n+1)$ such that*

$$\|\Delta_{n+1}(x)\| = \beta.$$

Proof. When $x(j)$ ($0 \leqslant j \leqslant n$) are given, $\Delta_{n+1}(x)$ depends only on $x(n+1)$. Hence, we denote $\Delta_{n+1}(x)$ by $\Delta_{n+1}(x, x(n+1))$.

From Property 1 follows $\lim_{x(n+1) \to \infty} \|\Delta_{n+1}(x, x(n+1))\| = \infty$. Consequently, there exists $x_1(n+1)$ such that $\|\Delta_{n+1}(x, x_1(n+1))\| \geqslant \beta$.

From Property 6, we know that there exists $x_0(n+1)$ such that $\|\Delta_{n+}(x, x_0(n+1))\| = \|\Delta_n(x)\|$.

According to Property 7, $\|\Delta_{n+1}(x, x(n+1))\|$ is a continuous function of $x(n+1)$. Therefore, there exists $x_2(n+1) \in \{x(n+1): |x(n+1)| \leqslant \max(|x_0(n+1)|, |x_1(n+1)|)\}$ such that $\|\Delta_{n+1}(x, x_2(n+1))\| = \beta$. Theorem 1 is proved.

II. The Property of the Pure Phase Sequence

In this section, using convolution-type matrix we investigate the properties of the pure phase sequence.

Lemma 1. *Let $g = (g(0), g(1), \cdots, g(t), \cdots)$ be a complex number sequence. For*

$$\|\Delta_n(g)\| \leqslant \beta, \quad \beta > 0, \ n \geqslant 0, \qquad (2.1)$$

the necessary and sufficient condition is as follows:

$$|G(e^{-i\omega})| \leqslant \beta \ (a.\ e.). \qquad (2.2)$$

Proof. Necessity. From (2.1) and Property 1 in Sec. I, it follows that

$$\sum_{t=0}^{\infty} |g(t)|^2 \leqslant \beta.$$

Thus $g \in l_2$. Hence there exists the frequence function $G(e^{-i\omega})$ of g.

We assume that (2.2) does not hold.

Let

$$E_1 = \{\omega: |G(e^{-i\omega})| > \beta\},$$

$$E_2 = \{\omega: |G(e^{-i\omega})| \leqslant \beta\},$$

$$\alpha > \int_{E_2} (\beta^2 - |G(e^{-i\omega})|^2) d\omega \Big/ \int_{E_1} (|G(e^{-i\omega})|^2 - \beta^2) d\omega,$$

$$|X(e^{-i\omega})|^2 = \begin{cases} \alpha, & \omega \in E_1, \\ 1, & \omega \in E_2, \end{cases}$$

then

$$\frac{1}{2\pi} \int_{-\pi}^{\pi} (|G(e^{-i\omega})|^2 - \beta^2) |X(e^{-i\omega})|^2 d\omega > 0. \qquad (2.3)$$

Since $|X(e^{-i\omega})|^2$ and $\ln |X(e^{-i\omega})|$ are integrable, there exists $x \in l_2$ such that $|X(e^{-i\omega})|$ is the amplitude spectrum of x[1,2]. According to the Parseval equality, (2.3) can be rewritten as

$$\sum_{t=0}^{\infty} |g(t) * x(t)|^2 > \beta^2 \sum_{t=0}^{\infty} |x(t)|^2.$$

We can always select n_0 such that

$$\sum_{t=0}^{n_0} |g(t) * x(t)|^2 > \beta^2 \sum_{t=0}^{n_0} |x(t)|^2,$$

that is,

$$\|\Delta_{n_0}(g) x_{n_0}\|^2 > \beta^2 \|x_{n_0}\|^2.$$

It follows that

$$\|\Delta_{n_0}(g)\| > \beta.$$

However, this is contrary to (2.1). Necessity is proved.

Sufficiency. By (2.2), we have $g \in l_2$. If $y = (y(0), \cdots, y(n), 0, \cdots)$, then from (2.2) we have

$$\|\Delta_n(g)y_n\|^2 = \sum_{t=0}^{n} |g(t)*y(t)|^2$$
$$\leqslant \frac{1}{2\pi}\int_{-\pi}^{\pi} |G(e^{-i\omega})|^2 |Y(e^{-i\omega})|^2 d\omega$$
$$\leqslant \beta^2 \|y_n\|^2.$$

From this (2.1) holds. The lemma is proved.

Theorem 2. *Let $g = (g(0), \cdots, g(n), \cdots)$ be a complex sequence. In order that g may be a pure phase sequence it is necessary and sufficient that*

$$\begin{cases} \|\Delta_n(g)\| \leqslant 1, \quad n \geqslant 0, & (2.4) \\ \sum_{t=0}^{\infty} |g(t)|^2 = 1. & (2.5) \end{cases}$$

Proof. Necessity. Let $y \in l_2$. Applying Theorem 2.5 of [1], we know

$$\|\Delta_n(g)y_n\|^2 = \sum_{t=0}^{n} |g(t)*y(t)|^2 \leqslant \sum_{t=0}^{n} |y(t)|^2 = \|y_n\|^2. \quad (2.6)$$

Hence, (2.4) holds. (2.5) always holds.

Sufficiency. From (2.4), we can obtain (2.6). According to Theorem 2.5 of [1], from (2.5) and (2.6) it follows that g is a pure phase sequence. Theorem 2 is proved.

Of course we can ask if

$$\|\Delta_n(g)\| \to 1, \quad n \to \infty, \quad (2.7)$$

whether g is a pure phase sequence. The following Theorem discusses this problem.

Theorem 3. *Let $g = (g(0), \cdots, g(t), \cdots)$ be a complex sequence. In order that (2.7) may hold it is necessary and sufficient that*

$$|G(e^{-i\omega})| \leqslant 1 \quad (a.\ e.), \quad (2.8)$$
$$\mu\{\omega: \alpha \leqslant |G(e^{-i\omega})| \leqslant 1\} > 0, \quad 0 < \alpha < 1, \quad (2.9)$$

where μ is a Lebesgue measure.

Proof. Necessity. Property 2 of Sec. I and (2.7) imply that

$$\|\Delta_n(g)\| \leqslant 1, \quad n \geqslant 0. \quad (2.10)$$

According to Lemma 1, (2.8) holds.

We now proceed to prove (2.9). Assume that (2.9) does not hold, then there is α_o such that $0 < \alpha_o < 1$ and

$$|G(e^{-i\omega})| \leqslant \alpha_o \quad (a.\ e.).$$

It follows from Lemma 1 that
$$\|\Delta_n(g)\| \leqslant \alpha_0 < 1, \quad n \geqslant 0,$$
Which is contrary to (2.7). Therefore, (2.9) holds.

Sufficiency. According to Lemma 1 and from (2.8), (2.10) holds.

Let
$$E_a = \{\omega : \alpha \leqslant |G(e^{-i\omega})| \leqslant 1\},$$
$$|X(e^{-i\omega})|^2 = \begin{cases} 1, & \omega \in E_a, \\ \beta, & \omega \in \overline{E}_a, \end{cases}$$

where $\beta > 0$ satisfies
$$\frac{\varepsilon}{1-\varepsilon} \mu(E_a) > \beta \mu(\overline{E}_a), \tag{2.11}$$
here ε is assigned such that $0 < \varepsilon < 1$.

By (2.11) we have
$$\mu(E_a) > (1-\varepsilon)\mu(E_a) + (1-\varepsilon)\beta\mu(\overline{E}_a),$$
and can rewrite it as
$$\int_{E_a} |X(e^{-i\omega})|^2 d\omega > (1-\varepsilon) \int_{-\pi}^{\pi} |X(e^{-i\omega})|^2 d\omega.$$

Hence we know
$$\frac{1}{2\pi} \int_{-\pi}^{\pi} |G(e^{-i\omega})|^2 \cdot |X(e^{-i\omega})|^2 d\omega \geqslant \frac{\alpha^2}{2\pi} \int_{E_a} |X(e^{-i\omega})|^2 d\omega$$
$$\geqslant \alpha^2 (1-\varepsilon) \frac{1}{2\pi} \int_{-\pi}^{\pi} |X(e^{-i\omega})|^2 d\omega.$$

Using the method of the proof of Lemma 1, we can know there exists n_0 such that
$$\|\Delta_{n_0}(g) x_{n_0}\|^2 \geqslant \alpha^2(1-\varepsilon) \|x_{n_0}\|^2.$$

This means
$$\|\Delta_{n_0}(g)\| \geqslant \alpha(1-\varepsilon)^{1/2}. \tag{2.12}$$

We can assign α and ε such that $|1-\alpha|$ and ε are arbitrarily small, it follows from (2.12) and Property 2 of Sec. II that (2.7) holds. The theorem is proved.

Theorem 3 leads to: For a complex sequence g, to have $\|\Delta_n(g)\| \to \beta (\beta > 0)$, it is necessary and sufficient that
$$\begin{cases} |G(e^{-i\omega})| \leqslant \beta, \\ \mu\left\{\omega : \alpha \leqslant \frac{|G(e^{-i\omega})|}{\beta} \leqslant 1\right\} > 0, \quad 0 < \alpha < 1. \end{cases}$$

Lemma 2. Let $x=(x(0), \cdots, x(t), \cdots)$ be a complex sequence, $\|\Delta_{n+k}(x)\| = 1 (k \geqslant 0)$, then x is a pure phase sequence and the z-transformation $X(z)$ of x can be expressed as

$$X(z) = \frac{\sum_{t=0}^{n} y(t) z^t}{\sum_{t=0}^{n} b(t) z^t}. \tag{2.13}$$

When $x(0) \neq 0$, $X(z)$ can be denoted by

$$X(z) = e^{i\theta} \prod_{j=1}^{l} \frac{z - \alpha_j}{1 - \bar{\alpha}_j z}, \tag{2.14}$$

where θ is a real number, $|\alpha_j| < 1$, $l \leqslant n$.

Proof. From Lemma 1 we have $|X(e^{-i\omega})| \leqslant 1$ (a. e.). With the aid of Property 3 in Sec. I, there exists a minimum phase sequence $b_n = (1, b(1), \cdots, b(n))'$ satisfying $\|\Delta_n(x) b_n\| = \|b_n\|$. Set $b = (1, b(1), \cdots, b(n), 0, 0, \cdots)$ and $y = x * b$. Property 4 yields that $y(t) = 0 (t > n)$. Consequently, $\|y\| = \|y_n\| = \|\Delta_n(x) b_n\| = \|b_n\| = \|b\|$. Applying the Parseval Theorem, we obtain

$$\int_{-\pi}^{\pi} |X(e^{-i\omega})|^2 |B(e^{-i\omega})|^2 d\omega = \int_{-\pi}^{\pi} |B(e^{-i\omega})|^2 d\omega.$$

Since $|X(e^{-i\omega})| \leqslant 1$ (a. e.), we have $|X(e^{-i\omega})| = 1$ (a. e.). Hence, x is a pure phase sequence. According to Theorem 2.4 of [1], (2.13) and (2.14) hold. Lemma 2 is proved.

Theorem 4. Let $x(0), x(1), \cdots, x(n)$ be $n+1$ complex numbers. In order that there may exist a pure phase sequence g satisfying

$$g(t) = x(t), \quad 0 \leqslant t \leqslant n,$$

it is necessary and sufficient that

$$\|\Delta_n(x)\| \leqslant 1.$$

Proof. Necessity. Applying Theorem 2, we have

$$\|\Delta_n(x)\| = \|\Delta_n(g)\| \leqslant 1.$$

Sufficiency. From Theorem 1 it follows that there exist $x(n+1), x(n+2), \cdots$ such that $\|\Delta_{n+k}(x)\| = 1 (k \geqslant 1)$. According to Lemma 2, $g = (x(0), \cdots, x(n), x(n+1), \cdots)$ is a pure phase sequence. The theorem is proved.

III. The Consistency Problem

The consistency problem in the complex space l_2 is as follows: given $h =$

$(h(0), \cdots, h(t), \cdots) \in l_2$ and $n+1$ complex numbers $x(0), x(1), \cdots, x(n)$, the problem whether there exist $x(n+1), x(n+2), \cdots$ such that $x=(x(0), \cdots, x(n), x(n+1), \cdots)$ satisfies $\gamma_{xx} = \gamma_{hh}$.

In the context of frequence spectrum, the consistency problem is as follows: given $x(0), x(1), \cdots, x(n)$ and $|H(e^{-i\omega})|$ where $|H(e^{-i\omega})|$ satisfies that $|H(e^{-i\omega})|^2$ and $\ln |H(e^{-i\omega})|$ are integrable, the problem whether there exist $x(n+1), x(n+2), \cdots$ such that $x=(x(0), x(1), \cdots, x(n), x(n+1), \cdots) \in l_2$ and $|X(e^{-i\omega})| = |H(e^{-i\omega})|$.

For reasons of expediency, we set

$$R(h, x_n) = \{y: y \in l_2, \gamma_{yy} = \gamma_{hh}, y(t) = x(t), 0 \leq t \leq n\}, \quad (3.1)$$

$$\tilde{H}(z) = \exp\left\{\frac{1}{2\pi}\int_{-\pi}^{\pi} \ln |H(e^{-i\omega})| \frac{e^{-i\omega}+z}{e^{-i\omega}-z} d\omega\right\}$$

$$= \sum_{t=0}^{\infty} \tilde{h}(t) z^t, \quad (3.2)$$

$$\tilde{x}_n = \begin{pmatrix} \tilde{x}(0) \\ \tilde{x}(1) \\ \vdots \\ \tilde{x}(n) \end{pmatrix} = \begin{pmatrix} \tilde{h}(0) & & & 0 \\ \tilde{h}(1) & \tilde{h}(0) & & \\ \vdots & \vdots & \ddots & \\ \tilde{h}(n) & \tilde{h}(n-1) & \cdots & \tilde{h}(0) \end{pmatrix}^{-1} \begin{pmatrix} x(0) \\ x(1) \\ \vdots \\ x(n) \end{pmatrix}. \quad (3.3)$$

By using the above symbols, the consistency problem is that given $x(0), \cdots, x(n)$ and $h \in l_2$ we will ask whether $R(h, x_n)$ is empty.

Theorem 5. *Given $x(0), x(1), \cdots, x(n)$ and $h \in l_2 (h \neq 0)$, and to make $R(h, x_n)$ not empty, it is necessary and sufficient that*

$$\|\Delta_n(\tilde{x})\| \leq 1. \quad (3.4)$$

Proof. Necessity. Let $y \in R(h, x_n)$. It follows from [1] or [2] that there exists a pure phase sequence g such that $g * \tilde{h} = y$. From this we have

$$\Delta_n(\tilde{h}) g_n = x_n. \quad (3.5)$$

Hence

$$g_n = [\Delta_n(\tilde{h})]^{-1} x_n = \tilde{x}_n. \quad (3.6)$$

As g is a pure phase sequence, it is noted from Theorem 2 and (3.6) that (3.4) holds.

Sufficiency. According to Theorem 4, it follows from (3.4) that there exists a pure phase sequence g such that (3.6) holds. Therefore (3.5) holds.

Set $y = g * \tilde{h}$, then $y \in R(h, x_n)$. Hence, $R(h, x_n)$ is not empty. The theorem is proved.

We now discuss the error relation in the consistency problem.

Theorem 6. *Given* $x(0), x(1), \cdots, x(n)$ *and* $h \in l_2 (h \neq 0)$ *which satisfy* $\|\Delta_n(\tilde{x})\| \leqslant 1$, *then*

$$\sup_{\substack{y^{(1)}, y^{(2)} \in R(h, x_n) \\ t \geqslant 0}} |y^{(1)}(t) - y^{(2)}(t)| \leqslant c[(1 - \|\Delta_n(\tilde{x})\|^2)\gamma_{hh}(0)]^{1/2}. \quad (3.7)$$

$$\sup_{y^{(1)}, y^{(2)} \in R(h, x_n)} \|y^{(1)} - y^{(2)}\| \geqslant [2(1 - \|\Delta_n(\tilde{x})\|^2)\gamma_{hh}(0)]^{1/2}, \quad (3.8)$$

where c is a positive constant depending on $\tilde{x}_n$.

Proof. With the aid of Lemma 2, we know that for $\tilde{x}_n$ there exists $\tilde{x}$ such that

$$\|\Delta_n(\tilde{x})\| \cdot e^{i\theta} \prod_{j=1}^{l} \frac{z - a_j}{1 - \bar{a}_j z} = \sum_{t=0}^{\infty} \tilde{x}(t) z^t. \quad (3.9)$$

Refer to Lemma 2 for the sense of parameters.

Let

$$B(z) = \sum_{t=0}^{l} b(t) z^t = \prod_{j=1}^{l} (1 - \bar{a}_j z). \quad (3.10)$$

Let $y \in R(h, x_n)$, then $y = g * \tilde{h}$ and g satisfies (3.6). It follows from (3.6) and (3.9) that

$$\|b * (g - \tilde{x})\|^2 = \|b * g - b * \tilde{x}\|^2$$

$$= \sum_{t=n+1}^{\infty} |b(t) * g(t)|^2 = \sum_{t=0}^{\infty} |b(t) * g(t)|^2 - \sum_{t=0}^{n} |b(t) * g(t)|^2$$

$$= \sum_{t=0}^{l} |b(t)|^2 - \|\Delta_n(\tilde{x})\| \sum_{t=0}^{l} |b(t)|^2. \quad (3.11)$$

Setting

$$d = \min_{\omega \in [-\pi, \pi]} |B(e^{-i\omega})|,$$

we have

$$\|g - \tilde{x}\|^2 = \frac{1}{2\pi} \int_{-\pi}^{\pi} |G(e^{-i\omega}) - \tilde{X}(e^{-i\omega})|^2 d\omega$$

$$\leqslant \frac{1}{d^2} \cdot \frac{1}{2\pi} \int_{-\pi}^{\pi} |B(e^{-i\omega})|^2 \cdot |G(e^{-i\omega}) - \tilde{X}(e^{-i\omega})|^2 d\omega$$

$$= \frac{1}{d^2} \|b * (g - \tilde{x})\|^2. \quad (3.12)$$

Let
$$c = 2\frac{\left[\sum_{t=0}^{l}|b(t)|^2\right]^{1/2}}{d}. \quad (3.13)$$

By (3.12) and (3.13) we get
$$\|g - \tilde{x}\|^2 \leqslant \frac{c^2}{4}(1 - \|\Delta_n(\tilde{x})\|^2). \quad (3.14)$$

We have
$$|y^{(1)}(t) - y^{(2)}(t)| = \left|\sum_{s=0}^{t}(g^{(1)}(s) - g^{(2)}(s))\tilde{h}(t-s)\right|$$
$$\leqslant \|g^{(1)} - g^{(2)}\| \cdot \|\tilde{h}\|$$
$$\leqslant (\|g^{(1)} - \tilde{x}\| + \|g^{(2)} - \tilde{x}\|)\|\tilde{h}\|$$
$$\leqslant c[(1 - \|\Delta_n(\tilde{x})\|^2)\gamma_{hh}(0)]^{1/2}. \quad (3.15)$$

From this we obtain (3.7).

Set
$$\lambda = \|\Delta_n(\tilde{x})\|,$$
$$\hat{x} = \frac{1}{\lambda}\tilde{x},$$
$$P^{(l)}(z) = \sum_{t=0}^{\infty} p^{(l)}(t)z^t = \frac{\lambda - z^l}{1 - \lambda z^l},$$
$$g^{(l)} = p^{(l)} * \hat{x}.$$

When $l \geqslant n$, $g_n^{(l)} = \tilde{x}_n$. Hence, $y^{(l)} = g^{(l)} * \tilde{h} = p^{(l)} * (\hat{x} * \tilde{h}) \in R(h, x_n)$.

Given $\varepsilon > 0$, we can mark off $N_1 \geqslant n$ so that
$$\sum_{t=N_1}^{\infty} |\hat{x}(t) * \tilde{h}(t)|^2 < \varepsilon^2.$$

Set
$$(\hat{x} * \tilde{h})^{(N_1)}(t) = \begin{cases} (\hat{x} * \tilde{h})(t), & t < N_1, \\ 0, & t \geqslant N_1. \end{cases}$$

Taking $l_1 = N_1$, we can mark off $N_2 > N_1$ such that
$$\sum_{t=N_2}^{\infty} |p^{(l_1)}(t) * (\hat{x} * \tilde{h})^{(N_1)}(t)|^2 < \varepsilon^2.$$

Taking $l_2 = N_2$, we have
$$\|y^{(l_1)} - y^{(l_2)}\| \geqslant \|g^{(l_1)} * (\hat{x} * \tilde{h})^{(N_1)} - g^{(l_2)} * (\hat{x} * \tilde{h})^{(N_1)}\| - 2\varepsilon$$

$$\geqslant \| (g^{(l_1)} * (\hat{x} * \tilde{h})^{(N_1)})^{(N_2)} - g^{(l_2)} * (\hat{x} * \tilde{h})^{(N_1)} \| - 3\varepsilon$$
$$= [\| (g^{(l_1)} * (\hat{x} * \tilde{h})^{(N_1)})^{(N_2)} - \lambda(\hat{x} * \tilde{h})^{(N_1)} \|^2$$
$$+ \| g^{(l_2)} * (\hat{x} * \tilde{h})^{(N_1)} - \lambda(\hat{x} * \tilde{h})^{(N_1)} \|^2]^{1/2} - 3\varepsilon$$
$$\geqslant [(1-\lambda^2)(\| \tilde{h} \|^2 - 2\varepsilon) + (1-\lambda^2)(\| \tilde{h} \|^2 - \varepsilon)]^{1/2} - 3\varepsilon.$$
(3.16)

As ε is arbitrarily assigned, by (3.16) we obtain (3.8). Thus the theorem is proved.

Applying (3.14) and the method of deriving (3.15), we can obtain the following corollary.

Corollary 1. *Under the supposition of Theorem 6, when $y \in R(h, x_n)$, we observe*

$$|y(t) - \tilde{x}(t) * \tilde{h}(t)| \leqslant \frac{c}{2} [(1 - \| \Delta_n(\tilde{x}) \|^2) \gamma_{hh}(0)]^{1/2}, \quad t \geqslant 0, \quad (3.17)$$

where c and $\tilde{x}(t)$ are determined by (3.13) *and* (3.9) *respectively.*

Theorem 6 and Corollary 1 give the restrictive condition of elements in $R(h, x_n)$. When $\| \Delta_n(\tilde{x}) \| = 1$, it follows from (3.7) that $R(h, x_n)$ has only one element. When $\| \Delta_n(\tilde{x}) \| \leqslant \frac{1}{\sqrt{2}}$, the norm of difference of two elements in $R(h, x_n)$ could be so great that it is not as smaller as $[\gamma_{hh}(0)]^{1/2}$.

We now give a simple corollary of Theorem 5.

Let $x(0), x(1), \cdots, x(n)$ be $n+1$ complex numbers, $h \in l_2 (h \neq 0)$.
Set
$$\mathscr{D} = \{q : q \text{ is a complex number such that } R(h, qx_n) \text{ is not empty}\},$$
$$R_\rho(h, x_n) = \left\{ y : y \in l_2, \frac{\gamma_{yy}}{\gamma_{yy}(0)} = \frac{\gamma_{hh}}{\gamma_{hh}(0)}, y(t) = x(t), 0 \leqslant t \leqslant n \right\},$$
$$R'_\rho(h, x_n) = \{y : y \in l_2 \text{ and there exists } q \neq 0 \text{ such that } qy \in R(h, qx_n)\}.$$
It is evident that
$$R_\rho(h, x_n) = R'_\rho(h, x_n).$$

Corollary 2.
$$\mathscr{D} = \{q : |q| \leqslant 1/ \| \Delta_n(\tilde{x}) \| \},$$
$$\min \{ \| y \| : y \in R'_\rho(h, x_n)\} = \| \Delta_n(\tilde{x}) \| \sqrt{\gamma_{hh}(0)}.$$

Here we discussed the consistency problem in the complex space l_2. As to the problem about the classes of related functions in the real space l_2, refer to [3].

References

[1] 舒立华,数学学报,**17**(1974),20—27.
[2] 程乾生,信号数字处理的数学原理,石油工业出版社,北京,1979.
[3] 周性伟,张箴,相关函数类中的一些分析性质,中国科学,1981,5:513—520.

原文载于 Scientia Sinica (Series A), Vol. XXV, No. 2, February 1982, 125—137.

Parameter Estimation in Exponential Models

Cheng Qiansheng

Institute of Mathematics, Peking University

Abstract: In this paper the problems of parameter estimation and order determination of an exponential (EX) model are studied in the time domain. In order to estimate the parameters, the parameter equations of an EX model are given in terms of the autocorrelation function, which is similar to the Yule-Walker equations of an autoregressive moving-average model. Estimates of parameters are obtained with the aid of the parameter equations and theorems are proved relating the convergence rate and asymptotic distribution of the estimates. We present two kinds of methods for estimating the order and prove that the estimates of the order are consistent.

Keywords: exponential model process; parameter estimation

1. Introduction

Let x_t be a weakly stationary time series with mean zero and autocorrelation function $r_x(t) = E x_{t+s} x_s$ and with an absolutely continuous spectrum function. It is well known that, when the spectral density function $S(\omega)$ of x_t satisfies $\log S(\omega) \in L(-\pi, \pi)$, $S(\omega)$ can be expressed as

$$S(\omega) = \exp\left(\sum_{l=-\infty}^{\infty} \lambda_l e^{-il\omega}\right).$$

If the spectral density function has the form

$$S(\omega) = \exp\left(\sum_{l=-p}^{p} \lambda_l e^{-il\omega}\right), \quad \lambda_p \neq 0, \tag{1}$$

then x_t is said to be an exponential model, denoted by $EX(p)$. p is called the order of the exponential model and λ_l are its parameters. The relation between $r_x(t)$ and $S(\omega)$ is

$$r_x(t) = \frac{1}{2\pi}\int_{-\pi}^{\pi} S(\omega) e^{it\omega} d\omega \tag{2}$$

or
$$\sum_t r_x(t)e^{-it\omega} = S(\omega). \tag{3}$$

When x_t is a real time series, parameters λ_t are real and $\lambda_t = \lambda_{-t}$. Thus (1) can be rewritten as

$$S(\omega) = \sigma^2 \exp\left\{\sum_{l=1}^{p} 2\lambda_l \cos(l\omega)\right\}. \tag{4}$$

For simplicity, throughout this paper we assume that x_t is a real time series.

Cheng (1985) has shown that the EX model is a type of maximum entropy spectrum model and has given some basic relations. Bloomfield (1973) and Dzhaparidze and Yaglom (1983) have studied the frequency domain method of parameter estimation in terms of the periodogram. The purpose of this paper is to investigate parameter estimation and order determination of the EX model in the time domain with the aid of autocorrelation.

The parameter equations of an EX model are presented in Section 2 and a property of the coefficient matrix of the parameter equations is given in Theorem 1. In Section 3 we discuss estimation of the parameters λ_l and σ^2 and the convergence rate of the estimates. In Section 4 we investigate the asymptotic distribution of the estimates. Section 5 gives two kinds of methods for estimating the order and proves the consistency. In Section 6 we give the proof of Theorem 1.

2. Parameter Equations of the Exponential Model

Taking the derivatives of both sides of Equations (3) and (1), we obtain

$$\sum_{t=-\infty}^{\infty} tr_x(t)e^{-it\omega} = \left\{\sum_{t=-\infty}^{\infty} r_x(t)e^{-it\omega}\right\}\left(\sum_{l=-p}^{p} l\lambda_l e^{-il\omega}\right). \tag{5}$$

By comparison of coefficients we obtain

$$\sum_{l=-p}^{p} l\lambda_l \{r_x(t-l) - r_x(t+l)\} = tr_x(t) \quad (-\infty < t < \infty).$$

For a real time series we have

$$\sum_{l=1}^{p} l\lambda_l \{r_x(t-l) - r_x(t+l)\} = tr_x(t) \quad (t \geq 1). \tag{6}$$

In practice, we consider the set of linear equations

$$\sum_{l=1}^{L} h_l \{r_x(t-l) - r_x(t+l)\} = \mathrm{tr}_x(t) \quad (1 \leqslant t \leqslant M), \qquad (7)$$

where $M \geqslant L \geqslant p$. Formula (7) is the parameter equation of the EX model and is similar to the Yule-Walker equations of the autoregressive movingaverage (ARMA) model.

The coefficient matrix of (7) is denoted by
$$Q_{M \times L} = (r_x(t-l) - r_x(t+l))_{1 \leqslant t \leqslant M, \; 1 \leqslant l \leqslant L} \qquad (8)$$

$$= \begin{bmatrix} r_x(0)-r_x(2) & r_x(-1)-r_x(3) & \cdots & r_x(1-L)-r_x(1+L) \\ \vdots & \vdots & & \vdots \\ r_x(M-1)-r_x(M+1) & r_x(M-2)-r_x(M+2) & \cdots & r_x(M-L)-r_x(M+L) \end{bmatrix}.$$
$$(9)$$

A property of the matrix $Q_{M \times L}$ is very important for solving Equation (7). We have the following theorem.

Theorem 1. *For positive integers of L and M,*
$$\mathrm{rank}\,(Q_{M \times L}) = \min(M, L). \qquad (10)$$

The proof of Theorem 1 will be given in Section 6.

3. Estimation of $\lambda_l (1 \leqslant l \leqslant p)$ and σ^2

We assume that N observations $x_1, x_2, \cdots, x_N$ are available. Set
$$\hat{r}_x(t) = \frac{1}{N} \sum_{s=1}^{N-t} x_{t+s} x_s, \quad (0 \leqslant t \leqslant N-1),$$
$$\hat{r}_x(-t) = \hat{r}_x(t) \quad (0 \leqslant t \leqslant N-1). \qquad (11)$$

We assume that p is known and that integers M and L satisfy $M \geqslant L \geqslant p$. Set
$$\hat{Q}_{M \times L} = (\hat{r}_x(t-l) - \hat{r}_x(t+l))_{1 \leqslant t \leqslant M, 1 \leqslant l \leqslant L}, \qquad (12)$$
$$\hat{h} = (\hat{h}_1, \cdots, \hat{h}_L)^{\mathrm{T}}, \qquad (13)$$
$$\hat{g} = (\hat{r}_x(1), 2\hat{r}_x(2), \cdots, M\hat{r}_x(M))^{\mathrm{T}}. \qquad (14)$$

Here $\hat{h}$ is the solution of the following set of linear equations:
$$\hat{Q}_{M \times L} \hat{h} = \hat{g}. \qquad (15)$$

Estimates of λ_l and σ^2 are given by the expressions
$$\hat{\lambda}_l = \hat{h}_l / l \quad (1 \leqslant l \leqslant p), \qquad (16)$$
$$\hat{\sigma}^2 = \hat{r}_x(0) \Big/ \Big[\frac{1}{2\pi} \int_{-\pi}^{\pi} \exp\Big\{ \sum_{l=1}^{p} 2\hat{\lambda}_l \cos(l\omega) \Big\} d\omega \Big]. \qquad (17)$$

We now discuss the convergence properties of the estimators. We know that x_t can be expressed as

$$x_t = w_t * u_t = \sum_{s=0}^{\infty} w_s u_{t-s}, \tag{18}$$

where u_t is a white noise with mean zero and $Eu_t^2 = \sigma^2$. The Z-transform of w_t is

$$W(Z) = \sum_{t=0}^{\infty} w_t Z^t = \exp\left(\sum_{l=1}^{p} \lambda_l Z^l\right). \tag{19}$$

Throughout the following discussion we further assume that the u_t are independent and identically distributed and $Eu_t^4 < \infty$. Because $W(Z)$ is an entire function, the following lemma holds.

Lemma 1. *There exist two positive numbers B and b such that*

$$|w_t| \leqslant Be^{-bt} \quad (t \geqslant 0). \tag{20}$$

Employing Lemma 1 and the method used in An Hong-Zhi et al. (1982), we have the following lemma.

Lemma 2.

$$\max_{0 \leqslant t \leqslant g_N} |\hat{r}_x(t) - r_x(t)| = O(\varepsilon_N) \text{ almost surely (a. s.)}, \tag{21}$$

where

$$\varepsilon_N = (\log\log N/N)^{1/2}, \tag{22}$$

$$g_N = O\{(\log N)^a\} \text{ for some } a > 0. \tag{23}$$

We now turn to the following set of linear equations:

$$Q_{M \times L} h = g. \tag{24}$$

Here

$$h = (h_1, \cdots, h_L)^T, \tag{25}$$

$$g = (r_x(1), 2r_x(2), \cdots, Mr_x(M))^T. \tag{26}$$

It follows from Theorem 1 that Equation (24) has an unique solution. According to (7), the solution is

$$h = (h_1, \cdots, h_p, \cdots, h_L)^T$$
$$= (\lambda_1, 2\lambda_2, \cdots, p\lambda_p, 0, \cdots, 0)^T. \tag{27}$$

Label the columns in the matrix $Q_{M \times L}$ as α_i ($i = 1, 2, \cdots, L$). Applying the Gram-Schmidt orthonormalization procedure to the columns yields

$$\beta_1 = \alpha_1 / \|\alpha_1\|,$$

$$\tilde{\alpha}_i = \alpha_i - \sum_{j=1}^{i-1} (\alpha_i, \beta_j) \beta_j, \tag{28}$$

$$\beta_i = \tilde{\alpha}_i / \|\tilde{\alpha}_i\| \quad (2 \leqslant i \leqslant L),$$

where $(\alpha_i, \beta_j) = \alpha_i^T \beta_j$ and $\|\alpha_1\| = (\alpha_1^T \alpha_1)^{1/2}$.

We can easily obtain

$$h_p = (g, \beta_p) / \|\tilde{\alpha}_p\|,$$

$$h_k = \left\{ (g, \beta_k) - \sum_{i=k+1}^{p} h_i (\alpha_i, \beta_k) \right\} \Big/ \|\tilde{\alpha}_k\| \quad (k = p-1, p-2, \cdots, 1). \quad (29)$$

Here $(h_1, \cdots, h_p, 0, \cdots, 0)$ is the solution of equation (24). In order to find the convergence property of λ_l, we now introduce the following lemma.

Lemma 3. *Let $\hat{\theta}_l$ be the estimators of parameters $\theta_l (l=1,2,\cdots,L)$ and let L be a function of sample size N. Let $\hat{\theta}_l$ satisfy*

$$\sup_{1 \leqslant l \leqslant L} |\hat{\theta}_l - \theta_l| = O(e_N) \text{ a. s.}, \quad (30)$$

where

$$e_N \to 0, \quad L^2 e_N \to 0, \quad \text{when } N \to \infty. \quad (31)$$

Assume that $f(\theta_1, \cdots, \theta_L)$ is a continuous function with second partial derivatives in a neighbourhood of the point $(\theta_1, \cdots, \theta_L)$ and satisfies

$$\sum_{l=1}^{L} \left| \frac{\partial}{\partial \theta_l} f \right| < c_1 < \infty, \quad (32)$$

$$\left| \frac{\partial^2}{\partial \theta_k \partial \theta_l} f \right| < c_2 < \infty. \quad (33)$$

Then

$$f(\hat{\theta}_1, \cdots, \hat{\theta}_L) - f(\theta_1, \cdots, \theta_L) = O(e_N) \text{ a. s.} \quad (34)$$

Proof. We have

$$f(\hat{\theta}_1, \cdots, \hat{\theta}_L) - f(\theta_1, \cdots, \theta_L) = \sum_{l=1}^{L} \frac{\partial f}{\partial \theta_l} (\hat{\theta}_l - \theta_l)$$

$$+ \frac{1}{2} \left\{ \sum_{l=1}^{L} (\hat{\theta}_l - \theta_l) \frac{\partial}{\partial \theta_l} \right\}^2 f\{\theta_1 + \lambda(\hat{\theta}_1 - \theta_1), \cdots, \theta_L + \lambda(\hat{\theta}_L - \theta_L)\}$$

$$\doteq \text{I} + \text{II}, \quad (35)$$

where $0 < \lambda < 1$.

By (30) and (32), we have

$$|\text{I}| \leqslant c_1 \sup_{1 \leqslant l \leqslant L} |\hat{\theta}_l - \theta_l| = O(e_N) \text{ a. s.} \quad (36)$$

Equations (30) and (33) yield

$$|\mathrm{II}| \leqslant \frac{1}{2}c_2 L^2 \Big(\sup_{1\leqslant l\leqslant L}|\hat{\theta}_l-\theta_l|\Big)^2 = \frac{1}{2}c_2 L^2 O(e_N)O(e_N) \leqslant O(e_N) \text{ a. s.} \quad (37)$$

It follows from (35), (36) and (37) that (34) holds. Lemma 3 is proved. ∎

Theorem 2. *For the estimators $\hat{\lambda}_l$ and $\hat{\sigma}^2$ in* (16) *and* (17),

$$\sup_{1\leqslant l\leqslant p}|\hat{\lambda}_l-\lambda_l|=O(\varepsilon_N) \text{ a. s.}, \quad (38)$$

$$\hat{\sigma}^2-\sigma^2=O(\varepsilon_N) \text{ a. s.}, \quad (39)$$

here $\varepsilon_N=(\log\log N/N)^{1/2}$.

Proof. With $\hat{r}_x(t)$ replacing the true autocorrelation $r_x(t)$, Equation (24) becomes Equation (15) and we get $\hat{\alpha}_i, \hat{\tilde{\alpha}}_i$ and $\hat{\beta}_i$ corresponding to $\alpha_i, \tilde{\alpha}_i$ and β_i in (28). Similar to (29), we have

$$\begin{aligned}\hat{h}_p &= (\hat{g},\hat{\beta}_p)/\|\hat{\alpha}_p\|, \\ \hat{h}_k &= \Big\{(\hat{g},\hat{\beta}_k)-\sum_{i=k+1}^{p}\hat{h}_i(\hat{\alpha}_i,\hat{\beta}_k)\Big\}\Big/\|\hat{\tilde{\alpha}}_k\| \quad (k=p-1,p-2,\cdots,1).\end{aligned} \quad (40)$$

Note that Theorem 1 implies

$$\|\tilde{\alpha}_i\|>0 \quad (1\leqslant i\leqslant L). \quad (41)$$

Here $\|\tilde{\alpha}_1\|=\|\alpha_1\|$. It follows from Lemmas 2 and 3 that

$$\begin{aligned}(\hat{g},\hat{\beta}_k)-(g,\beta_k)&=O(\varepsilon_N) \text{ a. s.}, \\ (\hat{\alpha}_i,\hat{\beta}_k)-(\alpha_i,\beta_k)&=O(\varepsilon_N) \text{ a. s.} \quad (1\leqslant i,k\leqslant L), \\ \|\hat{\tilde{\alpha}}_k\|-\|\tilde{\alpha}_k\|&=O(\varepsilon_N) \text{ a. s.}\end{aligned} \quad (42)$$

Using Lemmas 2 and 3 again, from (40) we get

$$\hat{h}_l-h_l=O(\varepsilon_N) \text{ a. s.} \quad (1\leqslant l\leqslant p). \quad (43)$$

By (16), (27) and (43), it follows that

$$\hat{\lambda}_l-\lambda_l=O(\varepsilon_N) \text{ a. s.} \quad (1\leqslant l\leqslant p). \quad (44)$$

From (4) we get

$$\sigma^2 = r_x(0)\Big/\Big[\frac{1}{2\pi}\int_{-\pi}^{\pi}\exp\Big\{\sum_{l=1}^{p}2\lambda_l\cos(l\omega)\Big\}d\omega\Big]. \quad (45)$$

By (45), (17), (44) and Lemma 2, according to Lemma 3, we know that (44) holds. Theorem 2 is proved. ∎

We point out that from (19) we have

$$\frac{1}{2\pi}\int_{-\pi}^{\pi}\exp\Big\{\sum_{l=1}^{p}2\lambda_l\cos(l\omega)\Big\}d\omega = \sum_{t=0}^{\infty}w_t^2 \quad (46)$$

and
$$w_t = 0 \quad (t<0),$$
$$w_0 = 1, \qquad (47)$$
$$w_t = \frac{1}{t}\sum_{l=0}^{p} l\lambda_l w_{t-l} \quad (t \geqslant 1)$$

(see Cheng, 1985). With $\hat{\lambda}_l$ replacing the true parameters λ_l in (47), we get $\hat{w}_t$ and

$$\frac{1}{2\pi}\int_{-\pi}^{\pi} \exp\left\{\sum_{l=1}^{p} 2\hat{\lambda}_l \cos(l\omega)\right\} d\omega \doteq \sum_{t=0}^{T} \hat{w}_t^2. \qquad (48)$$

Here T is a large positive integer. Equation (48) can be used to estimate σ^2.

4. Asymptotic Distribution

To establish the asymptotic distribution theorems for the estimates $\hat{\lambda}_l$ and $\hat{\sigma}^2$, we need the following lemma which follows directly from Hannan (1970, Chapter Ⅳ).

Lemma 4. *For any integer $k \geqslant 0$, the joint distribution of the quantities*
$$\{N^{1/2}[\hat{r}_x(t) - r_x(t)], \; t = 0, 1, \cdots, k\}$$
converges to the multivariate normal distribution with mean zero and covariance matrix $V = (v_{ij})$, a $(k+1)\times(k+1)$ positive definite matrix.

It is easily verified that
$$\lim_{N\to\infty} NE\{\hat{r}_x(i) - r_x(i)\}\{\hat{r}_x(j) - r_x(j)\} = v_{ij}$$
$$= \sum_{m=-\infty}^{\infty} \{r_x(m+i)r_x(m+j) + r_x(m+i)r_x(m-j)\}$$
$$+ r_x(i)r_x(j)\frac{c_4(u_0)}{\sigma^4}, \qquad (49)$$

where $c_4(u_0)$ is the fourth cumulant of u_0 which is equal to $Eu_0^4 - 3\sigma^4$.

Consider Equations (15) and (24). Let G be the diagonal matrix
$$G = \mathrm{diag}\{1, 1/2, \cdots, 1/L\}. \qquad (50)$$
Thus
$$\hat{\lambda} = (\hat{\lambda}_1, \cdots, \hat{\lambda}_L)^T = G\hat{h},$$
$$\lambda = (\lambda_1, \cdots, \lambda_L)^T = Gh.$$

Notice that, for $l > p, \lambda_l = 0$.

Theorem 3. *For fixed integers M, L, $M \geq L \geq p$, the joint distribution of the quantities*

$$\{N^{1/2}(\hat{\lambda}_l - \lambda_l), l = 1, 2, \cdots, L\}$$

converges to the multivariate normal distribution with zero mean and covariance matrix Σ:

$$\Sigma = G(Q_{M \times L}^T Q_{M \times L})^{-1} Q_{M \times L}^T W Q_{M \times L} (Q_{M \times L}^T Q_{M \times L})^{-1} G, \tag{51}$$

where $W = (w_{st})$ is an $M \times M$ matrix,

$$w_{st} = \sum_{m=-\infty}^{\infty} m^2 \{r_x(m+s) r_x(m+t) - r_x(m+s) r_x(m-t)\}, \tag{52}$$

and $Q_{M \times L}$ is defined by (9).

Proof. Equations (15) and (24) yield

$$(\hat{Q}_{M \times L} - Q_{M \times L})(\hat{h} - h) + Q_{M \times L}(\hat{h} - h) = f,$$

where

$$f = (f_1, \cdots, f_M)^T = (\hat{g} - g) - (\hat{Q}_{M \times L} - Q_{M \times L}) h.$$

It follows from Lemma 2 and Theorem 2 that

$$N^{1/2}(\hat{Q}_{M \times L} - Q_{M \times L})(\hat{h} - h) = N^{1/2} O(\varepsilon_N) O(\varepsilon_N) \to 0 \text{ a. s.}$$

Therefore, $N^{1/2} Q_{M \times L}(\hat{h} - h)$ and $N^{1/2} f$ have asymptotically the same distribution.

Set

$$u_t = (-h_L, -h_{L-1}, \cdots, -h_1, t, h_1, \cdots, h_L)^T,$$

$$v_t = (\hat{r}_x(t-L) - r_x(t-L), \hat{r}_x(t-L+1) - r_x(t-L+1), \cdots, \hat{r}_x(t+l) - r_x(t+L))^T.$$

Thus the element f_t of f can be expressed as

$$f_t = u_t^T v_t \quad (1 \leq t \leq M). \tag{53}$$

By (49),

$$\lim_{N \to 0} N E v_s v_t^T = \lim_{N \to 0} \{N E\{\hat{r}_x(s+i) - r_x(s+i)\} \times \{\hat{r}_x(t+j) - r_x(t+j)\}]_{-L \leq i, j \leq L} = \mathcal{L},$$

where

$$\mathcal{L} = \Big[\sum_{m=-\infty}^{\infty} \{r_x(m+s+i) r_x(m+t+j) + r_x(m+s+i) r_x(m-t-j)\}$$

$$+ r_x(s+i) r_x(t+j) \frac{c_4(u_0)}{\sigma^4} \Big]_{-L \leq i, j \leq L}.$$

Notice that Equation (5) implies

$$u_s^T (r_x(m+s-L), \cdots, r_x(m+s+l))^T = s r_x(m+s) - (m+s) r_x(m+s)$$

$$= -mr_x(m+s),$$
$$u_t^T(r_x(m-t+L),\cdots,r_x(m-t-L))^T = (m-t)r_x(m-t)+\text{tr}_x(m-t)$$
$$= mr_x(m-t).$$

Then we have
$$u_s^T L u_t = w_{st},$$
where w_{st} is defined by (52).

Hence
$$EN^{1/2}f(N^{1/2}F)^T = (u_s^T NEv_s v_t^T u_t)_{1\leqslant s,t\leqslant M}$$
$$\to (u_s^T L u_t)_{1\leqslant s,t\leqslant M} = W \quad (N\to\infty). \quad (54)$$

From (53), (54) and Lemma 4, the joint distribution of the vector $N^{1/2}f$ converges to the multivariate normal distribution with zero mean and covariance matrix W. Since $N^{1/2}Q_{M\times N}(\hat{h}-h) = N^{1/2}Q_{M\times N}G^{-1}(\hat{\lambda}-\lambda)$ and $N^{1/2}f$ have asymptotically the same distribution, the proof is thus complete. ∎

Consider next the asymptotic distribution of the estimator $\hat{\sigma}^2$.

Lemma 5. *Suppose that the conditions of Lemma 3 hold and, in addition,*
$$N^{1/2}L^2 e_N^2 \to 0 \text{ when } N\to\infty.$$

Then the quantity
$$N^{1/2}(f(\hat{\theta}_1,\cdots,\hat{\theta}_L)-f(\theta_1,\cdots,\theta_L))$$

and the quantity
$$N^{1/2}\sum_{l=1}^{L}\frac{\partial f}{\partial \theta_l}(\hat{\theta}_l-\theta_l)$$

have asymptotically the same distribution.

The proof of Lemma 5 is similar to that of Theorem 3. Since
$$\sigma^2 = r_x(0) \Big/ \frac{1}{2\pi}\int_{-\pi}^{\pi}\exp\Big\{\sum_{l=1}^{p}2\lambda_l\cos(l\omega)\Big\},$$
we have
$$\frac{\partial \sigma^2}{\partial r_x(0)} = \frac{\sigma^2}{r_x(0)},$$
$$\frac{\partial \sigma^2}{\partial \lambda_l} = 2r_x(l)\frac{\sigma^2}{r_x(0)}.$$

Let

$$a = \frac{\sigma^2}{r_x(0)}(1, 2r_x(1), \cdots, 2r_x(p))^T, \qquad (55)$$

$$\hat{b} = N^{1/2}(\hat{r}_x(0) - r_x(0), \hat{\lambda}_1 - \lambda_1, \cdots, \hat{\lambda}_p - \lambda_p)^T. \qquad (56)$$

From Lemma 5, $N^{1/2}(\hat{\sigma}^2 - \sigma^2)$ and $a^T \hat{b}$ have the same asymptotic distribution.

In the same way as in the proof of Theorem 3, it is easily verified that the joint distribution of the vector

$$N^{1/2}(\hat{r}_x(0) - r_x(0), \hat{\lambda}_1 - \lambda_1, \cdots, \hat{\lambda}_L - \lambda_L)^T \qquad (57)$$

converges to the multivariate normal distribution with zero mean and covariance matrix D:

$$D = \begin{bmatrix} 1 & 0 \\ 0 & G(Q_{M\times L}^T Q_{M\times L})^{-1} Q_{M\times L}^T \end{bmatrix} C \begin{bmatrix} 1 & 0 \\ 0 & G(Q_{M\times L}^T Q_{M\times L})^{-1} Q_{M\times L}^T \end{bmatrix}^T, \qquad (58)$$

where $C = (c_{ij})$ is an $(L+1) \times (L+1)$ matrix,

$$c_{11} = 2 \sum_{m=-\infty}^{\infty} r_x^2(m) + \frac{c_4(u_0)}{\sigma^4},$$

$$c_{1i} = c_{i1} = -2 \sum_{m=-\infty}^{\infty} m r_x(m+i) r_x(m) \quad (2 \leqslant i \leqslant L+1),$$

$$c_{ij} = \sigma_{i-1, j-1} \quad (i \geqslant 2, j \geqslant 2),$$

and where σ_{ij} are the entries of the matrix (51).

Comparing (56) with (57), we obtain the following theorem.

Theorem 4. *The distribution of $N^{1/2}(\hat{\sigma}^2 - \sigma^2)$ converges to the normal distribution with zero mean and covariance*

$$a^T \tilde{D} a, \qquad (59)$$

where a is defined by (55) and $\tilde{D} = (\tilde{d}_{ij})$ is a $(p+1) \times (p+1)$ matrix,

$$\tilde{d}_{ij} = d_{ij} \quad (1 \leqslant i, j \leqslant p+1),$$

and where the d_{ij} are the entries of the matrix (58).

5. Estimation of Order

Let the time series x_t contain no white noise and the maximum possible EM order be L. The maximum can be roughly deduced from a *priori* knowledge. We consider formulas (24)—(28). Set

$$g_0 = g / \|g\|, \qquad (60)$$

$$\mu_j = \sum_{m=1}^{j}(g_0,\beta_m)^2 \quad (1\leq j\leq L). \tag{61}$$

We obviously have
$$\mu_j<1,\quad j<p;\quad \mu_j=1,\quad j\geq p. \tag{62}$$
That is
$$\sum_{m=1}^{p}(g_0,\beta_m)^2 = 1,\quad (g_0,\beta_p)^2>0,\quad (g_0,\beta_j)=1,\quad j>p. \tag{63}$$

With $\hat{r}_x(t)$ replacing the true autocorrelation $r_x(t)$, let
$$\begin{aligned}\hat{g}_0 &= \hat{g}/\|\hat{g}\|,\\ \hat{\mu}_j &= \sum_{m=1}^{j}(\hat{g}_0,\beta_m)^2 \quad (1\leq j\leq L)\end{aligned} \tag{64}$$

(see Section 3). In a way similar to the proof of Theorem 2, we have the following lemma.

Lemma 6.
$$\begin{aligned}(\hat{g}_0,\hat{\beta}_m)^2-(g_0,\beta_m)^2&=O(\varepsilon_N)\text{ a.s.},\\ \hat{\mu}_j-\mu_j&=O(\varepsilon_N)\text{ a.s.}\quad (1\leq j,m\leq L),\end{aligned} \tag{65}$$
where $\varepsilon_N=(\log\log N/N)^{1/2}$.

Let
$$e_N=c\left(\frac{\log\log N}{N}\right)^{1/2-v}, \tag{66}$$
where $c>0, 0<v<\frac{1}{2}$. We have
$$e_N\to 0,\quad O(\varepsilon_n)/e_N\to 0,\quad \text{when } N\to\infty. \tag{67}$$
Let
$$\text{QSC}(j)=1-\hat{\mu}_j+je_N. \tag{68}$$

Now we give the theorem of the order determination.

Theorem 5. *Let*
$$\hat{p}_1=\min_{1\leq i\leq L}\{i\,|\,1-\hat{\mu}_i<e_N\} \tag{69}$$
and let $\hat{p}_2$ satisfy
$$\text{QSC}(\hat{p}_2)=\min_{1\leq j\leq L}\text{QSC}(j). \tag{70}$$
Then
$$\hat{p}_1\to p\text{ a.s.}, \tag{71}$$

$$\hat{p}_2 \to p \text{ a. s.} \tag{72}$$

Proof. It follows from Lemma 6 and Equations (63) and (64) that

$$1-\hat{\mu}_i = \sum_{m=i+1}^{L}(g_0,\beta_m)^2 + O(\varepsilon_N). \tag{73}$$

According to (63) and (67), when N is sufficiently large we have

$$\begin{aligned}1-\hat{\mu}_i \geqslant (g_0\beta_p)^2 + O(\varepsilon_N) > e_N \text{ a. s.} \quad (i<p),\\ 1-\hat{\mu}_i = O(\varepsilon_N) < e_N \text{ a. s.} \quad (i\geqslant p).\end{aligned} \tag{74}$$

Thus (71) holds.

By (68) and (73), we obtain

$$\text{QSC}(j) = \sum_{i=j+1}^{L}(g_0,\beta_i)^2 + O(\varepsilon_N) + je_N, \tag{75}$$

$$\text{QSC}(p) = pe_N + O(\varepsilon_N). \tag{76}$$

Noting (67), (63), (75) and (76), for sufficiently large N we find that

$$\text{QSC}(j) = pe_N + (j-p)e_N + O(\varepsilon_N) > \text{QSC}(p) \text{ a. s.} \quad (j>p), \tag{77}$$

$$\text{QSC}(j) \geqslant (g_0,\beta_p)^2 + je_N + O(\varepsilon_N) > \text{QSC}(p) \text{ a. s.} \quad (j<p). \tag{78}$$

Expressions (77) and (78) yield (72). Theorem 5 is proved. ∎

Theorem 5 shows that the estimators $\hat{p}_1$ and $\hat{p}_2$ of order p are consistent.

6. Proof of Theorem 1

Now we prove Theorem 1 for any autocorrelation function $r_x(t)$. Let

$$\begin{aligned}S(\omega) &= \sum_t r_x(t)e^{-it\omega},\\ E &= \{\omega: S(\omega) \neq 0\}.\end{aligned} \tag{79}$$

Since

$$\frac{1}{2\pi}\int_{-\pi}^{\pi} S(\omega)d\omega = r_x(0) > 0,$$

we obviously have

$$\mu(E) > 0, \tag{80}$$

where μ denotes Lebesgue measure. Set

$$b_t = \frac{1}{2\pi}\int_{-\pi}^{\pi}\{S(\omega)\}^{1/2} e^{it\omega}d\omega \quad (t=0,\pm 1,\cdots). \tag{81}$$

Then

$$B(\omega) \triangleq \sum_t b_t e^{-it\pi} = \{S(\omega)\}^{1/2},$$

$$r_x(t) = b_t * b_{-t} = \sum_s b_{s+t} b_s. \tag{82}$$

In particular

$$r_x(t-l) = \sum_s b_{s+t} b_{s+l},$$

$$r_x(t+l) = \sum_s b_{s+t} b_{s-l}. \tag{83}$$

Let

$$a = (a_1, a_2, \cdots, a_L)^T, \tag{84}$$

$$A(\omega) = \sum_{t=1}^{L} a_t e^{-it\omega}, \tag{85}$$

$$a_t = \begin{cases} a_t, & 1 \leqslant t \leqslant L, \\ 0, & \text{otherwise}, \end{cases} \tag{86}$$

$$g = b_t * a_{-a} = \sum_{s=1}^{L} a_s b_{t+s}, \tag{87}$$

$$h_t = b_t * a_t = \sum_{s=1}^{L} a_s b_{t-s}, \tag{88}$$

$$\mathrm{I} = \sum_{t,s=1}^{L} a_t a_s r_x(t-s), \tag{89}$$

$$\mathrm{II} = \sum_{t,s=1}^{L} a_t a_s r_x(t+s). \tag{90}$$

Substituting (83) for $r_x(t-s)$ and $r_x(t+s)$ in (89) and (90) yields

$$\mathrm{I} = \sum_s g_s^2, \quad \mathrm{II} = \sum_s g_s h_s. \tag{91}$$

By (87) and (88), the spectral functions $G(\omega)$ and $H(\omega)$ of g_t and h_t are

$$G(\omega) = B(\omega)\overline{A(\omega)}, \quad H(\omega) = B(\omega) A(\omega) \tag{92}$$

respectively. Thus

$$|G(\omega)| = |H(\omega)|.$$

According to the Parseval equality we have

$$\sum_s g_s^2 = \sum_s h_s^2. \tag{93}$$

By using the Schwartz inequality and (85) and (87) it follows that

$$|\text{II}| \leqslant \Big(\sum_s g_s^2 \sum_s h_s^2\Big)^{1/2} = \text{I}. \qquad (94)$$

We shall prove that the equality in (94) does not hold. We assume that the equality holds. Thus $g_s = h_s$. Then $G(\omega) = H(\omega)$. By (92) and (82), we get

$$\overline{A(\omega)} = A(\omega), \quad \omega \in E. \qquad (95)$$

Here E is a nonzero measure set (see (80)). Hence

$$e^{iL\omega}\overline{A(\omega)} - e^{iL\omega}A(\omega) = 0, \quad \omega \in E. \qquad (96)$$

However, when $a^T = (a_1, \cdots, a_L) \neq (0, \cdots, 0)$, the polynomial

$$p(Z) = Z^L \sum_{t=1}^{L} a_t Z^{-t} - Z^L \sum_{t=1}^{L} a_t Z^t$$

has at most $2M$ zeros. It is contrary to (90). Hence we have

$$|\text{II}| = \text{I}. \qquad (97)$$

We now discuss $Q_{L \times L}$. From (9) we get

$$a^T Q_{L \times L} a = \text{I} - \text{II}. \qquad (98)$$

It follows from (97) and (98) that

$$a^T Q_{L \times L} a > 0. \qquad (99)$$

Hence

$$\text{rank}(Q_{L \times L}) = L. \qquad (100)$$

Set

$$K = \min(M, L).$$

It is evident that

$$\text{rank}(Q_{K \times K}) \leqslant \text{rank}(Q_{M \times L}) \leqslant K.$$

Therefore, it follows from (100) that

$$\text{rank}(Q_{M \times L}) = \min(M, L).$$

Theorem 1 is proved. ∎

References

[1] An, H. Z., Chen, Z. G. and Hannan, E. J. (1982) Autocorrelations, autoregressions and autoregressive approximations. *Ann. Statist.* 10, 926—36.

[2] Bloomfield, P. (1973) An exponential model for the spectrum of a scalar time series. *Biometrka* 60, 217—26.

[3] Cheng, Q. S. (1985) Z-transform models and data extrapolation formulas in the maximum entropy methods of power spectral analysis. *Kexue Tongbao* (*Science Bulletin*) 30, 436—40.

[4] Dzhaparidze, K. O. and Yaglom, A. M. (1983) Spectrum parameter estimation in time series analysis. *Dev. Statist.* 4, 1—96.

[5] Hannan, E. J. (1970) *Multiple Time Series*. New York: Wiley.

原文载于 Journal of Time Series Analysis, Vol. 12, No. 1, January 1991, 27—40.

On Time-Reversibility of Linear Processes

Cheng Qiansheng

School of Mathematics Sciences, Peking University

Summary: A necessary and sufficient condition for time-reversibility of stationary linear processes is given which, contrary to previous results, does not require existence of moments of order higher than two.

Some keywords: non-Gaussian linear process; time-reversibility

1. Introduction

Let x_t be a stationary linear process:

$$x_t = w_t * u_t = \sum_{s \in Z} w_s u_{t-s}, \tag{1.1}$$

where u_t is a sequence of independent and identically distributed random variables with $E(u_t) = 0$, $E(u_t^2) = \sigma^2$, w_t is a square-summable sequence of constants, and Z is the set of all integers. The relationship between w_t and the spectral density $S_x(\omega)$ of x_t is

$$S_x(\omega) = |W(\omega)|^2 E(u_t^2), \tag{1.2}$$

where $W(\omega)$ and $S_x(\omega)$ are the Fourier transforms of w_t and $E(x_{t+s} x_s)$ respectively. In this paper, we will assume that

$$S_x(\omega) = |W(\omega)|^2 E(u_t^2) \neq 0 \tag{1.3}$$

almost everywhere.

A stationary process x_t is time-reversible if, for any positive integer n and any $t_1, \cdots, t_n \in Z$, $(x_{t_1}, \cdots, x_{t_n})$ and $(x_{-t_1}, \cdots, x_{-t_n})$ have the same joint probability distributions.

Weiss(1975) shows that, for causal autoregressive moving-average linear processes, time-reversibility is essentially restricted to Gaussian processes under certain conditions. Hallin, Lefevre & Puri(1988) extend this result to general linear processes under some conditions including the existence of moments of all orders.

In this paper, a necessary and sufficient condition for time-reversibility of stationary linear processes is given which, contrary to previous results, does not require existence of moments of order higher than two.

In § 2, we prove a basic lemma. In § 3, we provide a basic theorem which gives a necessary and sufficient condition for distributional equivalence of two linear processes without requiring the existence of higher-order moments. The relation between distributional equivalence and time-reversibility is investigated. The time-reversibility theorem is given as a consequence of the basic theorem.

2. The Basic Lemma

Lemma 1. *Let u_t and v_t, for $t \in Z$, be two independent and identically distributed processes with zero means and finite variances, and let*

$$v_t = h_t * u_t = \sum_{s \in Z} h_s u_{t-s}, \tag{2.1}$$

where h_t is a constant sequence. If v_t is non-Gaussian, then $h_t = a\delta_{t-t_0}$, where a is a nonzero constant, t_0 is an integer, $\delta_0 = 1$ and $\delta_t = 0$ for $t \neq 0$.

Proof. Since u_t and v_t are independent and identically distributed, we have

$$E(u_t u_s) = E(u_0^2 \delta_{t-s}), \quad E(v_t v_s) = E(v_0^2 \delta_{t-s}). \tag{2.2}$$

From (2.1), we get

$$E(v_0^2 \delta_t) = E(v_t v_0) = E\left(\sum_s h_s u_{t-s} \sum_l h_l u_{-l}\right) = \sum_s \sum_l h_s h_l E(u_{t-s} u_{-l})$$
$$= \sum_s \sum_l h_s h_l E(u_0^2 \delta_{t-s+l}) = E(u_0^2) \sum_s h_s h_{-(t-s)} = E(u_0^2) h_t * h_{-t};$$

that is,

$$h_t * h_{-t} = b\delta_t, \tag{2.3}$$

where $b = E(v_0^2/Eu_0^2)$.

It follow from (2.1) and (2.3) that

$$u_t = b^{-1} h_{-t} * v_t = b^{-1} \sum_l h_{-l} v_{t-l}. \tag{2.4}$$

Let $f(\lambda)$ and $g(\lambda)$ denote the characteristic functions of v_t and u_t respectively. By independence, from (2.1) and (2.4) we get

$$f(\lambda) = \prod_s g(h_s \lambda), \quad g(\lambda) = \prod_t f(b^{-1} h_{-t} \lambda) \tag{2.5}$$

and thereby

$$f(\lambda) = \prod_s \prod_t f(b^{-1}h_s h_{-t}\lambda). \tag{2.6}$$

Applying Theorem 5.6.1 of Kagan, Linnik & Rao (1973) yields that there exists (t_0, t_1) such that

$$h_{t_0} h_{-t_1} \neq 0, \quad h_s h_{-l} = 0, \quad (s,l) \neq (t_0, t_1).$$

That means $h_t = 0$, for $t \neq t_0$. This shows that $h_t = a\delta_{t-t_0}$ and the proof is complete.

3. The Basic Theorem and Time-Reversibility

Two processes x_t and y_t are said to be distributionally equivalent if, for any positive integer n and any $t_1, t_2, \cdots, t_n \in Z$, $(x_{t_1}, \cdots, x_{t_n})$ and $(y_{t_1}, \cdots, y_{t_n})$ have the same joint probability distribution.

Theorem 1. *Let x_t and y_t be two non-Gaussian linear processes:*

$$x_t = w_t * u_t, \quad y_t = w'_t * u'_t \quad (t \in \mathbf{Z}), \tag{3.1}$$

where u_t and u'_t are independent and identically distributed with zero means and finite variances, and w_t and w'_t are square-summable sequences satisfying (1.3). In order for x_t and y_t to be distributionally equivalent, it is necessary and sufficient that

(i) $w'_t = \dfrac{1}{a} w_{t+t_0},$ \hfill (3.2)

(ii) *u'_t and au_{t-t_0} have the same distributions,*

where a is a nonzero constant and t_0 is an integer.

Proof. Necessity. Let H_x and H_y denote two Hilbert subspaces generated by x_t and y_t respectively. As a result of the equivalence of x_t and y_t under the correspondent relation $x_t \leftrightarrow y_t$, H_x and H_y are isometrically isomorphic. Then there exists $v_t \in H_y$ corresponding to $u_t \in H_x$. As a result of the equivalence and isometrical isomorphism, v_t is independent and identically distributed with the same distribution as u_t, and $y_t = w_t * v_t$.

Then $y_t = w_t * v_t = w'_t * u'_t$. By formula (3.5) of Cheng (1990), we have

$$u'_t = c_t * v_t = \sum_s c_s v_{t-s}.$$

By Lemma 1, it follows that

$$u'_t = a v_{t-t_0}, \quad w'_t = \frac{1}{a} w_{t+t_0}.$$

Hence, necessity holds. Sufficiency is obvious and the theorem is proved.

The following corollary is immediate.

Corollary 1. Let x_t and y_t be two stationary linear processes:
$$x_t = w_t * u_t, \quad y_t = w'_t * u'_t \quad (t \in Z),$$
where u_t and u'_t are two sequences of independent and identically distributed random variables with zero means and finite variances, and w_t and w'_t are square-summable sequences satisfying (1.3). Then x_t and y_t are distributionally equivalent if and only if either

(a) x_t and y_t are Gaussian, and their spectral density functions $S_x(\omega)$ and $S_y(\omega)$ are identical,

or

(b) $w'_t = a^{-1} w_{t+t_0}$, and u'_t and au_{t-t_0} have the same distribution, where a is a nonzero constant and t_0 is an integer.

We now turn to discuss time-reversibility, which is in fact a special case of distributional equivalence. For a linear process $x_t = w_t * u_t$, we have $x_{-t} = w_{-t} * u_{-t}$. For an independent and identically distributed series u_t, u_{-t} is independent and identically distributed too. If we set $y_t = x_{-t} = w_{-t} * u_{-t}$, by Theorem 1 the following theorem is immediate.

Theorem 2. Let x_t be a non-Gaussian linear process, $x_t = w_t * u_t$, where the u_t are independent and identically distributed, and w_t is a square-summable sequence satisfying (1.3). Then x_t is time-reversible if and only if

(i) $w_t = a w_{t_0 - t}$, \hfill (3.3)

(ii) u_t and au_t have the same distributions,

where a is a nonzero constant and t_0 is an integer.

Note that $a = 1$ or -1, because $E(u_t^2) = E(a^2 u_t^2)$. It follows from (3.3) that a is unique.

Theorem 2 can be rewritten in the following form.

Corollary 2. Let $x_t = w_t * u_t$, where $t \in Z$, u_t is independent and identically distributed with zero means and finite variances, and w_t is square-summable and satisfies (1.3). Then x_t is time-reversible if and only if either (a) x_t is Gaussian, or (b) conditions (i) and (ii) of Theorem 2 hold.

Now we discuss the general solution of (3.3) when w_t is real. Let $W(\omega)$ be the Fourier transform of w_t:

$$W(\omega) = A(\omega)\exp\{i\theta(\omega)\} = \sum_t w_t e^{-t\omega}, \tag{3.4}$$

where $A(\omega)$ and $\theta(\omega)$ are the amplitude spectrum and the phase spectrum of w_t respectively, which satisfy

$$A(-\omega) = A(\omega), \quad \theta(-\omega) = -\theta(\omega). \tag{3.5}$$

From (3.3) it follows that

$$\exp\{-i\theta(\omega)\} = a \exp\{it_0\omega + i\theta(\omega)\}. \tag{3.6}$$

Note that $a = \exp\{i\pi(a-1)/2\}$. Then the general solution of (3.6) is

$$\theta(\omega) = \frac{1}{2}\{2\pi(1-a) + t_0\omega\} + \pi k(\omega) \quad (k(\omega) \in Z). \tag{3.7}$$

Hence, the general solution of (3.3) is $W(\omega) = A(\omega)\exp\{i\theta(\omega)\}$, where $A(\omega)$ is square integrable in $[-\pi, \pi]$ and positive almost everywhere, and $\theta(\omega)$ satisfies (3.5) and (3.7).

For causal sequences w_t, that is $w_t = 0$ for $t < u$, from (3.3) we have

$$w_t = \begin{cases} 0 & (t < 0 \text{ or } t > t_0), \\ aw_{t_0-t} & (0 \leq t \leq t_0), \end{cases}$$

where t_0 is a unique nonnegative integer. Thus, x_t is a special moving average model.

References

[1] Cheng, Q. S. (1990). Maximum standardized cumulant deconvolution of non-Gaussian linear processes. *Ann. Statist.* **18**, 1774—83.

[2] Hallin, M., Lefevre, C. & Puri, M. L. (1988). On time-reversibility and the uniqueness of moving average representations for non-Gaussian stationary time series. *Biometrika* **75**, 170—1.

[3] Kagan, A. M., Linnik, Y. V. & Rao, C. R. (1973). *Characterization Problems in Mathematical Statistics*. New York: Wiley.

[4] Weiss. G. (1975). Time-reversibility of linear stochastic processes. *J. Appl. Prob.* **12**, 831—6.

原文载于 Biometrika(1999), Vol. 86, No. 2, 483—486.

Almost Sure Convergence Analysis of Mixed Time Averages and kth-Order Cyclic Statistics

Li Hongwei, Cheng Qiansheng

Abstract: The strong law of large numbers for general processes is established under certain conditions and the convergence rate of time averages is given. These results are used to obtain the convergence rates and the almost sure convergence properties of mixed time averages and kth-order cyclic statistics.

Index Terms: almost sure convergence; convergence rate; cyclic-cumulant; cyclic-moment; mixed time average

I. Introduction

For stationary processes, consistency (in the mean-square sense) of third- and fourth-order sample moment and cumulant estimators was established respectively in [9], [10], and [6]. Earlier results on the asymptotic properties of sample averages of not necessarily stationary processes were provided by Parzen [8]. Dandawaté and Giannakis [3] generalized Parzen's analysis of "asymptotically stationary" processes to mixtures of deterministic, stationary, nonstationary, and complex-valued time series. Under some mixing conditions, they discussed the convergence in the mean-square sense for sample averages of such mixtures. Almost sure convergence of sample (cross-)correlations for linear processes and covariance estimators for cyclostationary processes have been reported respectively in [7] and [5]. For general processes, the almost sure convergence of any order sample moments and cumulants was left as an open problem. It is well known that the convergence rate is potentially useful in practice and implies almost sure convergence. But up to now it has not been studied for general processes.

In Section II of this correspondence, we establish the strong law of large numbers for complex-valued processes and give the convergence rate. Using these results, we obtain the convergence rates and the almost sure convergence properties of mixed time averages and kth-order cyclic statistics in Section III. Finally, conclusions are drawn in Section IV.

II. Strong Law of Large Numbers

Let $f(t)$ be a discrete-time complex process (deterministic or random). Our mixing condition can be stated as follows.

Assumption 1.

$$\sum_{\xi_0=-\infty}^{\infty} \sup_t |\operatorname{cov}\{f(t), f^*(t+\xi_0)\}| = C_0 < +\infty, \tag{1}$$

where * denotes comples conjugation and C_0 represents a constant.

Mixing conditions of this form were used in [2], [3], [9], and [10]. For stationary processes, (1) means that the second-order moments are absolutely summable. It is a common condition which is frequently used in spectral theory of random processes [6], [10]. There is a large class of processes satisfying Assumption 1 which includes linear processes, amplitude modulation (AM) signals, and mixed processes [3]. Following (1), the mixing condition also holds for nonlinear processes of the Volterra type with absolutely summable kernels.

Now we state and prove the strong law of large numbers for comples-valued processes.

Theorem 1. If $f(t)$ is a discrete-time complex process which satisfies Assumption 1, then the following holds:

$$E \left| \frac{1}{T} \sum_{t=0}^{T-1} f(t) - \frac{1}{T} \sum_{t=0}^{T-1} Ef(t) \right|^2 \leq \frac{C_0}{T}, \tag{2}$$

and for $\gamma > \frac{3}{4}$,

$$\left| \frac{1}{T} \sum_{t=0}^{T-1} f(t) - \frac{1}{T} \sum_{t=0}^{T-1} Ef(t) \right| \stackrel{a.s.}{=} o\left(\frac{1}{T^{1-\gamma}}\right), \tag{3}$$

where a. s. represents almost sure convergence.

Proof. From Assumption 1, we have

$$E\left|\frac{1}{T}\sum_{t=0}^{T-1}f(t)-\frac{1}{T}\sum_{t=0}^{T-1}Ef(t)\right|^2$$

$$\leq \frac{1}{T^2}\sum_{t_1,t_2=0}^{T-1}E[f(t_1)-Ef(t_1)][f(t_2)-Ef(t_2)]^*$$

$$\leq \frac{1}{T^2}\sum_{\xi_0=-(T-1)}^{T-1}\sum_{t=-\min\{\xi_0,0\}}^{T-1-\max\{\xi_0,0\}}|\text{cov}\{f(t),f^*(t+\xi_0)\}|$$

$$\leq \frac{1}{T}\sum_{\xi_0=-(T-1)}^{T-1}\sup_t|\text{cov}\{f(t),f^*(t+\xi_0)\}|$$

$$\leq \frac{C_0}{T}, \tag{4}$$

which is (2).

For $\gamma \geq \frac{1}{2}$, it follows from (4) that

$$E\left\{\frac{1}{T^{2\gamma}}\left|\sum_{t=0}^{T-1}[f(t)-Ef(t)]\right|^2\right\}\leq \frac{C_0}{T^{2\gamma-1}}<+\infty. \tag{5}$$

Let

$$S(T) \triangleq T^{-\gamma}\left|\sum_{t=0}^{T-1}[f(t)-Ef(t)]\right|. \tag{6}$$

For $\gamma>3/4$, there is a β which satisfies $\beta>1$ and

$$\gamma-\frac{1}{2}>\frac{1}{2\beta}>\frac{1}{4}. \tag{7}$$

It follows from (7) that $1<\beta<2$.

A. When $T=[N^\beta]$ (denotes the integer part of N^β), by (5) we have

$$E|S(T)|^2\leq C_1 N^{\beta(1-2\gamma)}, \tag{8}$$

where C_1 represents a constant. By (7), $\beta(1-2\gamma)<-1$. Hence,

$$\sum_{N=1}^{+\infty}E|S([N^\beta])|^2<+\infty. \tag{9}$$

It follows from the Chebyshev inequality and the Borel-Cantelli lemma [7, pp. 45-49], [11, p. 10] that

$$\lim_{N\to\infty}S([N^\beta])\stackrel{\text{a.s.}}{=}0. \tag{10}$$

B. For general T, let $T_N\leq T\leq T_{N+1}$, where $T_N=[N^\beta]$, we have

$$|S(T)-T^{-\gamma}[N^\beta]^\gamma S([N^\beta])|\leq [N^\beta]^{-\gamma}\sum_{t=[N^\beta]+1}^{[(N+1)^\beta]}|[f(t)-Ef(t)]| \tag{11}$$

and
$$E \sup_{T_N \leqslant T \leqslant T_{N+1}} | S(T) - T^{-\gamma}[N^\beta]^\gamma S([N^\beta]) |^2$$
$$\leqslant [N^\beta]^{-2\gamma} \sum_{t_1, t_2 = [N^\beta]+1}^{[(N+1)^\beta]} E\{| f(t_1) - Ef(t_1) || f(t_2) - Ef(t_2) |\}.$$
(12)

By Assumption 1, we have, for arbitrary t_1, t_2
$$E\{| f(t_1) - Ef(t_1) || f(t_2) - Ef(t_2) |\}$$
$$\leqslant \frac{1}{2} \{\text{cov}\{f(t_1), f^*(t_1)\} + \text{cov}\{f(t_2), f^*(t_2)\}\}$$
$$\leqslant C_0. \quad (13)$$

From (12) and (13), we obtain
$$E \sup_{T_N \leqslant T \leqslant T_{N+1}} | S(T) - T^{-\gamma}[N^\beta]^\gamma S([N^\beta]) |^2$$
$$\leqslant C_0 [N^\beta]^{-2\gamma} ([(N+1)^\beta] - [N^\beta])|^2$$
$$\leqslant C_2 N^{-2\gamma\beta + 2\beta - 2}, \quad (14)$$

where C_2 represents a constant and the second inequality follows from the fact that
$$[(N+1)^\beta] - [N^\beta] = O(N^{\beta-1}).$$

It follows from (7) that $\beta < 2$. For $\gamma > 3/4$
$$-2\gamma\beta + 2\beta - 2 < -2 \times \frac{3}{4}\beta + 2\beta - 2 = \frac{\beta}{2} - 2 < -1. \quad (15)$$

By (14) and (15), we obtain
$$\sum_{N=1}^{+\infty} E \sup_{T_N \leqslant T \leqslant T_{N+1}} | S(T) - T^{-\gamma}[N^\beta]^\gamma S([N^\beta]) |^2 \leqslant +\infty. \quad (16)$$

It follows from the Chebyshev inequality and the Borel-Cantelli lemma that
$$\lim_{N \to \infty} \sup_{T_N \leqslant T \leqslant T_{N+1}} | S(T) - T^{-\gamma}[N^\beta]^\gamma S([N^\beta]) | \stackrel{a.s.}{=} 0. \quad (17)$$

By (10), (17) and the estimate
$$S(T) \leqslant \sup_{T_N \leqslant T \leqslant T_{N+1}} | S(T) - T^{-\gamma}[N^\beta]^\gamma S([N^\beta]) | + T^{-\gamma}[N^\beta]^\gamma S([N^\beta]).$$
we have, for $\gamma > 3/4$
$$S(T) = T^{-\gamma} \left| \sum_{t=0}^{T-1} f(t) - E \sum_{t=0}^{T-1} f(t) \right| \stackrel{a.s.}{\longrightarrow} 0. \quad (18)$$

From (6) and (18), we obtain (3). Theorem 1 is proved. □

Theorem 1 gives the convergence rate and implies the almost sure convergence of sample averages. Theorem 1 is very useful. It can be used to study the convergence rate of mixed time averages.

III. Almost Sure Convergence of Mixed Time Averages and Cyclic Moments and Cumulants

We now use Theorem 1 to derive the almost sure convergence of mixed time averages and cyclic statistics. The related symbols, definitions, and statements can be found in [3].

Let $\{x_m(t)\}_{m=0}^k$ be $(k+1)$ deterministic or random complex signals. Our mixing conditions on $\{x_m(t)\}_{m=0}^k$ are as follows.

Assumption 2. $\forall m \in \mathbf{Z}$,

$$\sum_{\substack{\xi_1 \cdots \xi_m = -\infty \\ <\infty}}^{\infty} \sup_t |\operatorname{cum}\{x_{n_1}(t), x_{n_1}(t+\xi_1), \cdots, x_{n_m}(t+\xi_m)\}|, \quad (19)$$

where

$$x_n(t) \in \{x_0(t), x_0^*(t), \cdots, x_k(t), x_k^*(t)\}.$$

This assumption is different from [3, Assumption 1.1]. There is an omission (or typo) in [3, Assumption 1.1]. In the assumptions of [3], the authors only stated [3, eq. (2)]

$$\sum_{\xi_1 \cdots \xi_m = -\infty}^{\infty} \sup_t |\xi_l \operatorname{cum}\{x_{n_1}(t), x_{n_2}(t+\xi_1), \cdots, x_{n_m}(t+\xi_m)\}| < \infty, l \in \{1, \cdots, m\}. \quad (20)$$

But in the derivation of their Theorem 2.1, they made use of [3, eqs. (80) and (81)]

$$\sum_{\xi_1 \cdots \xi_m = -\infty}^{\infty} \sup_t |\xi_l \operatorname{cum}\{x_{n_0}(t), x_{n_1}(t+\xi_1), \cdots x_{n_m}(t+\xi_m)\}| < \infty,$$

$$l \in \{0, 1, \cdots, m\}, \xi_0 = 1. \quad (21)$$

We give an example to show that the assumption (20) is insufficient for mean-square-sense convergence of mixed time averages.

Example. Let $x_0(t)$ be mutually independent real-valued random variables with $x_0(0) = 0$ and, for $t \neq 0$, $x_0(t)$ uniformly distributed in $\left[-\frac{|t|}{2}, \frac{|t|}{2}\right]$.

We have $Ex_0(t)=0$ and $Ex_0^2(t)=\dfrac{t^2}{12}$, $\forall m \in \mathbf{Z}$.

$$\text{cum}\{x_0(t), x_0(t+\xi_1), \cdots, x_0(t+\xi_m)\}$$
$$= \begin{cases} \text{cum}\{x_0(t), x_0(t), \cdots, x_0(t)\}, & \xi_1 = \cdots = \xi_m = 0, \\ 0, & \text{otherwise.} \end{cases}$$

Hence,

$$\sum_{\xi_1 \cdots \xi_m = -\infty}^{\infty} \sup_t | \xi_l \text{cum}\{x_0(t), x_0(t+\xi_1), \cdots, x_0(t+\xi_m)\} | = 0, \quad l \in \{1, \cdots, m\}$$

and $\forall \tau_0$,

$$\lim_{T \to \infty} \frac{1}{T} \sum_{t=0}^{T-1} Ex_0(t+\tau_0) = 0.$$

Thus $x_0(t)$ satisfies (20). Let

$$\hat{\mathcal{M}}_0^{(T)}(0) \triangleq \frac{1}{T} \sum_{t=0}^{T-1} x_0(t).$$

We have

$$\text{cum}\left\{ \frac{1}{T} \sum_{t=0}^{T-1} x_0(t), \frac{1}{T} \sum_{t=0}^{T-1} x_0(t) \right\} = \frac{1}{T^2} \sum_{t=0}^{T-1} Ex^2(t) = \frac{1}{72}\left(2 - \frac{1}{T}\right)(T-1).$$

This indicates that $\hat{\mathcal{M}}_0^{(T)}(0)$ is not mean-square-sense-consistent.

Under the assumption (21), [3, Theorem 2.1] is correct in both its claims and derivation. However, [3, Assumption 1.1] should be replaced by (21).

Comparing (19) with (21), it is obvious that Assumption 2 is a weaker condition than the corrected [3, Assumption 1.1].

The following lemma and its proof are similar to [3, Lemma 2.1].

Lemma 1. Let $\{x_l(t)\}_{l=0}^k$ be $(k+1)$ discrete-time complex processes (deterministic or random) and define the product processes of K_n order as

$$f_n(t; \underline{\tau}_n) \triangleq \prod_{l=0}^{K_n} x_{n,l}(t + \tau_{n,l}), \quad x_{n,l}(t) \in \{x_m(t)\}_{m=0}^k, \tag{22}$$

where $\underline{\tau}_n \triangleq (\tau_{n,0}, \cdots, \tau_n, K_n)$, $\tau_{n,l}$ are arbitrary but fixed lags. If Assumption 2 holds, then $\forall m$, and $\xi = (\xi_1, \cdots, \xi_m)$,

$$\sum_{\xi} \sup_t | \text{cum}\{f_0(t; \underline{\tau}_0), f_1(t+\xi_1; \underline{\tau}_1), \cdots, f_m(t+\xi_m; \underline{\tau}_m)\} | < \infty. \tag{23}$$

Moreover, (23) holds even when any of the f's are conjugated.

The following corollaries follow immediately from Theorem 1, Lemma 1 and the properties of cumulants [1, p. 19].

Corollary 1. Let $\{x_m(t)\}_{m=0}^k$ be a set of $(k+1)$ discrete-time complex processes (deterministic or random) satisfying Assumption 2. Then with
$$x_{n,l}(t) \in \{x_m(t)\}_{m=0}^k, \underline{\tau}_n \triangleq (\tau_{n,0}, \cdots, \tau_{n,K_n})$$
and
$$\hat{\mathscr{M}}_{K_n}^{(T)}(\underline{\tau}_n) \triangleq \frac{1}{T} \sum_{t=0}^{T-1} \prod_{l=0}^{K_n} x_{n,l}(t+\tau_{n,l}) = \frac{1}{T} \sum_{t=0}^{T-1} f_m(t; \underline{\tau}_n), \quad (24)$$
the following holds:
$$E|\hat{\mathscr{M}}_{K_n}^{(T)}(\underline{\tau}_n) - E\hat{\mathscr{M}}_{K_n}^{(T)}(\underline{\tau}_n)|^2 \leqslant \frac{C}{T} \quad (25)$$
for any arbitrary but fixed orders K_n and lags $\underline{\tau}_n$, where C represents a constant. Further, for $\gamma > \frac{3}{4}$
$$|\hat{\mathscr{M}}_{K_n}^{(T)}(\underline{\tau}_n) - E\hat{\mathscr{M}}_{K_n}^{(T)}(\underline{\tau}_n)| \stackrel{a.s.}{=} o\left(\frac{1}{T^{1-\gamma}}\right). \quad (26)$$

Corollary 2. Suppose
$$\lim_{T \to \infty} \frac{1}{T} \sum_{t=0}^{T-1} E\{x_0(t+\tau_0) \cdots x_k(t+\tau_k)\}$$
exists $\forall \tau_0, \cdots, \tau_k$. Under the hypotheses of Corollary 1, we have
$$\lim_{T \to \infty} \hat{\mathscr{M}}_{x_0 \cdots x_k}^{(T)}(\tau_0, \cdots, \tau_k) \stackrel{a.s.}{=} \lim_{T \to \infty} \frac{1}{T} \sum_{t=0}^{T-1} E\{x_0(t+\tau_0) \cdots x_k(t+\tau_k)\}$$
$$\triangleq \mathscr{M}_{x_0 \cdots x_k}(\tau_0, \cdots, \tau_k). \quad (27)$$

Corollary 3. Let $x(t)$ be a kth-order cyclostationary process which satisfies Assumption 2, with $x_{n,l} \in \{x(t), x^*(t)\}$. Then for $\gamma > \frac{3}{4}$, the sample cyclic-moment
$$\hat{\mathscr{M}}_{kx}^{(T)}(\alpha; \underline{\tau}) \triangleq \frac{1}{T} \sum_{t=0}^{T-1} x(t) x(t+\tau_1) \cdots x(t+\tau_{k-1}) e^{-j\alpha t} \quad (28)$$
satisfies,
$$E|\hat{\mathscr{M}}_{kx}^{(T)}(\alpha; \underline{\tau}) - E\hat{\mathscr{M}}_{kx}^{(T)}(\alpha; \underline{\tau})|^2 \leqslant \frac{C}{T}, \quad (29)$$
$$|\hat{\mathscr{M}}_{kx}^{(T)}(\alpha; \underline{\tau}) - E\hat{\mathscr{M}}_{kx}^{(T)}(\alpha; \underline{\tau})| \stackrel{a.s.}{=} o\left(\frac{1}{T^{1-\gamma}}\right), \quad (30)$$
where C is a constant. Further, If $\mathscr{M}_{kx}$ exists, then

$$\lim_{T\to\infty}\hat{\mathcal{M}}_{kx}^{(T)}(a;\underline{\tau}) \stackrel{a.s.}{=} \mathcal{M}_{kx}(a;\underline{\tau}). \quad (31)$$

From [3, eq. (36)], the estimator of the kth-order time-varying cumulant of $x(t)$ can be defined as

$$\hat{c}_{kx}^{(T)}(t;\underline{\tau}) \triangleq \sum_{v}(-1)^{p-1}(p-1)!\,\hat{m}_{v_1 x}^{(T)}(t;\underline{\tau}_{v_1})\cdots\hat{m}_{v_p x}^{(T)}(t;\underline{\tau}_{v_p}). \quad (32)$$

The almost sure convergence of the estimator of kth-order time varying moments [3, eq. (58)] follows readily from that of $\hat{\mathcal{M}}_{kx}^{(T)}$, provided that $a \in \mathcal{A}_k^m$ has a finite number of elements. The almost sure convergence of the estimator of kth-order time-varying cumulants follows readily from that of $\hat{m}_{kx}^{(T)}$. We summarize these observation in the following corollary.

Corollary 4. Under the hypotheses of Corollary 3, if $\mathcal{A}_k^m$ has a finite number of cycles and $\mathcal{M}_{kx}$ exists, then for $\gamma > \frac{3}{4}$

$$\sup_t |\hat{m}_{kx}^{(T)}(t;\underline{\tau}) - E\hat{m}_{kx}^{(T)}(t;\underline{\tau})| \stackrel{a.s.}{=} o\left(\frac{1}{T^{1-\gamma}}\right), \quad (33)$$

$$\sup_t |\hat{c}_{kx}^{(T)}(t;\underline{\tau}) - E\hat{c}_{kx}^{(T)}(t;\underline{\tau})| \stackrel{a.s.}{=} o\left(\frac{1}{T^{1-\gamma}}\right). \quad (34)$$

Further

$$\lim_{T\to\infty}\hat{m}_{kx}^{(T)}(t;\underline{\tau}) \stackrel{a.s.}{=} m_{kx}(t;\underline{\tau}), \quad (35)$$

$$\lim_{T\to\infty}\hat{c}_{kx}^{(T)}(t;\underline{\tau}) \stackrel{a.s.}{=} c_{kx}(t;\underline{\tau}). \quad (36)$$

Next, we consider the kth-order cyclic cumulant $\mathcal{C}_{kx}$. Equation (37) of [3] suggests the following estimator:

$$\hat{\mathcal{C}}_{kx}^{(T)}(a;\underline{\tau}) = \frac{1}{T}\sum_{t=0}^{T-1}\hat{c}_{kx}^{(T)}(t;\underline{\tau})e^{-jat}. \quad (37)$$

From

$$|\hat{\mathcal{C}}_{kx}^{(T)}(a;\underline{\tau}) - E\hat{\mathcal{C}}_{kx}^{(T)}(a;\underline{\tau})| \leq \sup_t |\hat{c}_{kx}^{(T)}(t;\underline{\tau}) - E\hat{c}_{kx}^{(T)}(t;\underline{\tau})| \quad (38)$$

and Corollary 4 we have the following.

Corollary 5. Under the hypotheses of Corollary 3, if $\mathcal{A}_k^m$ has a finite number of cycles and $\mathcal{M}_{kx}$ exists, then for $\gamma > \frac{3}{4}$

$$|\hat{\mathcal{C}}_{kx}^{(T)}(a;\underline{\tau}) - E\hat{\mathcal{C}}_{kx}^{(T)}(a;\underline{\tau})| \stackrel{a.s.}{=} o\left(\frac{1}{T^{1-\gamma}}\right). \quad (39)$$

Further

$$\lim_{T\to\infty}\hat{\mathcal{C}}_{kx}^{(T)}(\alpha;\underline{\tau}) \stackrel{a.s.}{=} \mathcal{C}_{kx}(\alpha;\underline{\tau}). \tag{40}$$

Applications of cyclic statistics can be found in [4].

IV. Conclusion

For general processes, we have established the strong law of large numbers and obtained the convergence rate under conditions based on absolute second-moment summability. As an application of these results, we have proved that the mixed time averages are almost surely convergent to their expectations and obtained the rate of almost sure convergence. Whe have also proved the almost sure convergence of estimators for time-varying and cyclic moments and cumulants of cyclostationary signals and given the convergence rate. Throughout our analysis, only mixing conditions based on absolute cumulant summability were employed. We emphasize that our convergence rate does not depend on the order of the sample average.

References

[1] D. R. Brillinger, *Time Series: Data Analysis and Theory*. San Francisco, CA: Holden-Day, 1981.

[2] D. R. Brillinger and M. Rosenblatt, "Asymptotic theory of estimates of k-th order spectra," in *Spectral Analysis of Time Series*, B. Harris, Ed. New York: Wiley, 1967, pp. 153—188.

[3] A. V. Dandawate and G. B. Giannakis, "Asymptotic theory of mixed time averages and kth-order cyclic-moment and cumulant statistics," *IEEE Trans. Inform. Theory*, vol. 41, pp. 216—232, 1995.

[4] G. B. Giannakis and G. Zhou, "Harmonics in multiplicative and additive noise: Parameter estimation using cyclic statistics," *IEEE Trans. Signal Processing*, vol. 43, pp. 2217—2221, 1995.

[5] H. L. Hurd and J. Leskow, "Strongly consistent and asymptotically normal estimation for the covariance of almost periodically correlated processes," *Statist. Decisions*, vol. 10, pp. 201—255, 1992.

[6] P. T. Kim, "Consistent estimation of the fourth-order cumulant spectra density," *J. Time Series Anal.*, vol. 12, no. 1, pp. 63—71, 1991.

[7] L. Ljumg, *System Identification: Theory for the User*. Englewood Cliffs, NJ: Prentice-Hall, 1987.

[8] E. Parzen, "Spectral analysis of asymptotically stationary time series," *Bull. Inst. Int. Statist.*, vol. 39, pp. 87—103, 1961.

[9] M. Rosenblatt, *Stationary Sequences and Random Fields*. Basel, Switzerland: Birkhauser, 1985.

[10] M. Rosenblatt and J. W. Van Ness, "Estimation of the bispectrum," *Ann. Math. Statist.*, vol. 36, pp. 1120—1136, 1965.

[11] W. F. Stout, *Almost Sure Convergence*. New York: Academic, 1974.

原文载于 IEEE Transactions on Information Theory, Vol. 43, No. 4, July 1997, 1265—1268.

模式识别与属性数学

有序样品聚类的相关序列法

程 乾 生

北京大学数学系和石油天然气研究中心

摘要:本文借助于广义相关系数(相关度),引入了相关序列的概念,并提出了有序样品聚类的相关序列法.应用该方法确定二叠—三叠系界线的实例表明,相关序列法是简单而有效的.

引言

有序样品的聚类有着广泛的应用.传统的方法是最优分割法或 Fisher 算法(文献[1]第 130—143 页或文献[2]). Fisher 算法在计算每一类样品的直径时没有考虑该类样品的有序性,而且所得结果没有给出更多细微的性质.

本文根据样品的有序性,并考虑到一个样品和它的下一个样品的相关度,提出了相关序列法. 如果 i_0 是一个分割点,即第 i_0 个样品属于一类,第 i_0+1 个样品属于另一类,那么这两个样品的相关度应该比较小,而且与同类附近样品的相关度相比,应有较大的差异. 下面讨论相关序列法并给出一个应用实例.

有序样品聚类的相关序列法

设有 m 个有序样品 $x_1,\cdots,x_m$,每一个样品都有 n 个观测值($n>1$). 因此,每一个样品都是一个 n 维向量. 对第 i 个样品 x_i,记做 $x_i=(x_{i1},\cdots,x_{in})$,$x_{ij}$ 表示第 i 个样品的第 j 个指标值,其中 $1\leqslant i\leqslant m, 1\leqslant j\leqslant n$.

用 $g_i=g(x_i,x_{i+1})$ 表示第 i 个样品 x_i 与第 $i+1$ 个样品 x_{i+1} 之间的相关度. g_i 必须满足

$$g_i=g(x_i,x_{i+1})\geqslant 0, \quad 1\leqslant i\leqslant m-1, \tag{1}$$

$$g_i=g(x_i,x_{i+1})=1, \quad \text{当 } x_i=x_{i+1} \text{ 时}. \tag{2}$$

g_i 通常可用广义相关系数来构造

$$g_i = \sum_{j=1}^{n} x'_{ij} x'_{i+1,j} \Big/ \Big[\sum_{j=1}^{n} (x_{ij})^2 \sum_{j=1}^{n} (x_{i+1,j})^2 \Big]^{1/2}, \tag{3}$$

其中

$$x'_{ij} = (x_{ij} - a_{ij})/b_{ij} \geqslant 0, \tag{4}$$

参数 a_{ij} 和 b_{ij} 要视具体问题来定. b_{ij} 的作用主要是消除不同指标测量尺度不同而带来的影响.

我们称 $(g_1, \cdots, g_{m-1})$ 为有序样品 $(x_1, \cdots, x_m)$ 的相关序列. 利用相关序列对 m 个有序样品进行分类的方法, 称为相关序列法. 用相关序列法进行分类, 采用的是逐次二分法. 关键的步骤是如何把 m 个样品分成两类.

设 i_1 为分割点, 即 $(x_1, \cdots, x_m)$ 可分成两类

$$(x_1, \cdots, x_{i_1}) \quad (x_{i_1+1}, \cdots, x_m). \tag{5}$$

在实际问题中, x_i 和 x_{i+1} 是两个不同的样品, 因此它们的相关度 g_i 可能大也可能小. 但是, 在分割点 i_1, 由于 x_{i_1} 和 x_{i_1+1} 属于不同的类(式(5)), x_{i_1} 和 x_{i_1+1} 的相关度 g_{i_1} 应比较小, 同时在 i_1 附近相关度的变化应比较大. 因此, 在相关系列 $(g_1, \cdots, g_{m-1})$ 的基础上, 可再考虑 $(g_1, \cdots, g_{m-1})$ 的差异序列 $(h_1, \cdots, h_{m-2})$

$$h_i = |g_i - g_{i+1}|, \quad i = 1, 2, \cdots, m-2. \tag{6}$$

取 $l(1 \leqslant l \leqslant m-2)$ 使

$$h_l = \max_{1 \leqslant i \leqslant m-2} h_i. \tag{7}$$

分割点 i_1 可由下式确定

$$\begin{cases} i_1 = l+1, & \text{当 } g_l > g_{l+1} \text{ 时,} \\ i_1 = l, & \text{当 } g_l < g_{l+1} \text{ 时.} \end{cases} \tag{8}$$

分割点 i_1 有两个特征: 在 i_1 上的相关度 g_{i_1} 比较小; 在 i_1 附近相关度变化大. 具体地说, h_{i_1} 或 h_{i_1+1} 至少有一个达最大值.

分割点 i_1 确定后, 有序样本 $(x_1, \cdots, x_m)$ 就分成两类: 第 1 类为 $(x_1, \cdots, x_{i_1})$; 第 2 类为 $(x_{i_1+1}, \cdots, x_m)$.

如果要把样本 $(x_1, \cdots, x_m)$ 分成三类, 除了保留分割点 i_1 外, 还需找出一个分割点 i_2. 确定 i_2 的方法如下: 按照上面的方法, 在第 1 类中确定分割点 i_{11}, 在第 2 类中确定分割点 i_{12}, 则 i_2 为

$$\begin{cases} i_2 = i_{11}, & \text{当 } g_{i_{11}} < g_{i_{12}} \text{ 时,} \\ i_2 = i_{12}, & \text{当 } g_{i_{11}} > g_{i_{12}} \text{ 时.} \end{cases} \tag{9}$$

按照上述方法,可把有序样品$(x_1,\cdots,x_m)$聚成k类,其中k为整数,且$2\leqslant k\leqslant m$. 把有序样品聚成k类,必有$k-1$个分割点,这些分割点称为k类分割点. 这里提出的相关序列法,其聚成的类具有谱系结构,即$k-1$类分割点也必定是k类分割点.

有序样品聚成多少类比较好,最好是结合实际问题来考虑. 仅从数据上考虑,相关序列$(g_1,\cdots,g_{m-1})$中,相关度比较大的点不应是分割点. 另外,可做聚类谱系图: 由1类聚2类,再聚3类,直至m类(每一个样品为一类). 这个谱系图可供研究者参考.

要指出的是,不论把样本聚成几类,由相关序列$(g_1,\cdots,g_{m-1})$和差异序列$(h_1,\cdots,h_{m-2})$就可以确定分割点.

应用实例

下面讨论在地层分析中确定二叠—三叠系界线的实例. 在江西省信丰县铁石口 P/T 界线剖面取 10 个层位,层序和序号见表 1. 每一层取一个样品,记为$x_i,1\leqslant i\leqslant 10$. 对每个样品进行分析测量,得到 34 个数据,第i个样品x_i的数据记为$x_i=(x_{i1},\cdots,x_{i34})$. 数据矩阵$[x_{ij}]_{10\times 34}$由表 1 给出. 希望通过数据分析找出 P/T 界线的分界位置.

在式(4)中,取

$$a_{ij}=0,\quad b_{ij}=\sum_{i=1}^{n}x_{ij},\quad 1\leqslant i\leqslant m,\quad 1\leqslant j\leqslant n, \tag{10}$$

其中$m=10,n=34$.

由式(3)计算的相关序列$(g_1,\cdots,g_9)$见表 2,由式(6)计算的差异序列$(h_1,\cdots,h_8)$见表 3. 由表 3 知,最大值为 7.5,据式(7),$l=5$. 由表 2 知,$g_l=g_5>g_6=g_{l+1}$,因此,由式(8)知,$i_1=6$. 序号$i_1=6$对应第 26 层,所以第 26 层应作为 P/T 界线的分界位置. 这一结论得到了古生物资料的支持,$(x_1,\cdots,x_6)$(即第 26 层到第 30 层)属下三叠统铁石口组;$(x_7,\cdots,x_{10})$(即第 22 层到第 25 层)属上二叠统长兴组.

表 1 铁石口 P/T 界线剖面数据

序号	1	2	3	4	5	6	7	8	9	10
层序	30	29b	29a	28	27	26	25	24	23	22
Si	5.9455	18.4992	19.9139	12.4973	15.6087	11.7735	3.8413	19.0820	1.5703	11.7030
Al	2.8212	6.2805	5.8671	3.6485	6.3017	4.4944	2.0135	6.5348	1.0685	4.9332
Fe	2.1182	1.4434	3.6617	3.2053	3.8389	2.4707	0.7233	2.0748	0.4808	3.6610
Ti	0.1570	0.2299	0.3182	0.2612	0.3276	0.2939	0.0634	0.2615	0.0446	0.4070
Mn	0.2560	0.2782	0.1214	0.1659	0.1303	0.0680	0.2190	0.0571	0.1284	0.0415
Mg	0.8940	0.9276	1.1508	1.0830	1.2684	0.8076	0.6516	1.4370	0.6744	1.1946
Ca	34.7758	14.8745	8.8727	13.8947	10.0181	2.7825	34.7758	2.2173	34.7758	0.6634
P	0.0708	0.0789	0.0893	0.0645	0.0939	0.0582	0.0540	0.1023	0.0448	0.1102
K	0.1713	0.1710	0.1671	0.1666	0.1425	0.1497	0.1361	0.1232	0.1413	0.1304
Ca/Mg	38.8991	16.0355	7.6318	12.8298	7.8982	3.4454	53.3699	1.5430	51.5655	0.5553
Sr/Ba	1.2429	1.3907	1.0295	2.3724	1.0040	0.3220	10.3688	0.3897	5.7297	0.3297
Sr	994.8000	423.0200	241.1000	470.2000	278.9000	38.1900	1237.0000	60.2400	932.8000	51.3600
Ba	800.9000	304.6000	234.2000	198.2000	277.1000	118.8000	119.3000	154.6000	162.8000	155.8000
Cr	40.7900	32.2600	66.8600	66.6300	68.4800	27.2500	22.5300	27.7500	21.1200	88.9200
Zr	318.7000	1501.0000	553.1000	451.1000	515.1000	1505.0000	188.5000	712.6000	124.4000	1386.0000
Zn	19.0500	44.3700	55.8900	64.9100	10.9800	8.9940	3.7710	37.3900	5.5380	13.1100
Nd	147.6000	128.1000	92.2700	107.5000	87.3600	71.7500	157.2000	49.7800	156.0000	43.7800
Th	9.7460	10.5100	15.8100	13.8100	16.4200	12.8300	4.9140	12.9000	4.1840	17.2600
Co	12.9400	14.0700	17.4800	15.8000	16.5800	16.0000	6.2770	11.5900	5.9550	16.3800
Ni	26.0500	27.9600	37.6900	36.4000	36.8600	30.7100	15.0100	29.3400	15.6300	72.8600
Be	1.2390	1.0620	2.3310	1.9770	2.3460	1.7110	0.5160	1.4020	0.3540	2.5230

续表

序号	1	2	3	4	5	6	7	8	9	10
层序	30	29b	29a	28	27	26	25	24	23	22
Pb	17.1800	27.8200	24.9700	19.7400	24.4000	24.8400	12.3100	25.6400	9.7010	22.6900
V	46.0700	57.8700	87.7600	74.6200	88.2200	41.4400	21.9600	48.8000	19.4200	124.9000
Cu	14.3600	26.7700	21.0800	18.3700	19.8700	21.0800	7.9280	15.3000	7.8340	36.7500
Ta	0.2794	0.2260	0.4164	0.3855	0.4296	0.3004	0.1210	0.2680	0.1027	0.4056
Sc	2.0690	3.8580	3.2990	2.4880	3.4940	2.6560	1.0620	1.5100	0.9500	2.0410
La	147.8000	117.4000	85.8700	111.6000	98.8100	46.7800	166.4000	29.2600	161.7000	17.4300
Ce	153.1000	182.3000	135.9000	136.8000	126.9000	135.5000	130.7000	71.0500	20.1000	73.8200
Y	38.8100	55.5000	31.9400	27.8000	34.3500	45.0300	40.2500	20.4500	17.3900	23.7200
Yb	2.8950	4.3920	3.0410	2.6400	2.8950	4.4280	2.7320	2.2750	1.3990	3.4430
As	0.8139	1.6490	1.6440	1.1480	1.7600	1.3690	0.5710	1.8320	0.3207	1.4300
Cd	0.0712	0.0787	0.0900	0.0750	0.0900	0.0900	0.0488	0.0900	0.0450	0.0862
Mo	0.3929	0.6631	0.6778	0.4961	0.6434	0.5599	0.2652	0.6852	0.1891	0.5525
ΣREE	490.2100	483.6900	349.0200	386.3400	344.3200	303.4900	497.5900	172.8100	456.6000	162.1900

表 2　相关序列 $(g_1,\cdots,g_9)$

i	1	2	3	4	5	6	7	8	9
g_i	0.839	0.877	0.907	0.910	0.911	0.836	0.778	0.745	0.705

表 3　差异序列 $(h_1,\cdots,h_8)$

i	1	2	3	4	5	6	7	8
$100\times h_i$	3.8	3.0	0.3	0.1	7.5	5.8	3.3	4.0

相关序列法不仅能用来研究 P/T 界线的分界位置，还可以给出其他的信息。由表 2 可以看出，从 g_1 到 g_5 比较大，这表明在下三叠统铁石口组内，每一层与相邻层的相关度比较大，岩石性质相对稳定；而从 g_7 到 g_9 的值相对来说要小些，表明在上二叠统长兴组内，岩石性质相对来说要不稳定些。

结论

(1) 有序样品聚类的相关序列法易于计算，很容易获得分割点，分类结果具有谱系结构。

(2) 相关序列法反映了有序样品的细微性质，有利于对有序样品分类的更深入的分析。

(3) 实例分析表明，相关序列法既简单又有效。相关序列法在地球物理、地质及其他领域有着广泛的应用前景。

本文数据由北京大学杨守仁教授提供，在此表示感谢。

参考文献

[1] Hartigan J A. *Clustering Algorithms*, John Wiley & Sons, 1975.

[2] Fisher W D. On grouping for maximum homogeneity. *J Am Stat Assoc*, 1958, 53, 789—798.

[3] 杨守仁，孙存礼。江西省信丰县铁石口地区二叠—三叠纪牙形石动物群的发现及其地质意义。北京大学学报（自然科学版），1990, 26(2): 243—256.

原文载于《石油地球物理勘探》，第 29 卷，第 1 期，1994 年 2 月，96—100.

一种新的样品聚类方法——差异序列法

程 乾 生

北京大学数学系

现有的有序样品的聚类方法叫 Fisher 算法[1,2]，Fisher 算法在计算每一类的直径时，没有考虑样品的有序性。针对样品的有序性，本文提出一种新的有序样品的聚类方法——差异序列法。先考虑每一个样品和它下一个样品之间的差异度，然后从整体上考虑差异度，选择出用于聚类的分割点。该方法计算简单，结果直观。最后我们用一个例子说明该方法的有效性。

1. 有序样品聚类的差异序列法

设有 m 个有序样品 $x_1, x_2, \cdots, x_m$。每一个样品都有 n 个指标观测值，对第 i 个样品 x_i，记作 $x_i = (x_{i1}, \cdots, x_{in})$，$x_{ij}$ 表示第 i 个样品的第 j 个指标观测值，其中 $1 \leqslant i \leqslant m, 1 \leqslant j \leqslant n$。

我们用非负数 $g_i = g(x_i, x_{i+1})$ 表示第 i 个样品 x_i 和第 $i+1$ 个样品 x_{i+1} 之间的差异，$i = 1, 2, \cdots, m-1$。当 $x_i = x_{i+1}$ 时，我们要求 $g_i = g(x_i, x_{i+1}) = 0$。通常，可取 g_i 为加权 l_p 模

$$g_i = \left[\sum_{j=1}^{n} w_j |x_{ij} - x_{i+1 j}|^p\right]^{1/p}, \quad i = 1, 2, \cdots, m-1, \tag{1}$$

其中 w_j 为权，$w_j \geqslant 0$。权 w_j 的作用主要是消除不同指标尺度的不同以及反映指标的重要性。也可取 g_i 为修正的广义相关系数

$$g_i = 1 - \frac{\sum_{j=1}^{n} x'_{ij} x'_{i+1 j}}{\sqrt{\sum_{j=1}^{n}(x'_{ij})^2 \sum_{j=1}^{n}(x'_{i+1 j})^2}}, \quad i = 1, 2, \cdots, m-1, \tag{2}$$

其中

$$x'_{ij} = \frac{x_{ij} - a_{ij}}{b_{ij}} \geqslant 0, \tag{3}$$

a_{ij} 和 b_{ij} 的选择要根据具体问题来定，a_{ij} 的作用使数据非负，b_{ij} 的作用类似于权

函数.

我们称 $g_i = g(x_i, x_{i+1})$ 为第 i 个样品与第 $i+1$ 个样品的差异度,称 $(g_1, g_2, \cdots, g_{m-1})$ 为样品的差异序列. $(g_1, g_2, \cdots, g_{m-1})$ 的差异度 $h_i = h(g_i, g_{i+1})$ 为样品的二次差异度,通常取为

$$h_i = |g_i - g_{i+1}|, \quad i = 1, 2, \cdots, m-2, \tag{4}$$

我们称 $(h_1, h_2, \cdots, h_{m-2})$ 为样品的二次差异序列.

把有序样品 $(x_1, x_2, \cdots, x_m)$ 聚成 $k(1 < k < m)$ 类的步骤是:首先确定 $k-1$ 个整数点 $i_1, \cdots, i_{k-1}$,它们满足 $1 \leqslant i_1 < i_2 < \cdots < i_{k-1} \leqslant m$;然后把样品聚成 k 类

$$(x_1, \cdots, x_{i_1})(x_{i_1+1}, \cdots, x_{i_2}) \cdots (x_{i_{k-2}+1}, \cdots, x_{i_{k-1}})(x_{i_{k-1}}, \cdots, x_m), \tag{5}$$

我们称 $i_1, \cdots, i_{k-1}$ 为 k 类分割点.

利用差异序列对有序样品进行分类的方法,称为差异序列法.基本思想是:在分割点处样品由一类变到另一类,因此在分割点处的样品差异度应该比较大;又由于样品是随机的,即使在同一类里,样品的差异度也是不同的,但是,在分割点附近,样品差异度变化应该比较大,也即样品的二次差异度应该比较大.差异序列法的步骤是:先把样品分成两类,在此基础上再把样品分成三类,直到 k 类.具体做法是:首先确定 2 类分割点 i_1.取 $l(1 \leqslant l \leqslant m-2)$ 使

$$h_l = \max_{1 \leqslant i \leqslant m-2} h_i, \tag{6}$$

i_1 由下式确定

$$\begin{cases} i_1 = l, & \text{当 } g_l > g_{l+1} \text{ 时}, \\ i_1 = l+1, & \text{当 } g_l < g_{l+1} \text{ 时}. \end{cases} \tag{7}$$

上式的意义是,在 l 和 $l+1$ 两个点中取差异度大的点为分割点 i_1.如果 l_1 和 l_2 ($l_1 \neq l_2$) 都满足 (6) 式,当 $\max(g_{l_1}, g_{l_1+1}) > \max(g_{l_2}, g_{l_2+1})$ 时,则由 l_1 按照 (7) 式确定 i_1.当 2 类分割点 i_1 确定后,就可把样品 $(x_1, x_2, \cdots, x_m)$ 分成两类 $(x_1, \cdots, x_{i_1})$ 和 $(x_{i_1+1}, \cdots, x_m)$.在此基础上,可把样品分成三类:对以上两类分别求最大二次差异度(见(6)),对这两个值中的最大值所在的类按以上方法再分成两类,这样就把样品分成了三类.依此下去,可把样品分成 k 类.我们注意,如果在一个类中只含两个样品,则这个类只有一个差异度,而没有二次差异度.在这种情况下,我们就比较各类的差异度,把最大差异度所在的类再分成两类.

由上可知,有序样品聚类的差异序列法具有谱系结构:$k-1$ 类分割点也必定是 k 类分割点.下面分析一个例子.

2. 例

我们考虑地层划分问题. 从上到下依次有 10 个地层，每一层取一个样品，依次记为 $x_1, x_2, \cdots, x_{10}$. 对每个样品进行分析测量，得到 34 个数据，第 i 个样品 x_i 的数据记为 $(x_{i1}, \cdots, x_{i34})$，数据矩阵 $[x_{ij}]_{10\times 34}$ 由表 1 给出. 在公式 (1) 中取 $p=1$，取权 w_j 为

$$w_j = \left[\sum_{i=1}^{m} x_{ij}\right]^{-1}, \quad j=1,2,\cdots,n, \text{其中 } m=10, n=34. \tag{8}$$

按公式 (1) 和 (8) 计算的差异序列 $(g_1, \cdots, g_9)$ 见表 2，按公式 (4) 计算的二次差异序列 $(h_1, \cdots, h_8)$ 见表 3.

由表 3 知，最大值 $h_5 = 1.63$ 比其他值大得多，这表明把样品分成两类是合理的. 按照 (6) 和 (7) 式，两类分割点为 $i_1 = 6$. 这一分割点是有地质意义的，它表明第 6 层层底是二叠—三叠系界线（即 P/T 界线）（见白俊峰和杨守仁的文章）. 这一结论得到古生物资料[3]的支持.

差异序列法不仅能用来研究 P/T 界线，还可获得更多的信息. 从表 2 可看出，从 g_1 到 g_5 比较小，从 g_7 到 g_9 比较大，这表明，在 $(x_1, \cdots, x_6)$ 这一类里，岩石性质相对稳定，在 $(x_7, \cdots, x_{10})$ 这一类里，岩石性质相对不稳定.

用差异序列法，样品 $(x_1, \cdots, x_{10})$ 的 3 类分割点为 $(6,9)$，4 类分割点为 $(6,9,2)$，5 类分割点为 $(6,9,2,7)$，6 类分割点为 $(6,9,2,7,8)$. 用 Fisher 算法（见白俊峰和杨守仁的文章），样品的 3 类分割点为 $(6,9)$，4 类分割点为 $(6,9,2)$，5 类分割点为 $(6,9,7,8)$，6 类分割点为 $(6,9,2,7,8)$. 两种方法的 2 类、3 类、4 类、6 类的分割点都是相同的，5 类分割点是不同的. 对 Fisher 算法，2 是 4 类分割点却不是 5 类分割点. 这在分类中是容易引起混淆的.

表 1 数据矩阵 $[x_{ij}]_{10\times 34}$

994.80	800.9	40.79	318.70	19.05	147.60	9.746	12.94	26.05	1.239
423.02	304.6	32.26	1501.0	44.37	128.10	10.510	14.07	27.96	1.062
241.10	234.2	66.86	553.10	55.89	92.27	15.810	17.48	37.69	2.331
470.20	198.2	66.63	451.10	64.91	107.50	13.810	15.80	36.40	1.977
278.90	277.1	68.48	515.10	10.98	87.36	16.420	16.58	38.86	2.346
38.19	118.8	27.25	1505.0	8.994	71.75	12.830	16.00	30.71	1.711
1237.0	119.3	22.53	188.50	3.771	157.20	4.9140	6.277	15.01	0.516
60.24	154.6	27.75	712.60	37.39	49.78	12.900	11.59	29.34	1.402
932.80	162.8	21.12	124.40	5.538	156.00	4.184	5.955	15.63	0.354
51.36	155.8	88.92	1386.0	13.11	43.78	17.260	16.38	72.86	2.523

17.18	5.9455	2.8212	2.1182	0.1570	0.2560	0.10	34.7758	0.0708	0.1713
27.82	18.4992	6.2805	1.4434	0.2299	0.2782	0.9276	14.8745	0.0789	0.1710
24.97	19.9139	5.8671	3.6617	0.3182	0.1214	1.1508	8.7827	0.0893	0.1671
19.74	12.4973	3.6485	3.2053	0.2612	0.1659	1.0830	13.8947	0.0645	0.1666
24.40	15.6087	6.3017	3.8389	0.3276	0.1303	1.2684	10.0181	0.0939	0.1425
24.84	11.7735	4.4944	2.4704	0.2939	0.0608	0.8076	2.7825	0.0582	0.1497
12.31	3.8413	2.0135	0.7238	0.0634	0.2190	0.6516	34.7758	0.0540	0.1361
25.64	19.0820	6.5348	2.0748	0.2615	0.0571	1.4370	2.2173	0.1023	0.1232
9.701	1.5703	1.0685	0.4808	0.0446	0.1284	0.6744	34.7758	0.0448	0.1413
22.69	11.7030	4.9332	3.6610	0.4070	0.0415	1.1946	0.6634	0.1102	0.1304
38.8991	1.2429	46.07	14.36	0.2794	2.069	147.80	153.10	38.81	2.895
16.0355	1.3907	57.87	26.77	0.2260	3.858	117.40	182.30	55.50	4.392
7.6318	1.0295	87.76	21.08	0.4164	3.299	185.87	135.90	31.94	3.041
12.8298	2.3724	74.62	18.37	0.3855	2.488	111.60	136.80	27.80	2.640
7.8982	1.0040	88.22	19.87	0.4296	3.494	192.81	126.90	34.35	2.895
3.4454	0.3220	41.44	21.08	0.3004	2.656	146.78	135.50	45.03	4.428
53.3699	10.3688	21.96	7.928	0.1210	1.062	166.40	130.70	40.25	2.732
1.5430	0.3897	48.80	15.30	0.2680	1.510	129.26	171.05	20.45	2.275
51.5655	5.7297	19.42	7.834	0.1027	0.950	161.70	120.10	17.39	1.399
0.5553	0.3297	124.90	36.75	0.4056	2.041	117.43	173.82	23.72	3.443
0.8139	0.0712	0.3929	490.21						
1.6490	0.0787	0.6631	483.69						
1.6440	0.0900	0.6778	349.02						
1.1480	0.0750	0.4961	386.34						
1.7600	0.0900	0.6434	344.32						
1.3690	0.0900	0.5599	303.49						
0.5710	0.0488	0.2652	497.59						
1.8320	0.0900	0.6852	172.81						
0.3207	0.0450	0.1891	456.60						
1.4300	0.0862	0.5525	162.19						

表2 $(x_1,\cdots,x_{10})$的差异序列

i	1	2	3	4	5	6	7	8	9
g_i	1.68	1.35	0.79	0.99	1.21	2.84	3.11	2.87	3.63

表3 $(x_1,\cdots,x_{10})$的二次差异序列

i	1	2	3	4	5	6	7	8
h_i	0.33	0.56	0.2	0.22	1.63	0.27	0.24	0.75

其他例子(包括文献[1]中的例子)也表明了差异序列法的优越性. 差异序列法还可用于无序样品的聚类和统计估计,这里就不做讨论了.

参考文献

[1] Hartigan, J. A., *Clustering Algorithms*, John Wiley & Sons, 1975.
[2] Fisher, W. D., *J. Am. Stat. Assoc.*, 1958, 53:789—798.
[3] 杨守仁,孙存礼,北京大学学报(自然科学版),1990,26(2):243.

原文载于《科学通报》,第39卷,第2期,1994年1月,97—99.

属性识别理论模型及其应用

程 乾 生

北京大学数学科学学院信息科学系

摘要:在属性测度空间和有序分割类概念的基础上,提出了属性识别准则,建立了属性识别理论模型,并讨论了在大气环境质量评价中的应用.

0. 引言

在实际中有大量问题,例如环境质量评价中的许多问题[1],都可归结为对定性描述的度量问题.为了讨论定性描述的度量问题和不同的定性描述之间的关系以及相应的度量之间的关系,笔者提出了属性集的概念,属性测度空间和有序分割类的概念[2].本文在此基础上,研究属性识别的准则、理论模型和应用.

本文正文共分 5 节.第 1 节讨论属性集、属性可测空间和有序分割类的概念.在第 2 节,针对不同的问题,提出 4 个属性识别准则.在第 3 节,研究一类属性识别模型、计算、识别和比较的方法.在第 4 节,通过一个实例讨论本文提出的方法在大气环境质量评价中的应用,并分析了模糊数学在应用中所产生的问题.第 5 节为结语.

1. 属性集和属性可测空间

1.1 属性集和属性集运算

设 X 为研究对象的全体,称为对象空间.F 为 X 中元素的某类属性,称为属性空间或最大属性集.例如,$X = \{$所有的大气样品$\}$,X 中的元素 x 表示某地区某时间某高度的大气样品.我们要研究大气污染情况,可以令 $F = \{$大气污染程度$\}$.属性空间 F 中的任何一种情况,都称为一个属性集.例如,$A = \{$清洁$\}$,$B = \{$轻污染$\}$,A 和 B 都是属性空间 F(污染程度)中的一种情况,所以,A 和 B 都是属性集,都可看成 F 的子集.

对于属性集,可以定义属性集运算.A 与 B 的和 $A \bigcup B$ 表示"或者有 A 属性，或者有 B 属性". A 与 B 的交 $A \cap B$ 表示"既有 A 属性，也有 B 属性". A 与 B 的差 $A - B$ 表示"有 A 属性而没有 B 属性". A 的余集 $\overline{A}$ 表示"不具有 A 属性". 属性集中的空集定义为"不具有任何属性"，记为 $\approx$. 例如, $A - A = \approx, A \cap \overline{A} = \approx$. 对于多个属性集 A_i 的和与交分别用符号 $\bigcup_i A_i$ 与 $\bigcap_i A_i$ 表示. $\bigcup_i A_i$ 表示"至少具有 A_i 中的一个属性", $\bigcap_i A_i$ 表示"具有所有 A_i 的属性".

关于通常集合运算的关系式，对属性集也一样成立，例如, $A - B = A \cap \overline{B}$, $\overline{A \cup B} = \overline{A} \cap \overline{B}$.

1.2 属性代数和属性 σ 代数

为了研究属性集和属性集之间的关系，我们必须研究属性集的集合 $\mathscr{B}$.

如果 $\mathscr{B}$ 满足以下 3 个条件：(1) $F \in \mathscr{B}$; (2) 如果 $A \in \mathscr{B}$, 则 $\overline{A} \in \mathscr{B}$; (3) 如果 $A \in \mathscr{B}, B \in \mathscr{B}$, 则 $A \cup B \in \mathscr{B}$, 那么，称 $\mathscr{B}$ 为属性代数.

属性代数的意义是：对属性代数中的属性集，对余运算、有限和运算、有限交运算皆封闭.

设 $\mathscr{B}$ 为属性代数，如果 $\mathscr{B}$ 还满足：对 $A_i \in \mathscr{B}, i = 1, 2, \cdots,$ 有 $\bigcup_i A_i \in \mathscr{B}$, 则称 $\mathscr{B}$ 为属性 σ 代数，称 $(F, \mathscr{B})$ 为属性可测空间. 从理论上看，属性代数和属性 σ 代数与概率论和测度论中的代数和 σ 代数是不同的[4,5]，因为在那里集合的运算是点集的运算，而属性集的运算是逻辑中的"或"、"并"、"非"等运算. 当然，更抽象的、更一般的 σ 代数，可在格和 Boole 代数的基础上直接定义. 关于格和 Boole 代数参见文献[3].

1.3 属性测度和属性测度空间

设 x 为 X 中的一个元素, A 为一个属性集，用"$x \in A$"表示"x 具有属性 A". "$x \in A$"仅是一种定性的描述，而更需要的是定量的刻画"x 具有属性 A"的程度. 用一个数来表示"$x \in A$"的程度，这个数记为 $\mu(x \in A)$ 或 $\mu_x(A)$, 称它为 $x \in A$ 的属性测度. 为方便起见，要求属性测度在 $[0, 1]$ 之内取值.

这里需要指出，用一个数 $\mu(x \in A)$ 反映一个事物 x 具有属性 A 的程度的做法是早已有之的. 例如，用分数衡量一个学生的学习能力就是这样. 设 x 表示一个学生, A 表示"某门课程学得好"，通过口试或笔试测量学生某门课学得好（即 $x \in A$）的程度，该程度用分数表示，采用百分制时分数取值在 $[0, 100]$ 之内，采用 5 分制时分数取值为 $1, 2, 3, 4, 5$. 当然，分数 $\mu(x \in A)$ 的取值都可转化为在 $[0, 1]$ 之内.

对不同的属性集都可以给出相应的属性测度,但是,这些属性测度不可以任意给,必须满足一定的规则.

设 $(F,\mathscr{B})$ 为属性可测空间. 称 μ_x 为 $(F,\mathscr{B})$ 上的属性测度,如果 μ_x 满足:
1) $\mu_x(A) \geqslant 0, \forall A \in \mathscr{B}$;
2) $\mu_x(F) = 1$;
3) 若 $A_i \in \mathscr{B}, A_i \cap A_j = \approxeq (i \neq j)$,则

$$\mu_x(\bigcup_{i=1}^{\infty} A_i) = \sum_{i=1}^{\infty} \mu_x(A_i).$$

上述第 3 条性质称为属性测度的可加性. 称 $(F,\mathscr{B},\mu_x)$ 为属性测度空间. 在这里需要指出,属性测度空间与传统的概率空间和测度空间是不同的. 传统的概率空间和测度空间是建立在点集基础上的[4,5],而属性集不是由元素或点组成的,它只是关于某种属性的一种定性描述. 模糊集仅是研究对象 X 到区间 $[0,1]$ 的一个映射[6],因此,属性集也不同于模糊集.

1.4 属性空间的分割和有序分割类

设 F 为 X 上某类属性空间,$C_1, C_2, \cdots, C_K$ 为属性空间中的 K 个属性集. 如果 $(C_1, C_2, \cdots, C_K)$ 满足

$$F = \bigcup_{i=1}^{K} C_i, \quad C_i \cap C_j = \approxeq (i \neq j),$$

则称 $(C_1, C_2, \cdots, C_K)$ 为属性空间 F 的分割.

现举一例. 设 F 为大气污染程度,把 F 分为 5 类:$C_1 = \{$清洁$\}$,$C_2 = \{$轻污染$\}$,$C_3 = \{$中污染$\}$,$C_4 = \{$重污染$\}$,$C_5 = \{$严重污染$\}$. 这 5 类彼此不相交,$(C_1, C_2, \cdots, C_5)$ 是属性空间 F 的分割.

在有些情况下,属性集和属性集之间是可"比较"的. 在上例中,可以认为污染程度越低越好或越低越"强". 因此,对有些属性集可建立"强"序或"弱"序. 当属性集 A 比属性集 B "强"时,记为 $A > B$,当 A 比 B "弱"时,记为 $A < B$.

当然,"强"与"弱"是相对的,从一个角度认为是"强"的,从相反的角度则认为是"弱"的. 如何确定"强"与"弱",由具体问题确定.

如果 $(C_1, C_2, \cdots, C_K)$ 为属性空间 F 的分割,并且 $C_1 > C_2 > \cdots > C_K$ 或者 $C_1 < C_2 < \cdots < C_K$,则称 $(C_1, C_2, \cdots, C_K)$ 为有序分割类.

在上面大气污染的例子中,可以认为污染程度越低越"强",因此,在该例中,$C_1 > C_2 > C_3 > C_4 > C_5$,于是,$(C_1, C_2, C_3, C_4, C_5)$ 为有序分割类.

2. 属性识别准则

2.1 问题

现在提出两个属性识别问题.

问题 1 设 $(C_1, C_2, \cdots, C_K)$ 是属性空间 F 的一个分割, x 属于 C_i 类的属性测度为 $\mu_x(C_i)$, 满足 $\sum_{i=1}^{K} \mu_x(C_i) = 1$. 已知 $\mu_x(C_i), 1 \leqslant i \leqslant K$, 如何判别 x 属于哪一类 C_i?

问题 2 设 $(C_1, C_2, \cdots, C_K)$ 是属性空间 F 的一个有序分割类, x 属于 C_i 类的属性测度为 $\mu_x(C_i)$, 满足 $\sum_{i=1}^{K} \mu_x(C_i) = 1$. 已知 $\mu_x(C_i), 1 \leqslant i \leqslant K$, 如何判别 x 属于哪一类 C_i?

2.2 最小代价准则

现在回答问题 1.

判别有对有错, 因此要考虑代价. 设属于 C_i 而判别为 C_j 的代价为 d_{ij}. β_j 表示判别为 C_j 的全部代价, β_j 为

$$\beta_j = \sum_{i=1}^{K} d_{ij} \mu_x(C_i).$$

最小代价准则 若

$$\beta_{j_0} = \min_{1 \leqslant j \leqslant K} \beta_j,$$

则认为 x 属于 C_{j_0} 类.

因为当 x 判别为 C_{j_0} 类时所付出的代价最小, 所以这种判别是合理的.

2.3 最大属性测度准则

现在继续回答问题 1. 考虑最小代价准则的一种最简单的情况. 若认为正确判别无需付出代价, 这时取 $d_{jj}=0$; 若认为错误判别的代价皆相同, 这时取 $d_{ij}=1 (i \neq j)$. 则有

$$\beta_j = \sum_{\substack{1 \leqslant i \leqslant K \\ i \neq j}} \mu_x(C_i) = 1 - \mu_x(C_j),$$

进而有

$$\beta_0 = \min_{1 \leqslant j \leqslant K} \beta_j = 1 - \max_{1 \leqslant j \leqslant K} \mu_x(C_j).$$

最大属性测度准则 若

$$\mu_x(C_{j_0}) = \max_{1 \leqslant j \leqslant K} \mu_x(C_j),$$

则认为 x 属于 C_{j_0} 类.

这个准则是在一组简单特殊的代价 d_{ij} 下获得的, 因此, 也称最大属性测度准则为简单最小代价准则. 在模糊数学中, 相应的准则称为最优从属原则[7].

2.4 置信度准则

现在回答问题 2.

对有序分割类, 上述的识别准则就不太合适了. 还是举大气污染的例子. 大气污染情况分为 5 类 C_1, C_2, C_3, C_4, C_5, 依次表示清洁、轻污染、中污染、重污染、严重污染. 设大气样品 x 的属性测度为 $(\mu_x(C_1), \mu_x(C_2), \mu_x(C_3), \mu_x(C_4), \mu_x(C_5)) = (0.1, 0.4, 0.2, 0.2, 0.1)$, 其中 0.4 最大. 按照最大属性测度准则, 认为 x 属于第 2 类 $C_2 = \{$轻污染$\}$. 但是, x 属于中污染、重污染和严重污染的属性测度之和为 0.5, 占整个属性测度总和的一半, 因此, 认为 x 为轻污染是不合理的. 为此, 这里提出置信度准则.

置信度准则 设 $(C_1, C_2, \cdots, C_K)$ 是属性空间 F 的有一个有序分割类, λ 为置信度. 若在 $C_1 > C_2 > \cdots C_K$ 时,

$$k_0 = \min\left\{k: \sum_{i=1}^{k} \mu_x(C_i) \geqslant \lambda, 1 \leqslant k \leqslant K\right\},$$

若在 $C_1 < C_2 < \cdots < C_K$ 时,

$$k_0 = \max\left\{k: \sum_{i=1}^{k} \mu_x(C_i) \geqslant \lambda, 1 \leqslant k \leqslant K\right\},$$

则认为 x 属于 C_{k_0} 类.

上述准则是从"强"的角度考虑的, 即认为越"强"越好, 而且"强"的类应占相当大的比例. λ 为置信度, 取值范围通常为 $0.5 < \lambda < 1$, 一般取 0.6 与 0.7 之间.

2.5 评分准则

现在讨论一个新的问题—— 对象比较问题.

问题 3 设 $(C_1, C_2, \cdots, C_K)$ 是属性空间 F 的一个有序分割类, x_1 的属性测度为 $\mu_{x_1}(C_i)$, x_2 的属性测度为 $\mu_{x_2}(C_i)$, 满足 $\sum_{i=1}^{K} \mu_{x_1}(C_i) = \sum_{i=1}^{K} \mu_{x_2}(C_i) = 1$. 已知 $u_{x_1}(C_i)$ 和 $\mu_{x_2}(C_i)$, $1 \leqslant i \leqslant K$, 如何比较 x_1 和 x_2?

由于属性集 C_i 之间有强弱的关系, 可以用分数表示属性集的强弱关系, 强属性集的分数比弱属性集的分数大. 设属性集 C_i 的分数为 n_i, 当 $C_1 < C_2 < \cdots$

$<C_K$ 时,有 $n_1<n_2<\cdots<n_K$;当 $C_1>C_2>\cdots>C_K$ 时,有 $n_1>n_2>\cdots>n_K$. 称

$$q_x = \sum_{i=1}^{K} n_i \mu_x(C_i)$$

为 x 的分数.

评分准则 若

$$q_{x_1} > q_{x_2},$$

则认为 x_1 比 x_2 强,记为 $x_1 > x_2$.

通常,对 $C_1>C_2>\cdots>C_K$ 的情形,取 $n_i = K+1-i$,它表示有序分割类 $(C_1,C_2,\cdots,C_K)$ 中类别的重要性是等间隔下降的;对 $C_1<C_2<\cdots<C_K$ 的情形,取 $n_i = i$,它表示有序分割类 $(C_1,C_2,\cdots,C_K)$ 中类别的重要性是等间隔上升的.

以上介绍了 4 种属性识别准则. 现在笔者用最小代价准则解释为什么北京大学规定对必修课和选修课,本科生在 60 分以上、研究生在 70 分以上才算及格. 设研究对象 X 为某班全体同学,属性空间 $F=\{$对某门课程的掌握程度$\}$. 这里"掌握"包含对知识的掌握程度和能力强弱两个方面. 对 F 进行两类分割 $(C_1,C_2), C_1=\{$掌握得好$\}, C_2=\{$掌握得不好$\}$. (C_1,C_2) 为有序分割类, $C_1 > C_2$. 学生 x 所得的分数为 $\mu_x(C_1)$,所丢失的分数为 $\mu_x(C_2)=1-\mu_x(C_1)$. 如何根据 $\mu_x(C_1)$ 来判断学生 x 是属于 $C_1=\{$掌握得好$\}$ 的那一类呢,也即如何判断学生是否及格呢? 我们考虑最小代价准则. 取代价 $d_{11}=d_{22}=0$,属于 C_1 类而判为 C_2 类的代价取为 $d_{12}=1$,属于 C_2 类而判为 C_1 类的代价取为 $d_{21}=\alpha$. 我们要求 $\alpha \geq 1$,即 $d_{21}>d_{12}$,它表示:我们宁可把"好"的判为"不好"的,也不要把"不好"的判为"好"的. 把以上代价代入公式,我们可算出判为 C_1 类的全部代价 β_1 和判为 C_2 类的全部代价 β_2:

$$\beta_1 = \alpha \mu_x(C_2), \quad \beta_2 = \mu_x(C_1).$$

按最小代价准则,若判为 C_1 类,则要求 $\beta_1 < \beta_2$,即 $\alpha \mu_x(C_2) < \mu_x(C_1)$. 由于 $\mu_x(C_2)=1-\mu_x(C_1)$,我们得到

$$\mu_x(C_1) > \lambda,$$

其中

$$\lambda = \frac{\alpha}{1+\alpha}.$$

上式表明,学生 x 的分数 $\mu_x(C_1)$ 只有大于 λ 时,才能认为达到及格水平. α 为"不好"判为"好"的代价参数,它必须大于或等于 1. 取不同的 α 值,就可以得到不同的及格水平值 λ,如下表:

α	1	1.5	2	2.34	2.5
λ	0.50	0.60	0.67	0.70	0.71

取 $\alpha=1.5$,表示 60 分为及格线,取 α 接近 2.5 时,70 分为及格线. 对学生要求越高,α 的值取得越大. 置信度准则实际上是把有序分割类归结为"好"与"不好"或"强"与"弱"两个类,因此,置信度 λ 的含义和上面的及格水平值 λ 是一样的. 上面的分析,不仅揭示了最小代价准则和置信度准则的关系,也揭示了置信度和代价的关系.

对于置信度准则和评分准则的应用,下面将在第 4 节通过具体例子加以说明.

3. 一类属性识别模型

3.1 已知指标分类标准的属性模式识别模型

在研究对象空间 X 取 n 个样品 $x_1,x_2,\cdots,x_n$. 对每个样品要测量 m 个指标 $I_1,I_2,\cdots,I_m$. 第 i 个样品 x_i 的第 j 个指标 I_j 的测量值为 x_{ij},因此,第 i 个样品 x_i 可以表示为一个向量 $x_i=(x_{i1},\cdots,x_{im}),1\leqslant i\leqslant n$.

设 F 为 X 上某类属性空间,$(C_1,C_2,\cdots,C_K)$ 为属性空间 F 的有序分割类,且满足 $C_1>C_2>\cdots>C_K$. 每个指标的分类标准已知,写成分类标准矩阵为

$$\begin{array}{c} \quad\quad C_1 \quad C_2 \quad \cdots \quad C_K \\ \begin{array}{c} I_1 \\ I_2 \\ \vdots \\ I_m \end{array} \begin{bmatrix} a_{11} & a_{12} & \cdots & a_{1K} \\ a_{21} & a_{22} & \cdots & a_{2K} \\ \vdots & \vdots & & \vdots \\ a_{m1} & a_{m2} & \cdots & a_{mK} \end{bmatrix}, \end{array}$$

其中 a_{jk} 满足 $a_{j1}<a_{j2}<\cdots<a_{jK}$ 或者 $a_{j1}>a_{j2}>\cdots>a_{jK}$.

问题 已知 $x_i=(x_{i1},\cdots,x_{im}),1\leqslant i\leqslant n$,和分类标准矩阵,如何判别 x_i 属于哪一类 C_K,如何比较 x_i?

下面进行分析,关键是计算样品的属性测度,然后按照置信度准则和评分准则就可以回答上述问题.

3.2 样品属性测度的计算

首先来计算第 i 个样品第 j 个指标值 x_{ij} 具有属性 C_k 的属性测度 $\mu_{ijk}=\mu(x_{ij}\in C_k)$. 不妨假定 $a_{j1}<a_{j2}<\cdots<a_{jK}$.

当 $x_{ij} \leqslant a_{j1}$ 时，取 $\mu_{ij1}=1, \mu_{ij2}=\cdots=\mu_{ijK}=0$.
当 $x_{ij} \geqslant a_{jK}$ 时，取 $\mu_{ijK}=1, \mu_{ij1}=\cdots=\mu_{ijK-1}=0$.
当 $a_{jl} \leqslant x_{ij} \leqslant a_{jl+1}$ 时，取

$$\mu_{ijl} = \frac{x_{ij}-a_{jl+1}}{a_{jl}-a_{jl+1}}, \quad \mu_{ijl+1}=\frac{x_{ij}-a_{jl}}{a_{jl+1}-a_{jl}},$$

$$\mu_{ijk}=0, \quad k<l \text{ 或 } k>l+1.$$

在知道第 i 个样品的各指标测量值的属性测度之后，现在计算第 i 个样品 x_i 的属性测度 $\mu_{ik} = \mu(x_i \in C_k)$. 指标共有 m 个，每个指标的重要性可能相同、也可能不相同. 因此，要考虑指标权 $(w_1, w_2, \cdots, w_m)$, $w_j \geqslant 0$, $\sum_{j=1}^{m} w_j = 1$. 由指标权可得到属性测度

$$\mu_{ik} = \mu(x_i \in C_k) = \sum_{j=1}^{m} w_j \mu_{ijk}, \quad 1 \leqslant i \leqslant n, 1 \leqslant k \leqslant K.$$

有了属性测度，就可以进行识别和比较分析.

按照置信度准则，对置信度 λ，计算

$$k_i = \min\left\{ k : \sum_{l=1}^{k} \mu_{x_i}(C_l) \geqslant \lambda, 1 \leqslant k \leqslant K \right\},$$

则认为 x_i 属于 C_{k_i} 类.

按照评分准则，计算

$$q_{x_i} = \sum_{l=1}^{K} nl \mu_{x_i}(C_l),$$

则可根据 q_{x_i} 的大小对 x_i 进行比较和排序.

在下节将讨论本节模型的应用.

4. 在大气环境质量评价中的应用

4.1 应用实例

现在讨论大气环境质量评价的一个实例. 按照国家大气环境质量标准 (G83095-82) 和规定，选取 SO_2、NO_x、TSP 和降尘为 4 个评价指标，把大气环境质量 (即污染程度) 分为 5 级或 5 类：$C_1 = \{清洁\}$，$C_2 = \{轻污染\}$，$C_3 = \{中污染\}$，$C_4 = \{重污染\}$，$C_5 = \{严重污染\}$. 分类标准矩阵为

$$\begin{array}{c} \quad\quad C_1 \quad\ C_2 \quad\ C_3 \quad\ C_4 \quad\ C_5 \\ \begin{array}{c}SO_2\\NO_x\\TSP\\降尘\end{array}\left[\begin{array}{ccccc}0.05 & 0.15 & 0.25 & 0.50 & 0.85\\0.05 & 0.10 & 0.15 & 0.30 & 0.50\\0.15 & 0.30 & 0.50 & 1.00 & 1.70\\3 & 13 & 22 & 44 & 75\end{array}\right].\end{array}$$

对宜宾市区进行 5 年大气监测,得到测量数据矩阵

$$\begin{array}{c}\quad\quad\quad SO_2 \ \ NO_x \ \ TSP \ \ 降尘\\\begin{array}{c}1985\ 年\\1986\ 年\\1987\ 年\\1988\ 年\\1989\ 年\end{array}\left[\begin{array}{cccc}0.37 & 0.03 & 0.59 & 15.9\\0.41 & 0.04 & 0.48 & 15.7\\0.61 & 0.06 & 0.62 & 11.0\\0.59 & 0.08 & 0.69 & 14.6\\0.40 & 0.06 & 0.39 & 9.9\end{array}\right].\end{array}$$

因为有 4 个指标,取指标权为 $(w_1, w_2, w_3, w_4) = (0.25, 0.25, 0.25, 0.25)$. 按照上一节的计算方法,可得到属性测度分布矩阵

$$\begin{array}{c}\quad\quad\quad C_1 \quad\ C_2 \quad\ C_3 \quad\ C_4 \quad\ C_5\\\begin{array}{c}1985\ 年\\1986\ 年\\1987\ 年\\1988\ 年\\1989\ 年\end{array}\left[\begin{array}{ccccc}0.25 & 0.17 & 0.42 & 0.16 & 0\\0.25 & 0.20 & 0.39 & 0.16 & 0\\0.25 & 0.25 & 0.19 & 0.23 & 0.08\\0.10 & 0.36 & 0.20 & 0.28 & 0.06\\0.28 & 0.36 & 0.21 & 0.15 & 0\end{array}\right].\end{array}$$

取置信度为 0.7,由置信度准则可判别各年大气质量的级别,1985 年为 3 级,1986 年为 3 级,1987 年为 4 级,1988 年为 4 级,1989 年为 3 级.

在应用评分准则时,取 $n_i = 6 - i, 1 \leqslant i \leqslant 5$,按公式

$$q_{x_i} = \sum_{l=1}^{5}(6-l)\mu_{x_i}(C_l)$$

计算各年的分数,得下表:

年份	1985	1986	1987	1988	1989
分数	3.51	3.54	3.36	3.16	3.77

从上表可知,按污染程度从轻到重来排序,这 5 年的次序是 1989、1986、1985、1987、1988.

对上面的实例,有的文献用模糊综合评判法(见文献[8])或模糊识别理论模型(见文献[9])进行分析. 利用模糊综合评判法进行大气环境质量评价时,会出现分级不清、结果不合理的问题[9]. 究其原因,模糊数学的取大取小运算损失了大批中间值的信息,而最大隶属度原则不适用于大气环境质量评价,因为环境质

量评价是有序分割类的识别问题,这在分析置信度准则时已经指出了.利用模糊识别理论模型进行大气环境质量评价时,存在两个问题:(1)所计算的隶属度已失去了最原始、最直观的含义.例如,1985年、1986年和1989年的4项指标值皆在1级与4级之间,而文献[9]计算的结果,属于5级的隶属度却不为0;(2)不能比较更细微的污染程度.例如,1987年和1988年的大气环境质量皆评判为4级,但是哪一个污染程度更大些呢?由文献[9]的评价结果表1得不出答案.究其原因,是没有考虑大气环境质量评价问题的特点,没有有序分割类的概念.

5. 结语

笔者提出了属性测度空间和有序分割类的概念,提出了4个属性识别准则:最小代价准则,最大属性测度准则,置信度准则,评分准则.在此基础上,建立了一类属性识别模型,并用于大气环境质量评价.本文所提出的理论和方法,可以用到更广泛的质量评价和其他有关的问题中.

参考文献

[1] 叶文虎,栾胜基. 环境质量评价学. 北京:高等教育出版社,1994.
[2] 程乾生. 属性集理论与模糊数学、模式识别、人工智能. 见:肖树铁,吴方主编. 中国工业与应用数学学会第三次大会文集. 北京:清华大学出版社,1994.
[3] 古德斯坦因著. 刘文,李忠儐译. 布尔代数. 北京:科学出版社,1978.
[4] Lo ve M. Probability Theory. New York:D Van Nostrand Company,1963.
[5] Halmos P R. Measure Theory,New York:D Van Nostrand Company,1950.
[6] Zadeh L A. Fuzzy Sets,Information and Control. 1965,8:338.
[7] 楼世博,孙章,陈化成. 模糊数学. 北京:科学出版社,1983.
[8] 余常贵,王文秀. 模糊数学在大气环境质量评价中的应用. 重庆环境科学,1991,13(1):28.
[9] 陈守煜,熊德琪. 城市大气环境质量评价模糊识别理论模型. 环境科学研究,1992,5(5):10.

原文载于《北京大学学报(自然科学版)》,第33卷,第1期,1997年1月,12—33.

属性层次模型 AHM——一种新的无结构决策方法

程 乾 生

北京大学数学科学学院信息科学系

摘要:提出了属性层次模型 AHM,一种无结构决策方法.在属性测度基础上,提出了相对属性测度和属性判断矩阵的概念.相对权重和合成权重很容易从属性判断矩阵获得.通过例子说明,属性层次模型是简单有效的.

0. 引言

对无结构决策问题,1977 年 Saaty 提出了层次分析法[1],该方法建立在测量模型之上,解决的关键问题是如何通过元素的两两比较确定元素的排序.本文在属性测度基础上,提出了一种新的无结构决策方法——属性层次模型 AHM.

1. 属性判断矩阵

设有 n 个元素 $u_1, u_2, \cdots, u_n$. 对准则 C,比较两个不同的元素 u_i 和 u_j ($i \neq j$),u_i 和 u_j 的对准则 C 的重要性分别记为 μ_{ij} 和 μ_{ji}. 按属性测度的要求,μ_{ij} 和 μ_{ji} 应满足:

$$\mu_{ij} \geq 0, \quad \mu_{ji} \geq 0, \quad \mu_{ij} + \mu_{ji} = 1. \tag{1}$$

元素 u_i 和自身的比较是没有意义的,我们规定:

$$u_{ii} = 0, \quad 1 \leq i \leq n. \tag{2}$$

定义 1 满足(1)和(2)的 μ_{ij} 称为相对属性测度,由 μ_{ij} 组成的矩阵 $(\mu_{ij})_{1 \leq i,j \leq n}$ 称为属性判断矩阵.

定义 2 若 $\mu_{ij} > \mu_{ji}$,则称 u_i 比 u_j 相对强,认为 $u_i > u_j$.

定义 3 属性判断矩阵 (μ_{ij}) 称为具有一致性,如果对任何 i, j, k 有 $u_i > u_j$,$u_j > u_k$,则 $u_i > u_k$.

令

$$g(x) = \begin{cases} 1, & x > 0.5, \\ 0, & x \leq 0.5, \end{cases} \tag{3}$$

$$I_i = \{j : g(\mu_{ij}) = 1, 1 \leqslant j \leqslant n\}. \tag{4}$$

定理 1 属性判断矩阵 (μ_{ij}) 具有一致性的必要充分条件是:对任何 i,当 I_i 非空时,

$$g(\mu_{ik}) - g\big(\sum_{j \in I_i} g(\mu_{jk})\big) \geqslant 0, \quad 1 \leqslant k \leqslant n. \tag{5}$$

定理 1 给出了属性判断矩阵一致性检验方法.

令

$$W_C(i) = \frac{2}{n(n-1)} \sum_{j=1}^{n} \mu_{ij},$$

我们称 $W_C = (W_C(1), W_C(2), \cdots, WC(n))^T$ 为相对属性权,其中 T 表示转置.

在层次分析法中[1,2],元素 u_i 和 u_j 的比较由相对比例标度 b_{ij} 给出.可由 b_{ij} 确定 μ_{ij},比如,可规定:

$$\mu_{ij} = \begin{cases} k/(k+1), & b_{ij} = k, \\ 1/(k+1), & b_{ij} = 1/k. \end{cases} \tag{6}$$

在准则 C 下所计算的相对属性测度和属性权可表示为:

C	u_1	u_2	$\cdots$	u_n	W_C
u_1	μ_{11}	μ_{12}	$\cdots$	μ_{1n}	$W_C(1)$
u_2	μ_{21}	μ_{22}	$\cdots$	μ_{2n}	$W_C(2)$
$\vdots$	$\vdots$	$\vdots$	$\vdots$	$\vdots$	$\vdots$
u_n	μ_{n1}	μ_{n2}	$\cdots$	μ_{nn}	$W_C(n)$

2. 属性层次模型

我们结合一个例子讨论属性层次模型.

例 1 在某闹市区一商场附近交通十分拥挤,为改善交通环境,提出了 3 种方案:修天桥,修地道,商场搬迁.问题是选择哪种方案为好?[3]

用属性层次模型进行决策,方法如下:

1) 建立递阶层次结构.

该结构分为 3 层.最高层为目标层,最低层为方案层,中间层为准则层,它包含为实现目标所涉及的中间环节,这些环节作为决策分析的准则.根据需要,中间层还可分成若干层.在层次结构中,目标、每一种方案、每一个准则,皆称为元素.每一个元素作为准则支配着与它有关的下一层元素.例 1 的层次结构见图 1.

2) 构造属性判断矩阵并计算相对属性权.

对最高层和中间层的每一元素,以它为准则,构造与它有关的下一层元素的属性判断矩阵,并计算相对属性权.构造与计算的方法见上节.在构造出属性判断矩阵之后,要用上节提出的方法进行一致性检验.

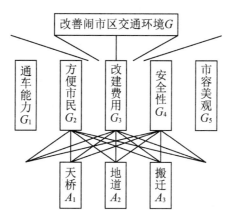

图 1 例 1 的层次结构图

对例1,按图1,中间层以上各元素的属性判断矩阵和相对属性权如下:

G	C_1	C_2	C_3	C_4	C_5	W_G
C_1	0	0.75	0.833	0.75	0.833	0.317
C_2	0.25	0	0.75	0.5	0.75	0.225
C_2	0.167	0.25	0	0.25	0.75	0.141
C_3	0.25	0.5	0.75	0	0.75	0.225
C_5	0.167	0.25	0.25	0.25	0	0.092

C_1	A_1	A_2	A_3	W_{C_1}	C_2	A_1	A_2	A_3	W_{C_2}
A_1	0	0.5	0.833	0.444	A_1	0	0.75	0.833	0.527
A_2	0.5	0	0.833	0.444	A_2	0.25	0	0.667	0.306
A_3	0.167	0.167	0	0.112	A_3	0.167	0.333	0	0.167

C_3	A_1	A_2	A_3	W_{C_3}
A_1	0	0.8	0.875	0.558
A_2	0.2	0	0.8	0.333
A_3	0.125	0.2	0	0.109

C_4	A_1	A_2	A_3	W_{C_4}
A_1	0	0.333	0.25	0.204
A_2	0.667	0	0.25	0.389
A_3	0.75	0.5	0	0.407

C_5	A_1	A_2	A_3	W_{C_5}
A_1	0	0.333	0.25	0.204
A_2	0.667	0	0.5	0.389
A_3	0.75	0.5	0	0.407

上面各属性判断矩阵, 皆按公式(6), 由文献[3]相应的判断矩阵转换而来. 经检验, 这些属性判断矩阵皆具有一致性.

3) 计算方案对系统目标的合成权重.

针对图 1, 合成权 W_G 的公式为

$$W_G = (W_{C_1} W_{C_2} \cdots W_{C_5}) W_G. \tag{7}$$

具体地, 上式为

$$W_G = \begin{pmatrix} 0.444 & 0.527 & 0.558 & 0.204 & 0.204 \\ 0.444 & 0.306 & 0.333 & 0.389 & 0.389 \\ 0.112 & 0.167 & 0.109 & 0.407 & 0.407 \end{pmatrix} \begin{pmatrix} 0.317 \\ 0.225 \\ 0.141 \\ 0.225 \\ 0.092 \end{pmatrix} = \begin{pmatrix} 0.403 \\ 0.38 \\ 0.217 \end{pmatrix}.$$

由此可知, 方案 A_1 (修建天桥) 的合成权重最大, 应做出选择方案 A_1 的决策. 这个结论与文献[3]用 AHP 所得出的结论是相同的.

对于多层结构系统, 可以按图 1 和公式(7)的模式由低层向高层逐层合成, 最终得到方案对目标的合成权重.

3. 定理 1 的证明

引理 1 $u_i > u_j$ 的必要充分条件是 $g(\mu_{ij}) = 1$.

证明 按定义 2, $u_i > u_j$ 是指 $\mu_{ij} > \mu_{ji}$. 由(1)知, $\mu_{ij} > \mu_{ji}$ 的充要条件是 $\mu_{ij} > 0.5$. 而由(3)知, $\mu_{ij} > 0.5$ 的充要条件是 $g(x) = 1$. 证毕.

定理 1 的证明 充分性 设 $u_{i_0} > u_{j_0}, u_{j_0} > u_{k_0}$. 由引理 1 知, $g(\mu_{i_0 j_0}) = 1$,

$g(\mu_{j_0 k_0}) = 1$. 由(4)知, I_{i_0} 非空, 而且 $j_0 \in I_{i_0}$. 由 g 函数的定义(3)式知, $g(\sum_{j \in I_{i_0}} g(\mu_{jk_0})) = 1$. 由(5)和(3)知, $g(\mu_{i_0 k_0}) = 1$. 由引理1知, $u_{i_0} > u_{k_0}$. 充分性成立.

必要性 反证法. 设 i_0 使 I_{i_0} 非空, 但是(5)式不成立. 这意味着, 存在 k_0 使

$$g(\mu_{i_0 k_0}) - g(\sum_{j \in I_{i_0}} g(\mu_{jk_0})) < 0.$$

由于 g 只取 0 或 1, 要使上式成立, 只有

$$g(\mu_{i_0 k_0}) = 0, \tag{8}$$

$$g(\sum_{j \in I_{i_0}} g(\mu_{jk_0})) = 1. \tag{9}$$

要使(9)成立, 至少得存在一个 j_0 使得 $j_0 \in I_{i_0}$, 同时 $g(\mu_{j_0 k_0}) = 1$. 由于 $j_0 \in I_{i_0}$, 于是 $g(\mu_{i_0 j_0}) = 1$. 由引理1知, $u_{i_0} > u_{j_0}$, $u_{j_0} > u_{k_0}$. 再由一致性定义3知, $u_{i_0} > u_{k_0}$, 因而 $g(\mu_{i_0 k_0}) = 1$. 而这与(8)矛盾. 必要性成立. 证毕.

参考文献

[1] Saaty T L. The Analytic Hierarchy Process. Pittsburgh : University of Pittsburgh, 1988.
[2] 王莲芬, 许树柏. 层次分析法引论. 北京: 中国人民大学出版社, 1990, 1—41.
[3] 程明熙. 决策理论和方法. 南京: 东南大学出版社, 1991, 228—233.
[4] 程乾生. 属性识别理论模型及其应用. 北京大学学报(自然科学版), 1997, 33(1):12.

原文载于《北京大学学报(自然科学版)》, 第 34 卷, 第 1 期, 1998 年 1 月, 10—13.

Fusion Prediction Based on the Attribute Clustering Network and the Radial Basis Function

Cheng Qiansheng, Wu Lianwen, Wang Shouzhang

Department of Information Science, School of Mathematical Sciences, Peking University

Abstract: A fusion prediction method is introduced on the basis of attribute clustering network and radial basis functions. An algorithm of quasi-self organization for developing the model for the fusion prediction is introduced. Some simulation results for chaotic time series are presented to show the performance of the method.

Keywords: chaotic time series; radial basis function; attribute clustering network; quasi-self organization; fusion prediction

In this note, we study how to model a nonlinear system and how to predict chaotic time series, i.e. how to construct a nonlinear map to denote a dynamical system with a series of iterative sequences in the phase space. If the map can be successfully established, it will be used for the prediction of the system.

The theoretical foundation of the above mentioned method is the reconstruction phase space theory[1]. For the observation function $s(t)$ of a determinate dynamical system, we can get a time series S_t with a scalar variable sampled at discrete points of time:

$$s_t = s(t \cdot \tau_s), \quad t=1,2,\cdots,N_s, \tag{1}$$

where τ_s denotes the sampling interval. By using time delay method, we can construct an orbit $x(t)$ in the m-dimensional Euclidean space from the time series s_t with a scalar variable:

$$x(t) = (s_t, s_{t+\tau_d}, \cdots, s_{t+(m-1)\tau_d}), \tag{2}$$

where m is the embedding dimension, τ_d the time delay. Normally, $\tau_d = k$, and k is an integer and the time delay as well. According to the Takens embedding

theorem[1], if $m \geq 2d+1$ (d is the dimension of the phase space in the original dynamical system), we will obtain a dynamical system $F: \mathbb{R}^m \to \mathbb{R}^m$ satisfying:
$$x(t+1) = F(x(t)). \quad (3)$$
From eq. (3), we can obtain another map $f: \mathbb{R}^m \to \mathbb{R}$ such that
$$s_{t+1+(m-1)\tau_d} = f(x(t)) = f(s_t, s_{t+\tau_d}, \cdots, s_{t+(m-1)\tau_d}). \quad (4)$$
Hence, with the known time series s_t, if we can construct an approximate model AF or Af of F or f satisfying eq. (3) or eq. (4), then a nonlinear prediction model of s_t will be obtained.

Up to now, there are several methods available for modeling and predicting of nonlinear time series. The main methods include the local linear model, the global model, the neural network model and the radial basis function model.

In this note, we use the fusion prediction model based on the attribute clustering network and the radial basis function to predict the time series. Firstly, we choose the center point set of the radial basis function with the attribute clustering network, then construct a model by means of the fusion prediction. Finally, test the performance of the model by sunspot data.

1. Nonlinear Prediction Model Based on Radial Basis Function and Attribute Clustering Network

As usual, the whole vector set $\{x(t)\}$ is divided into two groups: $\{x(t)\}_{t=1}^{N_L}$ is used to construct the model, which is called the learning set or the training set, and $\{x(t)\}_{t=N_L+1}^{N_L+N_T}$ is used to test our model, i.e. the testing set. To test fitting accuracy and predictive accuracy, we introduce normalized mean square errors. The fitting mean square error is used to denote fitting accuracy, which is defined as (for instance, one-step-ahead prediction)
$$\sigma_f^2 = \left\{ \sum_{t=1}^{N_L} (Af(x(t)) - s_{t+1+(m-1)\tau_d})^2 / N_L \right\} \Big/ \mathrm{Var}(s). \quad (5)$$
The predictive mean square error denoting predictive accuracy is defined as:
$$\sigma_p^2 = \left\{ \sum_{t=N_L+1}^{N_L+N_T} (Af(x(t)) - s_{t+1+(m-1)\tau_d})^2 / N_T \right\} \Big/ \mathrm{Var}(s). \quad (6)$$
where $\mathrm{Var}(s)$ denotes the variance of the time series. $\sigma_p^2 = 0$ means that the predictive value is the same as the real one, $\sigma_p^2 = 1$ shows that the predictive value

is not better than the mean one of $\{s_t\}$.

(ⅰ) Nonlinear prediction model based on radial basis function. Radial basis function model was defined as

$$Af(x(t)) = \sum_{j=1}^{N_c} \lambda_j \phi(\|x - c_j\|), \tag{7}$$

where $\phi(r): \mathbb{R}^+ \to \mathbb{R}$ is a radial basis function, it usually includes the following forms: $\phi(r) = r, r^3, r^2 \log r$ and $\exp(-r^2/\sigma^2)$. $\|\cdot\|$ denotes Euclidean norm, c_j ($j=1,2,\cdots,N_c$) the center points of the radial basis function, and $\lambda = (\lambda_1, \lambda_2, \cdots, \lambda_{N_c})'$ the unknown coefficients.

Casdagli[2] applied radial basis functions to chaotic time series modeling and prediction firstly, he used the function $Af(x(t))$ of eq. (7) to fit a series of points $(x(t), s_{t+1+(m-1)\tau_d})$, $(t=1,2,\cdots,N_L)$, such that $Af(x(t))$ satisfies the following form:

$$\min \sum_{t=1}^{N_L} (Af(x(t)) - s_{t+1+(m-1)\tau_d})^2. \tag{8}$$

Rewrite the above problem as the matrix form

$$b = A \cdot \lambda, \tag{9}$$

where $b = (s_{2+(m-1)\tau_d}, s_{3+(m-1)\tau_d}, \cdots, s_{N_L+1+(m-1)\tau_d})^T$, $A_{ij} = \phi(\|x(i) - c_j\|)$, $i=1,2,\cdots,N_L$, $j=1,2,\cdots,N_c$, the coefficient λ is determined with eq. (9), then the prediction function Af is obtained.

The training approach of the radial basis function network consists of two steps. In the first step we choose the center points $\{c_j\}$, in the second step we determine the weight vector λ such that the error between the real and prediction value is the least.

Casdagli[2] used the strict interpolation approach of the radial basis functions to construct the model, the center points $\{c_j\}$ are all points in the training set, then A is an $N_c \times N_c$ matrix. If the data points are all distinct, the matrix A is not singular and eq. (9) has a unique solution. However, the drawback of this technique is that high fitting accuracy can be reached for large amounts of data in the training set, so this technique has large memory requirements and is very time consuming. This technique will especially lead to overfitting for data with noise. If the time series is cyclical or approximately cyclical, usually the technique shows approximate singularity, and for some special radial basis functions, such as Gauss form, the technique leads to approximate singularity of A, the predictive accuracy will be very poor. In the second step, Broomhead suggested solving the problem of eq. (9) by using the least square method. Thus, when $N_L > N_c$, i. e. the number N_L of training samples is greater than the number N_c of center points, we get

$$\lambda = A^+ \cdot b, \quad (10)$$

where A^+ is Moore-Penrose pseudo-inverse of A.

In recent years, the methods of using the radial basis function for nonlinear prediction are based on the above ideas, but how to choose the center point set $\{c_j\}$ is open. In this note, we will choose the center point set of the radial basis function by means of attribute measure clustering network.

(ⅱ) Non-supervised clustering network algorithm based on attribute measure. Given the sample set $\{x(t)\}_{t=1}^{N_L}$, where $x(t)$ is reconstructed from the time series s_t with a scalar variable according to eq. (2). In the previous prediction, usually the center point set $\{c_j\}$ is beforehand chosen, and the importance of every index of $x(t)$ is assumed to be the same. However, in practice, we know that the effect of each index is different. Prof. Cheng[3] presented an attribute measure clustering algorithm (AKCN algorithm) for this problem, In this note, we will choose the center point set $\{c_j\}$ by AKCN algorithm, AKCN algorithm adopts the form of Kohonen self-organization, but has some special technique[3].

2.Quasi-self Organization Training Algorithm Based on Survival of the Fittest

We often obtain different results from different prediction models. If the number of the center point set $\{c_j\}$ and the embedding dimensions m are different in the radial basis function models, we will obtain different prediction models. In a sense, each prediction model is optimum under certain conditions. Hence, we will consider using fusion prediction to obtain good prediction accuracy.

We introduce the following rules: for the training set $\{x_i\}_{i=1}^n$, we will choose prediction model $f_j, j=1,2,\cdots,M$. Here we assume M is even, f_{ji} is the prediction value of f_j at $x_i, i=1,2,\cdots,n, j=1,2,\cdots,M, Y_i$ the real value, and the fusion prediction function f_c is obtained according to the following model:

$$f_c = \lambda_1 \cdot f_1 + \lambda_2 \cdot f_2 + \cdots + \lambda_M \cdot f_M. \quad (11)$$

We choose λ_j to minimize E,

$$E = \sum_{i=1}^{n} | f_c(x_i) - Y_i |^2, \qquad (12)$$

where λ_j satisfies the following condition:

$$\sum_{j=1}^{M} \lambda_j = 1. \qquad (13)$$

Now we introduce the following quasi-self organization algorithm to determine λ_j and f_j:

(i) Initially $\lambda_j (j=1,2,\cdots,M)$ is chosen randomly from the interval [0, 1], and satisfies eq. (13). Let T denote the total training number, $\eta(0)$ the initial value of the training parameter $\eta(t)$ $(0<\eta(t)<1)$, and Thv threshold value.

(ii) For a given prediction point x_i, f_{ci} is the fusion prediction value, Y_i the real value and f_{ji} the prediction value of model f_j.

(iii) Compute Euclidean norm d_{ji} by
$$d_{ji} = |f_{ji} - Y_i|^2, \quad j=1,2,\cdots,M.$$

(iv) Coupled Principle. Firstly, we arrange all d_{ji} $(j=1,2,\cdots,M)$ in the ascending order according to their values, secondly, we put all λ_j $(j=1,2,\cdots, M)$ into $M/2$ groups with two λ_j in each group according to the following rule: two λ_j corresponding to the maximum d_{ji} and the minimum d_j respectively will be grouped together, two λ_j corresponding to the second largest d_{ji} and the second smallest d_{ji} respectively will be grouped together, and so on, finally $M/2$ groups will be obtained.

(v) Mutually Complementary Principle. For λ_j in each group, the λ_j corresponding to the large d_{ji} subtracts AS, here $AS = \eta(t) * \lambda_j$, in reverse, another λ_j in the same group corresponding to the small d_{ji} adds AS, then they still satisfy eq. (13).

(vi) Give the next prediction point, then return to (iii), and so on, until deal with all prediction points.

(vii) Update the training parameter $\eta(t)$ as
$$\eta(t) = \eta(0)(1-t/T),$$
where t is the training time at present, T the total training number and $\eta(0)$ the initial value.

(viii) Set $t=t+1$, return to (ii) till $t=T$ or $\sum \{\Delta\lambda_j\}^2$ is less than Thv.

(ⅸ) After weight coefficient set $\{\lambda_j\}$ $(j=1,2,\cdots,M)$ is determined, we sort them by the ascending order, then cumulate λ_j from the minimum value to the maximum value until the sum reaches 10% of the total (10% is an experimental value, how to choose it depends on the practical problem and the experimental results) and delete them from the set. Finally, we obtain m models corresponding to the rest λ_j. By using the above algorithm repeatedly, we can obtain new weight coefficient sets, and then the algorithm will be ended in this way.

3. Analysis of the Above Algorithm

The algorithm adjusts $\{\lambda_j\}$ to the minimal direction from d_{ji}, the final space distribution of $\{\lambda_j\}$ can properly denote the probability of space distribution by the repeated training. The advantage of the algorithm is self-fitted and optimum training. There are two steps in the traning approach, in the first step, the rough training will approximately determine the weight vector λ_j, here $\eta(t)$ has a large value, in the second step, we obtain λ_j by means of the further training. In the training process, the weight vector is convergent.

We know that radial basis functions depend on the number of the center point set $\{c_j\}$ and the embedding dimensions m. Because our fusion model needs several radial basis function models, we must choose different numbers of the center point set $\{c_j\}$ and the embedding dimensions m to construct the following fusion-prediction model:

(ⅰ) Determine the center point set $\{c_j\}$ by AKCN algorithm, the center c_j of the prediction model is the final weight vector corresponding to the output neuron i;

(ⅱ) The error between the real value and the fitting value is minimal by adjusting $\{\lambda_j\}$ of eq. (7) in terms of the least square method;

(ⅲ) Repeat (ⅰ) and (ⅱ), several models with different number N_c of the center point set and the embedding dimension m will be obtained;

(ⅳ) Obtain fusion prediction model by quasi-self organization algorithm.

Now we would like to show how this fusion model performs with the prediction problem of sunspot data.

People have observed sunspots since the 17th century. There is an average record of the number of sunspot every year. Up to now we have approximately 300 samples. Sunspot date have been intensively investigated as standard modeling data in statistical science since Yule (1972). We will choose sunspot data to illustrate the performance of our model. The model is built by using the data of 1700—1920, i.e. the data of 1700—1920 are used as the training set, $N_L=218$. The data of 1921—1955 and those of 1956—1979 are the testing sets, they are respectively used to test the predictive ability of the model. We use one-step-ahead prediction for the model, i.e. $\tau_d=1$. The radial basis function model is based on eq. (7), here the radial basis function is the Gaussian form: $\phi(r)=\exp(-r^2/\sigma^2)$.

For the different N_c and m we might yield the different models and denote them as (m, N_c). We have obtained ten such models, which are $(3,10)$, $(3,15)$, $(3,20)$, $(3,24)$, $(3,30)$, $(4,16)$, $(4,22)$, $(4,26)$, $(5,14)$ and $(6,15)$, then the fusion prediction model is built by means of these models. Using fusion prediction model yields

$$\text{Var}_f(\text{train})=128.7,$$
$$MSE_{1921-1955}=155.6,$$
$$MSE_{1956-1979}=195.4,$$

where $\text{Var}_f=\sigma_f^2 * \text{Var}(s)$, $MSE=\sigma_p^2 * \text{Var}(s)$ (see eqs. (5) and (6)).

By using quasi-self organization algorithm, the iteration algorithm converges after the 285th operation in the first experiment, the value of λ_j is listed as follows:

Table 1 Parameter λ_j

Iterative number	λ_1	λ_2	λ_3	λ_4	λ_5	λ_6	λ_7	λ_8	λ_9	λ_{10}
1	0.07401	0.03099	0.02459	0.00811	0.02508	0.20094	0.10279	0.08560	0.18758	0.26026
41	0.07581	0.03348	0.03081	0.01295	0.03247	0.21536	0.09140	0.09397	0.19276	0.22094
81	0.06701	0.03816	0.04361	0.02407	0.04364	0.24021	0.07273	0.10793	0.21112	0.15146
121	0.04914	0.04356	0.06004	0.02628	0.04248	0.23856	0.06384	0.12261	0.22458	0.12888
161	0.04169	0.04494	0.07727	0.02504	0.04407	0.21504	0.04574	0.09224	0.24127	0.16564
201	0.04077	0.04774	0.08219	0.02644	0.04199	0.20546	0.03357	0.07733	0.26526	0.17919
241	0.03948	0.04863	0.08241	0.02678	0.04026	0.20385	0.03098	0.07331	0.27299	0.18127
281	0.03919	0.04876	0.08239	0.02682	0.03994	0.20363	0.03060	0.07274	0.27433	0.18156
282	0.03918	0.04876	0.08239	0.02682	0.03993	0.20363	0.03059	0.07273	0.27434	0.18156
283	0.03918	0.04876	0.08239	0.02682	0.03933	0.20363	0.03059	0.07273	0.27435	0.18157
284	0.03918	0.04877	0.08239	0.02682	0.03933	0.20363	0.03059	0.07273	0.27436	0.18157
285	0.03918	0.04877	0.08239	0.02682	0.03933	0.20363	0.03059	0.07272	0.27436	0.18157

The above table shows the change speed of parameter λ_j in our fusion model, we can see that its convergent speed is very fast.

In order to compare our method with others, some results of other methods[1] are given as follows:

(i) Threshold Autoregressive models:
$$Var_f(train) = 148.9,$$
$$MSE_{1921-1955} = 148.9,$$
$$MSE_{1956-1979} = 429.8.$$

(ii) Neural network (Weigend et al.):
$$Var_f(train) = 125.9,$$
$$MSE_{1921-1955} = 132.0,$$
$$MSE_{1956-1979} = 537.25.$$

(iii) Successive approximation of the radial basis function[4]:
$$Var_f(train) = 142.3,$$
$$MSE_{1921-1955} = 140.8.$$

(iv) Strict interpolation of the radial basis function:
$$Var_f(train) = 10^{-3},$$
$$MSE_{1921-1955} = 10^5.$$

(v) The radial basis function by Kohonen self-organization network[1]:
when $m=3, N_c=16$,
$$Var_f(train) = 147.0,$$
$$MSE_{1921-1955} = 136.7;$$
when $m=6, N_c=12$,
$$Var_f(train) = 129.1,$$
$$MSE_{1921-1955} = 292.9.$$

The above results indicate that the predictive result of our model is better than that of the other models. The fusion prediction model is suitable for chaotic time series, for the model has properly fused the several different models and overcome the defect of individual model.

---

[1] Wang Mingjin, Nonlinear prediction and chaotic time series, Ph. D. Thesis, Peking University, Beijing, 1997.

4. Conclusion

In this note, we study the problem of the nonlinear system fusion modeling and the prediction of chaotic time series. Even though the new prediction model is better than existing models, it might be improved further. Especially, it should be studied further how to fuse different models and promote advantage of each model better.

References

[1] Takens, F., Detecting strange attractors in turbulence, Dynamical Systems and Turbulence (eds. Rand, D. A., Young, L. S.) Warwick, 1980. Lecture Notes in Math., Vol 898, Berlin: Springer-Verlag, 1981, 366—381.

[2] Casdagli, M., Nonlinear prediction chaotic time series, Physica D. 1989, 35: 335.

[3] Cheng Qiansheng, Attribute pattern recognition and application. In Proceedings of the 4th Conference of Chinese Industry and Applied Mathematics, Shanghai: Fudan University Press, 1996, 27—32.

[4] He, S. D., Lapedes, A., Nonlinear modeling and prediction by successive approximation using radial basis functions, Physica D. 1993, 70: 289.

原文载于 Chinese Science Bulletin, Vol. 46, No. 9, May 2001, 789—792.

Analysis of the Weighting Exponent in the FCM

Yu Jian[1], Cheng Qiansheng[2], Huang Houkuan[1]

1. Department of Computer Science and Technology, Northern Jiaotong University;
2. Department of Information Science, School of Mathematics Science, Peking University

Abstract: The fuzzy c-means (FCM) algorithm is one of the most frequently used clustering algorithms. The weighting exponent m is a parameter that greatly influences the performance of the FCM. But there has been no theoretical basis for selecting the proper weighting exponent in the literature. In this paper, we develop a new theoretical approach to selecting the weighting exponent in the FCM. Based on this approach, we reveal the relation between the stability of the fixed points of the FCM and the data set itself. This relation provides the theoretical basis for selecting the weighting exponent in the FCM. The numerical experiments verify the effectiveness of our theoretical conclusion.

Index Terms: fixed point; fuzzy c-means; Hessian matrix; weighting exponent

I. Introduction

The fuzzy c-means algorithm (FCM) is a popular fuzzy clustering method. Many of its applications are indicated in [1]. One of the most important parameters in the FCM is the weighting exponent m. When m is close to one, the FCM approaches the hard c-means algorithm. When m approaches infinity, the only solution of the FCM will be the mass center of the data set. Therefore, the weighting exponent m plays an important role in the FCM algorithm.

Hence, choosing a suitable weighting exponent is very important when implementing the FCM. Up to now, there has been no theoretical basis for an optimal choice of m in the FCM just as many authors have pointed out (see [2]—[6], etc); namely, it is an open problem. However, many heuristic strategies are recommended in the literature. In 1976, a physical interpretation of the FCM algorithm when $m=2$ was given in [7]. Based on the performance of some cluster validity indices, Pal and Bezdek [5] have given heuristic guidelines regarding the best choice for m, suggesting that it is probably in the interval [1.5, 2.5]. Similar recommendations appear in [3], [8], and [9]. Most researchers have proposed $m=2$. But these recommendations are based on empirical studies and may not be appropriate for general data sets.

In this paper, our main goal is to give a theoretical and practical approach to selecting the weighting exponent in the FCM. In Section II, we recall the FCM and its solutions; some relevant results are also given. In Section III, a new local optimality test of the solutions of the FCM is proposed. Based on this test, we obtain theoretical rules of selecting the weighting exponent in the FCM, which show that a proper m depends on the data set itself. Section IV includes numerical experiments; the experimental results confirm the theoretical conclusion in this paper. Finally, in Section V, we summarize the results of this paper and make some remarks.

II. FCM Algorithm and its Solutions

Let $X=\{x_1, x_2, \cdots, x_n\} \subset R^s$ be a data set and $u=\{u_{ik}\}_{c \times n} \in M_{fcn}$ be a partition matrix, $v=\{v_1, v_2, \cdots, v_c\}$ be the cluster centers, $v_i \in R^s$; $\|x\| = \sqrt{x^T x}$ be an inner product norm; $1 < m < +\infty$ and $2 \leqslant c < n$, then the objective function of the FCM is defined as follows:

$$J_m(u,v) = \sum_{k=1}^{n} \sum_{i=1}^{c} (u_{ik})^m \|x_k - v_i\|^2, \tag{1}$$

where

$$M_{fcn} = \{u = [u_{ik}]_{c \times n} \mid \forall i, \forall k, u_{ik} \geqslant 0, \sum_{i=1}^{c} u_{ik} = 1, n > \sum_{k=1}^{n} u_{ik} > 0\}.$$

Let $G: u = [u_{ik}]_{c \times n} \in M_{fcn} \mapsto v = (v_1, v_2, \cdots v_c)^T \in R^{cs}$, where $v_i = \sum_{k=1}^{n} u_{ik}^m x_k / \sum_{k=1}^{n} u_{ik}^m (i=1,2,\cdots,c)$ i.e., $G(u) = v$; and set $F: v = (v_1, v_2, \cdots, v_c)^T \in R^{cs} \mapsto u = [u_{ik}]_{c \times n} \in M_{fcn}$, where $u_{ik} = \|x_k - v_i\|^{-2/(m-1)} / \sum_{j=1}^{c} \|x_k - v_j\|^{-2/(m-1)}$ ($i=1, 2, \cdots, c; k=1,2,\cdots, n$) i.e., $F(v) = u$. Then we can define a map: $T_m(u,v) = (\hat{u}, \hat{v})$, where $\hat{u} = F(v), \hat{v} = G(\hat{u})$.

$(u^{(l)}, v^{(l)})$, $l=1, 2, \cdots$, is called an iteration sequence of the FCM algorithm, if $(u^{(l)}, v^{(l)}) = T_m(u^{(l-1)}, v^{(l-1)})$, $\forall l \geqslant 1$, where $(u^{(0)}, v^{(0)})$ is any element of $M_{fcn} \times R^{cs}$. Set

$$\Omega = \left\{ (u^*, v^*) \in M_{fcn} \times R^{cs} \middle| \begin{array}{l} J_m(u^*, v^*) \leqslant J_m(u, v^*), \forall u \in M_{fcn}, u \neq u^* \\ \text{and } J_m(u^*, v^*) < J_m(u^*, v), \forall v \in R^{cs}, v \neq v^* \end{array} \right\}$$

In this paper, we always suppose that $\|x_k - v_i\| \neq 0$, $\forall k$, $\forall i$.

Theorem 2.1 (Bezdek *et al.*, [10]) Let $(F(v^{(0)}), v^{(0)})$ be the starting point of iteration with T_m and $v^{(0)} \in R^{cs}$, then the iteration sequence $T_m^{(1)}(u^{(0)}, v^{(0)})$, $l = 1, 2, \cdots$, either terminates at a point $(u^*, v^*) \in \Omega$, or there is a subsequence converging to a point in Ω.

Transparently, the point $(u^*, v^*) \in \Omega$, if and only if, (u^*, v^*) is the fixed point. In other words, $(u^*, v^*) \in \Omega \Leftrightarrow u^* = F(v^*)$ and $v^* = G(u^*)$. Let $\bar{x} = \sum_{k=1}^{n} x_k / n$. According to [1], the only solution of the FCM algorithm is the mass center of the data set $\bar{x}$ when m approaches infinity. In fact, the mass center of the data set $\bar{x}$ is a fixed point of the FCM algorithm for any $m > 1$ (see [11]).

Set $U^* = [1/c]_{c \times n}$, $D = [d_{ij}]_{c \times n}$, where

$$d_{ij} = \begin{cases} 1, & \text{if } j = 1 \text{ and } i = 1, \\ -1, & \text{if } j = 1 \text{ and } i = 2, \\ 0, & \text{otherwise,} \end{cases}$$

then set $U_\varepsilon = U^* + \varepsilon D$. In 1987, Tucker proved the following fact.

Theorem 2.2 (Tucker, [11]) If $\forall k$, $\forall i$, $\|x_k - v_i\|^2 > 0$ and $v^* = \bar{x}$, then for $n > 2$ the inequality $J_m(U_\varepsilon, G(U_\varepsilon)) < J_m(U^*, v^*)$ holds for every $\varepsilon > 0$ sufficiently small if and only if $m \leqslant n/(n-2)$.

Bezdek *et al.* [10] pointed out that Tucker's theorem contains the first theoretical result for avoiding distinguished values of m. By letting $m \geqslant n/(n-2)$, Bezdek *et al.* considered that (at least) the saddle points guaranteed by Theorem 2.2 would be avoided. This is the first theoretical result about m in [10], but it has little practical value.

III. Rules of Selecting Weighting Exponent

It is well known that the number of points in Ω may be so large that some points may be not the desired solution of the FCM. Generally, when the data set is clustered into c ($c > 1$) subsets, each subset is often expected to have a different prototype (or cluster center) than others. However, $(U^*, \bar{x})$ is a fixed point of the FCM. Particularly, the output of the FCM will be $\bar{x}$ with probability close to 1 if $\bar{x}$ is one stable solution of the FCM algorithm. In order to avoid such cases, $\bar{x}$ should not be a stable point of the FCM. Can we determine, however, whether or not $\bar{x}$ is a stable point of the FCM? In the following, we will address this problem.

In 1986, Sleim and Ismail [12] described the Hessian matrix H^u of $\psi_m(u) = \min_{v \in R^a} J_m(u,v)$.

Theorem 3.1 (Selim & Ismail, [12])

1) $\psi_m(u) = \sum_{k=1}^{n} \sum_{i=1}^{c} (u_{ik})^m \| x_k - v_i \|^2 ,$

where
$$v_i = \frac{\sum_{k=1}^{n} u_{ik}^m x_k}{\sum_{k=1}^{n} u_{ik}^m}, \tag{2}$$

2) $\dfrac{\partial \psi_m(u)}{\partial u_{ik}} = m u_{ik}^{m-1} \| x_k - v_i \|^2 , \tag{3}$

3) $\dfrac{\partial^2 \psi_m(u)}{\partial u_{ik} \partial u_{ir}} = -2m^2 \dfrac{u_{ik}^{m-1} u_{ir}^{m-1}}{\sum_{k=1}^{n} u_{ik}^m} (x_k - v_i)^T (x_r - v_i) \text{ if } k \neq r, \tag{4}$

$\dfrac{\partial^2 \psi_m(u)}{\partial u_{ik} \partial u_{ik}} = m u_{ik}^{m-2} \| x_k - v_i \|^2 \left\{ m - 1 - 2m \dfrac{u_{ik}^m}{\sum_{k=1}^{n} u_{ik}^m} \right\}, \tag{5}$

$\dfrac{\partial^2 \psi_m(u)}{\partial u_{ik} \partial u_{jr}} = 0, \text{ if } i \neq j. \tag{6}$

Therefore, the Hessian matrix H^u of $\psi_m(u)$ is given by $H^u_{cn \times cn} = diag(H^u_1, H^u_2, \cdots, H^u_c)$, where H^u_i is a $n \times n$ matrix whose components are given by (4) and (5). It is easy to show $H^u_i = m(m-1)(D^u_i - (2m/(m-1))G^u_i)$, where $D^u_i = diag(d^i_1, d^i_2, \cdots, d^i_n)$, $G^u_i = (g^{(i)}_{kr})_{n \times n}$, $d^i_k = u_{ik}^{m-2} \| x_k - v_i \|^2$, $g^{(i)}_{kr} = \left(u_{ik}^{m-1} u_{ir}^{m-1} / \sum_{l=1}^{n} u_{il}^m \right)(x_k - v_i)^T (x_r - v_i)$.

Theorem 3.2 (Selim & Ismail, [12]) A point (u,v) obtained by the FCM is a local minimum of $J_m(u,v)$ if $\psi_m(u)$ is convex at u.

Obviously, Theorem 3.2 is a sufficient condition for a local minimum of

$J_m(u,v)$, but not necessary one. But if $(u,v) \in \Omega$ is a local minimum of $J_m(u,v)$, the matrix H_u may be not positive semi-definite just as Kim et al. reported in [16]. Therefore, Kim et al. gave a new optimality test for fixed points of the fuzzy c-means algorithm as follows:

Theorem 3.3 (Kim, Bezdek, Hathaway, [13]) If $(u,v) \in \Omega$ is a local minimum of the objective function $J_m(u,v)$, then $L(u) = PH^uP$ is positive semi-definite, where $P = I_{cn \times cn} - (1/c) diag(O_1, O_2, \cdots, O_n)$, $O_i = [o^i_{pq}]_{c \times c}$, $\forall i$, $\forall p$, $\forall q$, $o^i_{pq} = 1$.

Clearly, the optimality test given by Kim et al. has the same order with the matrix H^u. In order to get a simpler criterion, we set $\xi_m(u) = J_m(u, G(u))$, where $u = [u_{ik}]_{cn}$ and $u_{ck} = 1 - \sum_{i=1}^{c-1} u_{ik}$. In other words, $\xi_m(u)$ is a function of $(c-1) \times n$ variables. By computation, we obtain Theorem 3.4.

Theorem 3.4

1) $\xi_m(u) = \sum_{k=1}^{n} \sum_{i=1}^{c} (u_{ik})^m \| x_k - v_i \|^2$,

where $v_i = \dfrac{\sum_{k=1}^{n} u_{ik}^m x_k}{\sum_{k=1}^{n} u_{ik}^m}$, and $u_{ck} = 1 - \sum_{i=1}^{c-1} u_{ik}$, (7)

2) $\dfrac{\partial \xi_m(u)}{\partial u_{ik}} = m u_{ik}^{m-1} \| x_k - v_i \|^2 - m u_{ck}^{m-1} \| x_k - v_c \|^2$, $1 \leq i \leq c-1$, (8)

3) $\dfrac{\partial^2 \xi_m(u)}{\partial u_{ik} \partial u_{ir}} = \dfrac{\partial^2 \psi_m(u)}{\partial u_{ik} \partial u_{ir}} + \dfrac{\partial^2 \psi_m(u)}{\partial u_{ck} \partial u_{cr}}$ if $k \neq r$ and $1 \leq i \leq c-1$, (9)

$\dfrac{\partial^2 \xi_m(u)}{\partial u_{ik} \partial u_{ik}} = \dfrac{\partial^2 \psi_m(u)}{\partial u_{ik} \partial u_{ik}} + \dfrac{\partial^2 \psi_m(u)}{\partial u_{ck} \partial u_{ck}}$ if $1 \leq i \leq c-1$, (10)

$\dfrac{\partial^2 \xi_m(u)}{\partial u_{ik} \partial u_{jr}} = \dfrac{\partial^2 \psi_m(u)}{\partial u_{ck} \partial u_{cr}}$ if $1 \leq i \neq j \leq c-1$ and $k \neq r$, (11)

$\dfrac{\partial^2 \xi_m(u)}{\partial u_{ik} \partial u_{jk}} = \dfrac{\partial^2 \psi_m(u)}{\partial u_{ck} \partial u_{ck}}$ if $1 \leq i \neq j \leq c-1$. (12)

Therefore, the Hessian matrix $H(\xi_m(u))$ of $\xi_m(u)$ is given by $H(\xi_m(u))_{(c-1)n \times (c-1)n} = diag(H^u_1, H^u_2, \cdots, H^u_{c-1}) + \Theta^u$, where $\Theta^u = (\theta_{ij})_{(c-1) \times (c-1)}$, $\theta_{ij} = H^u_c$. If $(u, G(u)) \in \Omega$, then $\partial \xi_m(u) / \partial u_{ik} = 0, 1 \leq i \leq c-1, \forall k$.

According to Theorem 3.4, the series expansion of $\xi_m(u)$ on the point $(u, G(u)) \in \Omega$ can be represented by

$$\xi_m(u+\varphi_u) = \xi_m(u) + \frac{1}{2}\sum_{i=1}^{c-1}\varphi_i^T H_i^u \varphi_i + \frac{1}{2}\Big(\sum_{i=1}^{c-1}\varphi_i\Big) H_c^u \Big(\sum_{i=1}^{c-1}\varphi_i\Big) + o(\|\varphi_u\|^2).$$
(13)

Certainly, we can decide whether or not a point $(u,v) \in \Omega$ is a strict local minimum of $J_m(u,v)$ by the matrix $H(\xi_m(u))$. The reason is as follows:

If $(u,v) \in \Omega$ and u is a strict local minimum of $\xi_m(u)$, then $\exists \varepsilon$ such that for all φ_u with $\|\varphi_u\| < \varepsilon$ and $\varphi_u \neq 0$, the inequality $J_m(u+\varphi_u, G(u+\varphi_u)) = \xi_m(u+\varphi_u) > \xi_m(u) = J_m(u, G(u))$ holds. According to [12], we know that $\forall \phi_v, J_m(u+\varphi_u, v+\phi_v) \geqslant J_m(u+\varphi_u, G(u+\varphi_u))$ holds, therefore, $J_m(u+\varphi_u, v+\phi_v) J_m(u, G(u))$. Since $(u,v) \in \Omega$ means $v = G(u)$, $(u,v) \in \Omega$ is a strict local minimum of $J_m(u,v)$. Conversely, if $(u,v) \in \Omega$ is a strict local minimum of $J_m(u,v)$, then $\exists \varepsilon, \delta$ s.t. $\forall \|\varphi_u\| < \varepsilon, \|\phi_v\| < \delta$, the inequality $J_m(u+\varphi_u, v+\phi_v) \geqslant J_m(u, G(u))$ holds. Since the function $(G(u))$ is continuous and $v = G(u)$, we can set $G(u+\varphi_u) = v + \phi_v$. It follows that $J_m(u+\varphi_u, G(u+\varphi_u)) > J_m(u, G(u))$. In other words, $\xi_m(u+\varphi_u) > \xi_m(u)$, i.e., u is a strict local minimum of $\xi_m(u)$. As for $(u,v) \in \Omega$, the matrix $H(\xi_m(u))$ is positive definite only if $(u,v) \in \Omega$ is a strict local minimum of $\xi_m(u)$. Therefore, the matrix $H(\xi_m(u))$ is positive definite only if $(u,v) \in \Omega$ is a strict local minimum of $J_m(u,v)$. Clearly, the matrix $H(\xi_m(u))$ is a simpler criterion than Theorem 3.3 on local optimality of fixed points of the FCM algorithm.

In the following, we will give the conditions under which $(U^*, \bar{x})$ is a stable point of the FCM.

Set $u = U^*$, $D_{U^*} = \text{diag}(\|x_1 - \bar{x}\|^2, \|x_2 - \bar{x}\|^2, \cdots, \|x_n - \bar{x}\|^2)$, $G_{U^*} = (g_{kr})_{n \times n}$, $F_{U^*} = (f_{kr}^{U^*})_{n \times n}$, $g_{kr} = (x_k - \bar{x})^T (x_r - \bar{x})/n$, $f_{kr}^{U^*} = 1/n(x_k - \bar{x}/\|x_k - \bar{x}\|)^T (x_r - \bar{x}/\|x_r - \bar{x}\|)$, then we have

$$\forall i, H_i^u = m(m-1)c^{2-m}\Big[D_{U^*} - \frac{2m}{(m-1)}G_{U^*}\Big]$$

$$= m(m-1)c^{2-m}(D_{U^*})^{1/2}\Big(I_{n \times n} - \frac{2m}{m-1}F_{U^*}\Big)(D_{U^*})^{1/2}.$$

Theorem 3.5 Let $\lambda_{\max}(F_{U^*})$ be the maximum eigenvalue of the matrix F_{U^*}. If $\lambda_{\max}(F_{U^*}) < 0.5$ and $m > (1/(1-2\lambda_{\max}(F_{U^*})))$, then $(U^*, \bar{x})$ is a strict local minimum of $J_m(u,v)$.

Proof. If $m > (1/(1-2\lambda_{\max}(F_{U^*})))$, then $1 - (2m/(m-1))\lambda_{\max}(F_{U^*}) > 0$. Therefore, $I_{n \times n} - (2m/(m-1))F_{U^*}$ is positive definite. By (13), we know that $H(\xi_m(u))$ is positive definite. According to the above analysis, $(U^*, \bar{x})$ is a strict local minimum of $J_m(u,v)$. QED.

Theorem 3.6 Let $\lambda_{\min}(F_{U^*})$ be the minimum eigenvalue of the matrix F_{U^*}, then $\lambda_{\min}(F_{U^*}) = 0$.

Proof. If $\lambda_{\min}(F_{U^*}) > 0, n > 2$, then $\lambda_{\min}(F_{U^*}) < 0.5$. Since $(U^*, \bar{x})$ is the fixed point of the FCM for any $m > 1$, we set $m < (1/(1-2\lambda_{\min}(F_{U^*})))$, then the matrix $H(\xi_m(u))$ is negative definite by (13). It implies that $\exists \varepsilon > 0, 0 < \|\varphi_u\| < \varepsilon, \xi_m(U^* + \varphi_u) < \xi_m(U^*)$.

So
$$J_m(U^*, \bar{x}) > J_m(U^* + \varphi_u, G(U^* + \varphi_u)). \tag{14}$$

Set $\varphi_u = \varepsilon/t \times E_1 - \varepsilon/t \times E_2$, where

$$E_1 = \lfloor e_{ik}^2 \rfloor_{c \times n}, e_{ik}^1 = \begin{cases} 1, \text{if } i = i_0 \\ 0, \text{otherwise} \end{cases}; \quad E_2 = \lfloor e_{ik}^2 \rfloor_{c \times n}, e_{ik}^2 = \begin{cases} 1, \text{if } i = j_0 \\ 0, \text{otherwise} \end{cases}; \quad t = 4.$$

According to [12], we know the following inequality holds
$$J_m(U^* + \varphi_u, G(U^* + \varphi_u)) \geq J_m(F(G(U^* + \varphi_u)), G(U^* + \varphi_u)). \tag{15}$$

It follows that $G(U^* + \varphi_u) = G(U^*)$. Therefore, we get $J_m(U^*, G(U^*)) > J_m(U^*, G(U^*))$ by (14) and (15). It is a contradiction. QED.

If $\lambda_{\max}(F_{U^*}) < 0.5$ and $m > (1/(1-2\lambda_{\max}(F_{U^*}))$, Theorem 3.5 tells us that U^* is a stable fixed point of the FCM. Moreover, U^* is not a stable fixed point of the FCM if $\lambda_{\max}(F_{U^*}) \geq 0.5$.

In order to simplify Theorem 3.5, we prove the following lemma.

Lemma 3.1 Let $\lambda_{\max}(F_{U^*})$ be the maximum eigenvalue of the matrix F_{U^*}, then $1 \geq \lambda_{\max}(F_{U^*}) \geq (1/\min(n-1,s))$.

Proof. Let $poly(F_{U^*})$ be the characterisitic polynomial of F_{U^*}, then the coefficient of the $(n-1)$th order term of the $poly(F_{U^*})$ is 1. Let $\lambda_1, \lambda_2, \cdots, \lambda_n$ be the eigenvalues of the matrix F_{U^*}. It is trivial to show that the matrix F_{U^*} is positive semi-definite. Hence, $\lambda_1 \geq 0, \lambda_2 \geq 0 \cdots, \lambda_n \geq 0$. If we set

$$H = \left[\frac{(x_1 -)\bar{x}}{\|x_1 - \bar{x}\|}, \frac{(x_2 - \bar{x})}{\|x_2 - \bar{x}\|}, \cdots, \frac{(x_n - \bar{x})}{\|x_n - \bar{x}\|} \right].$$

then $F_{U^*} = H^T H / n$. Therefore, the matrix F_{U^*} has at most s nonzero eigenvalues since the dimensionality of the point x_i is s. Obviously, we have $\sum_{i=1}^{n}\lambda_i =$ 1. By Theorem 3.6 and the principle of the pigeonhole, then it follows that $1 \geqslant \lambda_{\max}(F_{U^*}) \geqslant (1/\min(n-1,s))$. QED.

Thereom 3.7 Suppose $m < \min(s,n-1)/(\min(s,n-1)-2)$, then $(U^* \bar{x})$ is not a stable solution of the FCM, where $s \geqslant 3$.

Proof. If $m < \min(s,n-1)/(\min(s,n-1)-2)$, then it follows that $m < (1/(1-2\lambda_{\max}(F_{U^*})))$ by Lemma 3.1. By Theorem 3.5, it implies that the matrix $I_{n \times n} - (2m/(m-1))F_{u^*}$ is not positive definite. From (13), it follows that $H(\xi_m(u))$ is not positive definite. Thus, U^* is not a local minimum of $J_m(u, G(u))$. In other words, U^* is not a stable fixed point of the FCM. QED.

Obviously, Theorem 3.7 is more general and precise than Theorem 2.2.

According to the above analysis, we get two theoretical rules of selecting the weighting exponent as follows:

Rule α: $m \leqslant \dfrac{\min(s,n-1)}{\min(s,n-1)-2}$, if $\min(n-1,s) \geqslant 3$,

Rule β: $m \leqslant \dfrac{1}{1-2\lambda_{\max}(F_{U^*})}$, if $\lambda_{\max}(F_{U^*}) < 0.5$.

Obviously, Rule α is an approximation of Rule β. For most applications, the inequality $s < n$ holds, Rule α means $m \leqslant (s/(s-2))$, i.e., the dimension of the data set X plays an important role in the FCM. Sometimes, Rule α can be broken. Howerer, Rule β must be obeyed if $\lambda_{\max}(F_{U^*}) < 0.5$. Otherwise, the most probable output of the FCM will be $\bar{x}$ according to Theorem 3.5. Although $\bar{x}$ is meaningful for the data set, it is usually not expected to be the most probable solution of the FCM. When $\lambda_{\max}(F_{U^*}) \geqslant 0.5$, Rules α,β are invalid. In this case, how to select m depends on the user.

IV. Data Sets and Experimental Results

In this section, we will test partial results in this paper by numerical experiments. According to [14], the performance of the FCM can be evaluated by the non-fuzziness index $\text{NFI}(u,c) = (c/(n \times (c-1))) \sum_{i=1}^{c} \sum_{k=1}^{n} u_{ik}^2 - /(c-1)$. Since NFI

$(F(v),)=0 \Leftrightarrow v=\bar{x} \Leftrightarrow u=U^*$ and $0 \leqslant \text{NFI}(u,c) \leqslant 1$, it is available for determining whether or not $v=\bar{x}$. If $\text{NFI}(u,c) \neq 0$, the corresponding weighting exponent m in the FCM is considered valid on the data set X, otherwise, m is invalid for the data set X. In order to verify the theoretical conclusions in Section Ⅲ, we tested the FCM with different values of m and c on the datasets in Table I.

Table I Datasets Used in Section Ⅳ

Name of Dataset	No. of Samples	No. of features	No. of Classes	$\lambda_{max}(F_{Data})$	$\frac{1}{1-2\lambda_{max}(F_{Data})}$
Isolet1+2+3+4	6238	617	26	0.1889	1.6072
Isolet5	1559	617	26	0.1926	1.6265
Sonar	208	60	2	0.1949	1.6388
Vowel	990	10	11	0.2189	1.7787
PiamIndians Diabetes	768	8	2	0.2558	2.0475
Waveform	5000	21	3	0.3272	2.8935
Glass	214	9	6	0.3424	3.1726
IRIS	150	4	3	0.6652	$+\infty$

All the datasets in Table I are practical data sets and can be obtained from the UCI Repository of Machine Learning Databases [15]. As for any data set with n samples, we standardize this dataset by zero-mean normalization: if x_i stands for the ith point in these data, $x_{i,j}$ stands for the jth attribute of x_i, then

$$X_{ij} = (x_{ij} - \sum_{k=1}^{n} x_{kj}/n) \Big/ \sqrt{\sum_{k=1}^{n}(x_{kj} - \sum_{l=1}^{n} x_{lj}/n)^2/(n-1)}.$$

If the data is labeled, the error rate of clustering results by the FCM can be worked out when the correct number of clusters is given. By simple computation, we know that $\lambda_{max}(F_{IRIS})=0.06652$ (Without zero-mean normalization, $\lambda_{max}(F_{IRIS})=0.8079$). Therefore, there is no theoretically invalid weighting exponent of the FCM for IRIS. Fig. 1 clearly show that it is reasonable because the error rates of clustering results vary only slightly for different values of m on IRIS without zero-mean normalization.

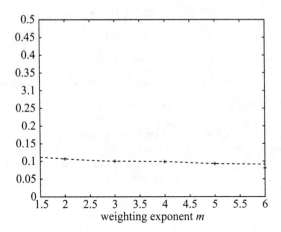

Fig. 1 Error rates of clustering results of IRIS with different m and $c=3$

As for Sonar, Vowel, Isolet5, Isolet1+2+3+4, PimaIndiansDiabetes, Waveform, and Glass, each of them has theoretical upper bound of the valid weighting exponent in the FCM. Moreover, the numerical experiments show that the heuristic rule $m=2$ is not appropriate for Sonar, Vowel, Isolet5, Isolet1+2+3+4 and tell us that the theoretical upper bound of the valid weighting exponent in the FCM given by Rule β is precise. In this paper, the experimental results on Sonar, Vowel aredrawn in Figs. 2 and 3, clearly show that $m=2$ is not appropriate for the FCM since the output of the FCM always seems to be $v_i=\bar{x}$ when $m=2$.

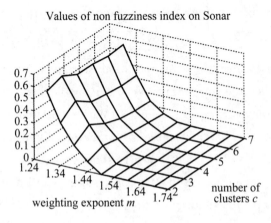

Fig. 2 Values of non fuzziness index on Sonar with various values of m and c

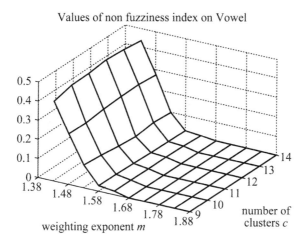

Fig. 3 Values of non fuzziness index on Vowel with various values of *m* and *c*

Moreover, we can give a theoretical explanation about Pal and Bezdek's heuristic suggestion about *m* according to our analysis. First, we describe the artificial data set-Normal_4 used in [5].

Normal_4 (Pal and Bezdek [5]): a sample of 800 points includes four cluster centers: $\mu_1, \mu_2, \mu_3, \mu_4$. Each cluster consists of 200 points and the points in the ith cluster obey the normal distribution $N(\mu_i, I_4)$, where $[\mu_1, \mu_2, \mu_3, \mu_4] = 3 * I_4$.

Using 2000 random samplings drawn from Normal_4, we get that the average value of $\lambda_{max}(F_{Normal\_4})$ is 0.3059, the standard deviation of $\lambda_{max}(F_{Norman\_4})$ is 0.0039 (Here, Normal_4 is not normalized). Therefore, the theoretical valid weighting exponent for a random Normal_4 should be not greater that 2.6 with approximately 50% probability. The results in Fig. 4 show that the sample from the distribution Normal_4 behaves according to our theoretical assertions. If we infer the valid weighting exponent only from IRIS and Normal_4, we can get the same conclusion as [5]. This gives a theoretical explanation about Pal and Bezdek's suggestion about *m*. In other words, the valid range of weighting exponent depends on the data set itself.

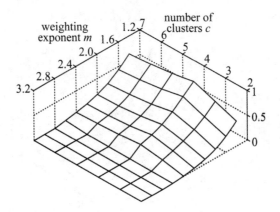

Fig. 4 Values of non fuzziness index on Normal_4 with various values of m and c

V. Conclusions and Discussions

In this paper, we give, not only a new local optimality test on fixed points of the FCM, but also present a new approach to selecting the weighting exponent in the FCM. Based on these results, we reveal the relation between the stability of the fixed points of the FCM and the data set X. When $\lambda_{\max}(F_{U^*}) \geqslant 0.5$, $(U^*, \bar{x})$ is never stable, but $(U^*, \bar{x})$ is stable if $m > (1/(1 - 2\lambda_{\max}(F_{U^*})))$ and $\lambda_{\max}(F_{U^*}) < 0.5$.

Usually, $(U^*, \bar{x})$ is not expected to be an acceptable solution of the FCM. Based on this assumption, a theoretical basis for the selection of the weighting exponent in the FCM is given by Rules α and β. Certainly, it can be argued that $(U^*, \bar{x})$ is one useful interpretation of cluster substructure in data even if $(U^*, \bar{x})$ is a stable point of the FCM, but $(U^*, \bar{x})$ is unacceptable, especially, as the most probable output of the FCM no matter what substructure the data set has. The theoretical analysis and numerical experiments show that this case really can happen if Rule β is broken. Furthermore, the numerical experiments show that Rule β produces better results than the traditional choice $m = 2$ or its neighbor.

If $\lambda_{\max}(F_{U^*}) \geqslant 0.5$, any $m > 1$ is a theoretical valid weighting exponent in the FCM according to our study. In fact, it has been reported by Choe and Jordan that the performance of the FCM is insensitive to the variation of m when two dimensional data is used to study m in [16]. By Lemma 3.1, $\lambda_{max}(F_{U^*}) \geqslant$ 0.5 always holds if the dimensionality of the data is 2. It implies that Choe and Jordan's discovery is a natural consequence of our analysis. Certainly, it is also consistent with our analysis of IRIS. However, our theoretical analysis and numerical experimental results show that $m = 2$ is not a reasonable heuristic guideline if $\lambda_{max}(F_{U^*}) < 0.25$.

References

[1] J. C. Bezdek, *Pattern Recognition With Fuzzy Objective Function Algorithms*. New York:Plenum,1981.

[2] A. Baraldi, P. Blonda, and F. Parmiggiani, "Model transitions in descending FLVQ," *IEEE Trans. Neural Networks*, vol. 9, pp. 724—737, Sept. 1998.

[3] M. J. Fadili, S. Ruan, D. Bloyet, and B. Mayoyer, "On the number of clusters and the fuzziness index for unsupervised FCA application to BOLD fMRI time series," *Med. Image Anal.*, vol. 5, pp. 55—67, Mar. 2001.

[4] I. Gath and A. B. Geva, "Unsupervised optimal fuzzy clustering," *IEEE Trans. Pattern Anal. Machine Intell.*, vol. 11, pp. 773—781, July 1989.

[5] N. R. Pal and J. C. Bezdek, "On cluster validity for the fuzzy c-mean model," *IEEE Trans. Fuzzy Syst.*, vol. 3, pp. 370—379, Aug. 1995.

[6] R. P. Li and M. Mukaidono, "A maximum entropy approach to fuzzy clustering," in *Proc. 4th IEEE Intern. Conf. Fuzzy Systems*, Yokohama, Japan, Mar. 20-24, 1995, pp. 2227—2232.

原文载于 IEEE Transactions on System, Man, and Cybernetics—PartB: Cybernetics, Vol. 34, No. 1, February 2004, 634—638.

附 录

程乾生教授生平简介

程乾生(1940—2010),汉族,安徽怀宁人,1940 年 5 月 7 日生于湖南乾城. 学习经历如下:

1946 年 3 月—7 月,在安徽省怀宁县五横乡月公冲小学上小学;

1947 年 9 月—1951 年 3 月转入安徽省安庆市高琦小学上小学;

1951 年 9 月—1953 年 3 月,在安徽省安庆市省立安庆初级中学上初中;

1953 年 4 月—1954 年 7 月转入安徽省芜湖市第三中学上初中;

1954 年 9 月—1957 年 7 月在安徽省芜湖市第一中学上高中;

1957 年 9 月考入北京大学数学力学系,1963 年 7 月以优异的成绩毕业于北京大学数学力学系概率统计专业.

1963 年 9 月至 2004 年 2 月先后在北京大学数学力学系、数学科学学院信息科学系担任助教、讲师、副教授、教授、博士生导师、北京大学数学研究所副所长.在此期间,应邀到奥地利维也纳大学、林兹大学、美国伊利诺伊大学、台湾淡江大学、香港中文大学进行访问和讲学.2004 年 2 月退休,但依然积极从事科研工作.程乾生教授于 2010 年 12 月 3 日在广西崇左讲学期间因猝死不幸逝世,享年 70 岁.程乾生教授还曾担任过北京大学石油天然气研究中心副主任,中国工业和应用数学学会常务理事兼学术委员会主任,中国电子学会中国仪器仪表学会信号处理专业委员会副主任委员等职.程乾生教授于 1994 年加入中国民主同盟,为北京大学民盟做了大量工作.他热心组织建设,积极参与盟组织的各项活动,参政议政,建言献策,关心年轻同志的发展.他为北京大学民盟的建设,特别是对数学支部的发展壮大做出重要贡献.

程乾生教授的研究兴趣很广泛,在数学理论发展和实际应用研究中均有重要建树.其主要研究领域包括信号处理、时间序列分析、模式识别、属性数学、决策与评价分析.在国内外学术刊物发表论文 200 余篇,出版专著 8 本.获省部级科技进步奖二项,获国家自然科学奖一项.他是最早研究非高斯线性过程的开拓型学者之一.他提出了多谱参数方法和最大规范累量反褶积方法,并给出非高斯线性过程的表示唯一性定理,被国外专家认为"解决了重要的非高斯线性时间序列模型的识别问题". 他在利用信息论研究熵谱估计方面, 做了一系列深入的工作, 给出了高斯平稳随机场最一般的熵率定义和相应的熵率公式, 建立了 ARMA 模型和指数模型 (EX 模型) 的最大熵谱模型, 完整地解决了指数模型时域估计的问题, 提出了最大联合熵谱估计方法. 他也是中国最早研究信号处理的学者之一. 他的著作《信号数字处理的数学原理》是国内第一本信号处理方面的专著, 广为传播, 被誉为中国信号处理学科的经典著作. 他对能量有限信号的结构和性质进行深入分析, 对能量延迟性质给出了完整而简洁的证明. 在高分辨率谱窗研究中, 对极大能量比的方法原理给出了理论分析和计算方法. 他提出了属性集理论, 建立了属性测度理论和信息窗理论, 并对属性集理论与模糊数学和人工智能的关系进行了研究, 产生了巨大的影响.

程乾生教授非常重视应用研究. 他认为应用研究必须做到"顶天立地". 所谓"顶天", 就是要瞄准最高水平, 立足超越, 构建新的理论和方法; 而"立地"则是要落实到解决实际问题中, 不能将应用仅仅作为研究的"背景"和"道具". 他作为主要参加者的石油地震勘探数字处理方法研究曾获 1978 年全国科学大会奖. 他作为主要作者之一所编著的《地震勘探数字技术》一书, 在地质勘探界有很大的影响. 他主持的信号处理和模式识别在地震勘探中的研究, 曾获国家科技攻关奖和地质矿产部科技成果奖. 他提出的波形相似检验法, 在天王星环的检测中起到重要作用.

程乾生教授积极投身到教学工作, 先后讲授过"高等数学"、"时间序列分析"、"数字信号处理"、"信息论与信号处理"等课程. 在教学中, 他能够与学生积极互动, 紧扣要点, 并重视思想和方法的讲解, 得到了学生的欢迎和好评. 他还致力于教材的建设, 所编写的教材《数字信号处理》被评为北京高等教育精品教材. 另外, 程乾生教授也是一位优秀的研究生导师. 他给予学生一种宽松的环境, 从思想和方法上进行学术指导, 并能够在生活上给予关心和帮助. 在他指导的数十名博士和硕士研究生中, 许多人已经成为各自领域中的专家和骨干.

程乾生教授严于律己, 积极进取, 始终保持着热情洋溢的工作和生活态度. 对于同事、同学、学生和朋友, 他尽最大努力给予关心和帮助; 对于集体和群众的事情, 他积极提出自己的观点和建议. 他时刻关心着北大数学院的发展, 热心于对年轻人的培养和鼓励, 为院系和学科的发展做出了重要的贡献.

程乾生论著目录

专著及译著(8本)
[1] 程乾生. 数字信号处理[M]. 2版. 北京:北京大学出版社,2010.

[2] 程乾生. 数字信号处理简明教程[M]. 北京:高等教育出版社,2007.

[3] 程乾生. 数字信号处理[M]. 北京:北京大学出版社,2003.

[4] 李宏伟,程乾生. 高阶统计量与随机信号分析[M]. 武汉:中国地质大学出版社,2002.

[5] 程乾生. 信号数字处理的数学原理[M]. 2版. 北京:石油工业出版社,1993.

[6] Johnson D A. 概率与机率[M]. 程乾生,译. 北京:科学出版社,1984.

[7] Silvia M T, Robinson E A. 油气勘探中地球物理时间序列的反褶积[M]. 甘章泉,程乾生,译. 北京:石油工业出版社,1982.

[8] 程乾生. 数字信号处理的数学原理[M]. 北京:石油工业出版社,1979.

杂志或会议论文(203篇)
2011年
[1] Wu L, Cheng Q S. An asymmetric adaptive classification method[J]. International journal of Wavelets, Multiresolution and Information processing,2011,9(1):169—179.

2010年
[2] Zhan H, Cheng Q S. A simple approach to valuing Asian rainbow options [J]. International Journal of Electronic Customer Relationship Management,2010,4(1):60—76.

2009年
[3] 程乾生. 无结构决策:层次分析法 AHP 和属性层次模型 AHM:决策科学与评价:中国系统工程学会决策科学专业委员会第八届学术年会,2009[C]. 2009:22—27.

2008年
[4] Yuan K, Zou J, Feng S, Duan C, Bao S, Cheng Q S. Roundness curve for classification of cell phase on microscopic image: Proceedings of the International Conference on Information Technology and Applications in Biomedi-

cine (ITAB),2008[C]. 2008:70—73.

[5] 詹惠蓉,程乾生.两个或多个几何平均价格的最小或最大值期权的定价[J].数学的实践与认识，2008,38(24):95—102.

2007 年

[6] 程乾生,吴柏林.模糊统计分析的数学原理及其应用[J].量化研究学刊,2007,1(1):84—106.

[7] Yan S,Hu Y,Xu D,Zhang H J,Zhang B, Cheng Q S. Nonlinear discriminant analysis on embedded manifold[J]. IEEE Transactions on Circuits and Systems for Video Technology,2007, 17(4):468—477.

[8] Sun X, Cheng Q S, Feng J. From Penalized Maximum Likelihood to Cluster Analysis:A Unified Probabilistic Framework of Clustering[J]. International Journal of Pattern Recognition and Artificial Intelligence,2007, 21(3):483—490.

2006 年

[9] Yan J,Liu N,Yang Q,Zhang B, Cheng Q S,Chen Z. Mining adaptive ratio rules from distributed data sources[J]. Data Mining and Knowledge Discovery,2006,12(2):249—273.

[10] Yan J,Zhou X,Yang Q,Liu N, Cheng Q S,Wong S T C. An effective system for optical microscopy cell image segmentation, tracking and cell phase identification:Proceedings of the IEEE Conference on Image Processing, 2006[C]. 2006:1917—1920.

[11] Yan J,Zhang B,Yan S,Liu N,Yang Q, Cheng Q S,Li H,Chen Z, Ma W Y. A scalable supervised algorithm for dimensionality reduction on streaming data[J]. Information Sciences,2006,17(14):2042—2065.

[12] Yan J,Zhang B,Liu N,Yan S, Cheng Q S,Fan W,Yang Q,Xi W, Chen Z. Effective and efficient dimensionality reduction for large-scale and streaming data preprocessing[J]. IEEE Transactions on Knowledge and Data Engineering,2006,18(3):320—333.

[13] Sun X, Cheng Q S. On Subspace Distance [J]. Lecture Notes in Conputer Science,2006,4142:81—89.

[14] Yu Y, Cheng Q S. Particle filters for maneuvering target tracking problem [J]. Signal Processing,2006, 86(1):195—203.

[15] Chen J,Zhang B,Shen D,Yang Q,Chen Z, Cheng Q S. Diverse topic

phrase extraction from text collection:Proceedings of the 15th International World Wide Web Conference (WWW), Edinburgh,UK,2006[C].

[16] 程乾生.非线性时间序列预测:基于数据结构的预测方法:全国第一届嵌入式技术联合学术会议,2006[C].2006:1－2.

[17] 吴光旭,程乾生.NGARCH 模型在证券投资风险分析中的应用[J].数学的实践与认识,2006,36(8):37－43.

[18] 孙喜晨,程乾生,封举富.关于矩阵2－范数的一个定理[J].数学的实践与认识,2006,36(6):300－301.

[19] 程乾生,武连文.时间序列的经验模态频率分解 EMFD[J].数学的实践与认识,2006,36(5):151－153.

[20] 杜建卫,程乾生,刘乃强.中药色谱指纹图谱的小波变换与分形[J].数学的实践与认识,2006,36(7):275－280.

[21] 孙喜晨,程乾生,封举富.子空间非相似性度量（WWF－SSD）的三角不等式[J].数学的实践与认识,2006,36(4):201－203.

[22] 程乾生.DNA 双螺旋是如何发现的[C].新观点新学说学术沙龙文集1:科学的本源,2006:57－59.

[23] 杜建卫,程乾生,刘乃强,林立.基于小波变换与分形的中药指纹图谱的特征提取[J].信号处理,2006,22(5):719－723.

2005 年

[24] Gao D, Ma J, Cheng Q S. A step by step optimization approach to independent component analysis [J]. Advances in Neural Networks-ISNN 2005,2005, 3496:11－22.

[25] Li S Z,Lu X G,Hou X,Peng X,Cheng Q S. Learning multiview face subspaces and facial pose estimation using independent component analysis [J]. IEEE Transactions on Image Processing,2005, 14(6):705－712.

[26] Yuan K,Wu L,Cheng Q S,Bao S,Chen C,Zhang H. A novel fuzzy C-means algorithm and its Application[J]. International Journal of Pattern Recognition and Artificial Intelligence,2005, 19(8):1059－1066.

[27] Ma J,Gao B,Wang Y,Cheng Q S. Conjugate and natural gradient rules for BYY harmony learning on Gaussian mixture with automated model selection[J]. International Journal of Pattern Recognition and Artificial Intelligence,2005,19(5):701－713.

[28] Lin M T,Zhang J L,Cheng Q S,Chen R. Independent particle filters

[J]. Journal of the American Statistical Association, 2005, 100(472): 1412 −1421.

[29] Lu Z, Cheng Q S, Ma J. A gradient BYY harmony learning algorithm on mixture of experts for curve detection[J]. Lecture Notes in Computer Science, 2005, 3578: 250−257.

[30] Zhang B, Yan J, Liu N, Cheng Q S, Chen Z, Ma W Y. Supervised semi-definite embedding for image manifolds: Proceedings of the IEEE International Conference on Multimedia and Expo (ICME), Beijing, China, April, 2005 [C].

[31] Gao D, Ma J, Cheng Q S. An alternative switching criterion for independent component analysis (ICA) [J]. Neurocomputing, 2005, 68: 267−272.

[32] He R, Cheng Q S, Wu L, Yuan K. New feature extraction in gene expression data for tumor classification [J]. Progress in Natural Science, 2005, 15 (9): 861−864.

[33] Yan J, Liu N, Zhang B, Yan S, Chen Z, Cheng Q S, Fan W, Ma W Y. OCFS: optimal orthogonal centroid feature selection for text categorization-Proceedings of the 28th Annual International ACM SIGIR Conference on Research and Development in Information Retrieva, ACM, 2005[C]. 2005: 122 −129.

[34] Yan J, Cheng Q S, Yang Q, Zhang B. An incremental subspace learning algorithm to categorize large scale text data[J]. Web Technologies Research and Development-APWeb, 2005. 2005, 3399: 52−63.

[35] Zhang M, Cheng Q S. An approach to VaR for capital markets with Gaussian mixture[J]. Applied mathematics and computation, 2005, 168(2): 1079−1085.

[36] 詹惠蓉,程乾生. 亚式期权在依赖时间的参数下的定价[J]. 管理科学学报, 2005, 7(6): 24−29.

[37] 刘敬伟,徐美芝,郑忠国,程乾生. 基于 DTW 的语音识别和说话人识别的特征选择[J]. 模式识别与人工智能, 2005, 18(1): 50−54.

[38] 詹惠蓉,程乾生. 拟蒙特卡罗法在亚洲期权定价中的应用[J]. 数学的实践与认识, 2005, 35(9): 20−27.

[39] 蔡惠萍,程乾生. 属性层次模型 AHM 在选股决策中的应用[J]. 数学的实践与认识, 2005, 35(3): 55−58.

[40] 周俊,杨静平,程士宏,程乾生. 二元混合型索赔分布的复合模型的递推方程[J]. 数学进展,2005,34(1):54—72.

[41] 武连文,程乾生. 关于动力学互相关因子指数的注记[J]. 物理学报,2005,54(7):3027—3028.

2004 年

[42] Gao B, Liu T Y, Cheng Q S, Ma W Y. A linear approximation based method for noise-robust and illumination-invariant image change detection[J]. Advances in Multimedia Information Processing-PCM 2004. 2004, 3333: 95—102.

[43] Yan S, He X, Hu Y, Zhang H J, Li M, Cheng Q S. Bayesian shape localization for face recognition using global and local textures [J]. IEEE Transactions on Circuits and Systems for Video Technology, 2004, 14(1): 102—113.

[44] Yan S, Zhang H, Hu Y, Zhang B, Cheng Q S. Discriminant analysis on embedded manifold[J]. Computer Vision-ECCV 2004. 2004, 3021: 121—132.

[45] Yan J, Zhang B, Yan S, Yang Q, Li H, Chen Z, Xi W, Fan W, Ma W Y, Cheng Q S. IMMC: incremental maximum margin criterion: Proceedings of the tenth ACM SIGKDD International Conference on Knowledge Discovery and Data Mining(KDD), New York, NY, USA ACM, 2004[C]. 2004: 725—730.

[46] Ma J, Gao B, Wang Y, Cheng Q S. Two Further Gradient BYY Learning Rules for Gaussian Mixture with Automated Model Selection Intelligent Data Engineering and Automated Learning-IDEAL 2004 [J]. Lecture Notes in Computer Science, 2004, 3177: 690—695.

[47] Wu J, Ma J, Cheng Q S. Further results on the asymptotic memory capacity of the generalized Hopfield network [J]. Neural Processing Letters, 2004, 20(1): 23—38.

[48] Zhang M H, Cheng Q S. Determine the number of components in a mixture model by the extended KS test[J]. Pattern Recognition Letters, 2004, 25(2): 211—216.

[49] Yu J, Cheng Q S, Huang H. Analysis of the weighting exponent in the FCM[J]. IEEE Transactions on Systems, Man, and Cybernetics, Part B:

Cybernetics. 2004,34(1):634—639.

[50] 詹惠蓉,程乾生. 亚洲期权价格的一个新的多元控制变量估计[J]. 北京大学学报(自然科学版),2004,40(1):5—11.

[51] 吴光旭,程乾生,潘家柱. 中国股票市场风险值的非参数估计[J]. 北京大学学报(自然科学版),2004,40(5):696—701.

2003 年

[52] Yan S,Liu C,Li S Z,Zhang H,Shum H Y,Cheng Q S. Face alignment using texture-constrained active shape models[J]. Image and Vision Computing,2003,21(1):69—75.

[53] Yan S,Hou X,Li S Z,Zhang H,Cheng Q S. Face alignment using view-based direct appearance models[J]. International journal of Imaging Systems and Technology,2003,13(1):106—112.

[54] Zhang M H,Cheng Q S. Gaussian mixture modelling to detect random walks in capital markets[J]. Mathematical and Computer Modelling,2003,38(5—6):503—508.

[55] Yu Y,Cheng Q S. MRF parameter estimation by an accelerated method[J]. Pattern Recognition Letters,2003,24(9):1251—1259.

[56] Yan S,Li M,Zhang H,Cheng Q S. Ranking prior likelihood distributions for bayesian shape localization framework:Proceedings of the Ninth IEEE International Conference on Computer Vision,2003[C]. 2003:51—58.

[57] Yu J,Shi H,Huang H,Sun X,Cheng Q S. Counterexamples to convergence theorem of maximum-entropy clustering algorithm[J]. Science in China Series F:Information Sciences,2003,46(5):321—326.

[58] 于剑,程乾生. 关于 FCM 算法中的权重指数 m 的一点注记[J]. 电子学报,2003,31(3):478—480.

[59] 刘敬伟,程乾生. 基于 AR 模型的基因芯片数据识别[J]. 生物数学学报,2003,18(3):328—332.

[60] 周俊,成世学,程乾生. 个体风险模型的 Poisson 复合模型近似[J]. 运筹学学报,2003,7(2):91—96.

[61] 于剑,石洪波,黄厚宽,孙喜晨,程乾生. 关于极大熵聚类算法的收敛性定理的反例[J]. 中国科学:E 辑,2003,33(6):531—535.

2002 年

[62] Cheng Q S,Fan Z T. The stability problem for fuzzy bidirectional as-

sociative memories[J]. Fuzzy Sets and Systems,2002,132(1):83—90.

[63] Yan S,Liu C,Li S Z,Zhang H,Shum H,Cheng Q S. Texture-constrained active shape models Proceedings of the First International Workshop on Generative-Model-Based Vision (with ECCV),2002[C]. 2002:107—113.

[64] Liu J,Cheng Q S,Zheng Z,Qian M. A DTW-based probability model for speaker feature analysis and data mining [J]. Pattern Recognition Letters, 2002,23(11):1271—1276.

[65] Li S,Zhang H,Yan S,Cheng Q S. Multi-view face alignment using direct appearance models: Proceedings of the Fifth IEEE International Conference on Automatic Face and Gesture Recognition,Washington,DC,USA,2002 [C]. 2002: 324— 329.

[66] Li S Z,Peng X H,Hou X W,Zhang H J,Cheng Q S. Multi-view face pose estimation based on supervised ISA learning: Proceedings of the Fifth IEEE International Conference on Automatic Face and Gesture Recognition, 2002 [C]. 2002:100—105.

[67] 刘敬伟,程乾生. 基于动态时间规划的基因芯片数据识别[J]. 北京大学学报(自然科学版),2002, 5(38):611—615.

[68] 贺仁亚,程乾生,孙喜晨. 属性均值聚类二叉树及其在人脸识别中的应用[J]. 北京大学学报(自然科学版),2002,38(5):616—621.

[69] 张明恒,程乾生. 金融资产收益分布的混合高斯分析[J]. 数学的实践与认识,2002, 32(3):416—421.

[70] 贺仁亚,程乾生. 一种新的分类方法[J]. 数学的实践与认识,2002,32(5):830—834.

[71] 于剑,程乾生. 模糊聚类方法中的最佳聚类数的搜索范围[J]. 中国科学:E辑,2002, 32(2):274—280.

2001 年

[72] Hou X,Li S Z,Zhang H J,Cheng Q S. Direct appearance models: Proceedings of the IEEE Computer Society Conference on Computer Vision and Pattern Recognition,2001[C]. 2001:828—833.

[73] Cheng Q S,Wu L,Wang S. Fusion prediction based on the attribute clustering network and the radial basis function[J]. Chinese Science Bulletin, 2001,46(9):789—792.

[74] Yu J,Cheng Q S,Huang H K. On weighting exponent of the fuzzy

c-means model: Proceedings of the sixth biennial International Conference for Young Computer Scientists (ICYCS), Hangzhou, 2001[C]. 2001: 631—633.

[75] Yu J, Cheng Q S. The upper bound of the optimal number of clusters in fuzzy clustering [J]. Science in China Series F: Information Sciences, 2001, 44(2): 119—125.

[76] 崔晓瑜,程乾生.非对称数字水印[J].北京大学学报(自然科学版),2001,37(5):618—622.

[77] 吴淑珍,程乾生.一种孤立词语音识别方法研究[J].北京大学学报(自然科学版),2001,37(1):69—72.

[78] 程乾生,周小波,孙喜晨.气候序列不连续点的小波分析和混合聚类分析[J].大气科学,2001,25(4):551—558.

[79] 于剑,程乾生.关于聚类有效性函数 $FP(u,c)$ 的研究[J].电子学报,2001,29(7):899—901.

[80] 国栋,程乾生.具有剪切的矢量压缩立体绘制算法[J].计算机辅助设计与图形学学报,2001,13(6):532—536.

[81] 贺仁亚,程乾生.一种用于认证的小波变换域的数字水印技术[J].计算机辅助设计与图形学学报,2001,13(9):812—815.

[82] 胡玉胜,涂序彦,崔晓瑜,程乾生.基于贝叶斯网络的不确定性知识的推理方法[J].计算机集成制造系统,2001,7(12):65—69.

[83] 于剑,程乾生,黄厚宽.基于粗集理论的知识库合并与分解[J].计算机科学,2001,28(5):17—18.

[84] 于剑,程乾生.关于聚类有效性函数熵公式 $HP(u,c)$ [J].模糊系统与数学,2001,15(2).

[85] 侯新文,程乾生.蛇形算法及其改进[J].数学的实践与认识,2001,31(2):202—205.

[86] 范周田,程乾生.模糊矩阵的特征向量[J].系统工程理论与实践,2001,21(1):45—52.

[87] 彭春华,程乾生.一种基于最小张树的属性聚类算法[J].系统工程理论与实践,2001,21(2):30—34.

2000 年

[88] Wang C, Cheng Q S. Attribute cluster network and fractal image compression: Proceedings of the 5th International Conference on Signal Processing, Beijing, China, 2000[C]. 2000: 1613—1616.

[89] Yu J, Cheng Q S. A note on rough set and non-measurable set[J]. Chinese Science Bulletin, 2000, 45(16):1456−1458.

[90] Cheng Q S, Zhou X, Sun X. Nonlinear fusion filters based on prediction and smoothing[J]. Chinese Science Bulletin, 2000, 45(18):1726−1728.

[91] 于剑,程乾生.模糊划分的一个新定义及其应用[J].北京大学学报(自然科学版),2000,36(5):619−623.

[92] 国栋,程乾生.一种基于神经网络的图像变形实验[J].计算机辅助设计与图形学学报,2000,12(6):455−458.

[93] 于剑,程乾生.粗集与不可测集[J].科学通报,2000,45(7):686−689.

[94] 程乾生,周小波,孙喜晨.基于预测和平滑的非线性融合滤波器[J].科学通报,2000,45(8):821−823.

[95] 程乾生,武连文,王守章.基于属性聚类网络和径向基函数的融合预测[J].科学通报,2000,45(11):1211−1216.

[96] 王明进,程乾生. Kohonen 自组织网络在混沌时间序列预测中的应用(Ⅱ)[J].系统工程理论与实践,2000,20(10):79−83.

[97] 周小波,陈志航,程乾生,朱迎善.预测混合中心平滑法去噪[J].系统工程理论与实践,2000,20(2):73−78.

1999 年

[98] Cheng Q S, Miscellanea. On time-reversibility of linear processes[J]. Biometrika, 1999, 86(2):483−486.

[99] 程乾生,王春梅.层次属性聚类网络与分形图像压缩:第九届全国信号处理学术年会(CCSP),1999[C].1999:404−407.

[100] 程乾生.复杂系统(网络)的属性模式识别网络分析法:第九届全国信号处理学术年会(CCSP)1999[C].1999:1−4.

[101] 熊春光,孙喜晨,程乾生.基于进化规划的属性均值 Kohonen 自组织网:第九届全国信号处理学术年会(CCSP)1999[C].1999:457−460.

[102] 陈志航,程乾生.基于隐马尔科夫模型的组合预测方法:第九届全国信号处理学术年会(CCSP)1999[C].1999:34−37.

[103] 程乾生,周小波.自适应预测法在去噪声中的应用[J].电子学报,1999,27(8):9−11.

[104] 王明进,程乾生.基于径向基函数的非线性预测模型[J].管理科学学报,1999,2(4):28−33.

[105] 彭春华,程乾生.基于属性理论的综合预报系统及其应用[J].系统工

程理论与实践,1999,19(5):83—88.

[106] 陈志航,程乾生.属性识别方法及其在期货价格预测中的应用[J].系统工程理论与实践,1999,19(6):90—93.

[107] 国栋,程乾生.基于最优采样模式的自适应光线投射法[J].中国图象图形学报:A辑,1999,4(10):843—848.

[108] 闫宇松,程乾生.可逆双正交小波变换在图象压缩中的应用[J].中国图象图形学报:A辑,1999,4(9):795—799.

[109] 王春梅,程乾生.算术编码在分形图象压缩中的应用[J].中国图象图形学报:A辑,1999,4(4):307—311.

1998年

[110] Cheng Q S, Zhou X. A new class of nonlinear filters:Proceedings of the Fourth International Conference on Signal Processing Proceedings,1998[C].1998:237—241.

[111] Li H, Cheng Q S, Yuan B. Strong laws of large numbers for two-dimensional processes:Proceedings of the Fourth International Conference on Signal Processing,1998[C].1998:43—46.

[112] Li H, Cheng Q S. Convergence properties of higher-order moment and Cumulant estimates[J]. Journal of Electronics (CHINA),1998,15(3):240—247.

[113] Li H, Cheng Q S. Eigenpolynomials of a 2-D harmonic signal[J]. Signal Processing Letters,1998,5(3):71—73.

[114] 高崇明,李益勋,程乾生,周小波.差异序列聚类算法在四膜虫遗传分析中的应用[J].北京大学学报(自然科学版),1998,34(6):765—769.

[115] 程乾生.属性层次模型AHM:一种新的无结构决策方法[J].北京大学学报(自然科学版),1998,34(1):10—14.

[116] 程乾生,周小波,朱迎善.气候突变的聚类分析[J].地球物理学报,1998,41(3):308—314.

[117] 李宏伟,程乾生.乘性和加性噪声中谐波恢复和循环统计量方法[J].电子学报,1998,26(7):105—111.

[118] 程乾生,李宏伟.调幅信号的参数循环累量估计[J].电子学报,1998,26(7):99—104.

[119] 李宏伟,程乾生.关于二维谐波的建模问题[J].电子学报,1998,26(4):19—23.

[120] 程乾生,王守章,武连文.基于遗传算法的多阶马氏链组合预测方法[J].管理科学学报,1998,1(4):26-33.

[121] 程乾生.属性数学:属性测度和属性统计[J].数学的实践与认识,1998,28(2):97-107.

[122] 李宏伟,程乾生.二维谐波的累量公式[J].通信学报,1998,19(2):59-62.

[123] 程乾生.属性均值聚类[J].系统工程理论与实践,1998,18(9):124-126.

[124] 程乾生,曾庆存,李大潜,薛识寿.在模式识别中的信息融合:中国工业与应用数学学会第五次大会论文集[C].北京:清华大学出版社,1998:278-280.

1997 年

[125] Li H, Cheng Q S. Almost sure convergence analysis of mixed time averages and kth-order cyclic statistics[J]. IEEE Transactions on Information Theory, 1997, 43(4):1265-1268.

[126] 程乾生.属性识别理论模型及其应用[J].北京大学学报(自然科学版),1997,33(1):12-20.

[127] 沈亮,程乾生.一种新的文字细化算法[J].模式识别与人工智能,1997,10(3):232-237.

[128] 程乾生.质量评价的属性数学模型和模糊数学模型[J].数理统计与管理,1997,16(6):18-23.

[129] 周小波,程乾生.自适应最优混合差异聚类算法[J].数学的实践与认识,1997,27(4):312-319.

[130] 王明进,程乾生.Kohonen自组织网络在混沌时间序列预测中的应用[J].系统工程理论与实践,1997,7(7):12-17.

[131] 程乾生.层次分析法AHP和属性层次模型AHMα[J].系统工程理论与实践,1997,17(11):25-28.

[132] 程乾生.属性集和属性综合评价系统[J].系统工程理论与实践,1997,17(9):1-8.

1996 年

[133] Cheng Q S, Li H. Parametric methods of cyclic-polyspectrum estimation for AM signals: Proceedings of the 3rd International Conference on Signal Processing, 1996[C]. 1996:19-22.

[134] Li H, Cheng Q S. Some estimation problems for harmonics in multiplicative and additive noise: Proceedings of the 3rd International Conference on Signal Processing, 1996[C]. 1996: 181—184.

[135] Cheng Q S, Chen R, Li T H. Simultaneous wavelet estimation and deconvolution of reflection seismic signals[J]. IEEE Transactions on Geoscience and Remote Sensing, 1996, 34(2): 377—384.

1990—1994 年

[136] 程乾生, 黄肖俊. 柯西滤波: 第五届全国信号处理学术年会(CCSP), 武汉, 1994.[C].

[137] 程乾生. 一种新的样品聚类方法: 差异序列法[J]. 科学通报, 1994, 39(2): 8—12.

[138] 程乾生. 褶积型矩阵和 H^2 函数在单位圆内零点的个数[J]. 科学通报, 1994, 39(2): 100—101.

[139] 程乾生. 有序样品聚类的相关序列法[J]. 石油地球物理勘探, 1994, 29(1): 96—100.

[140] 程乾生. 属性集理论与模糊数学、模式识别、人工智能: 中国工业与应用数学学会第三次大会, 1994[C]. 北京: 清华大学出版社, 1994: 6—14.

[141] 程乾生. 有序样品聚类的差异序列法[J]. 北京大学学报(自然科学版), 1993, 29(5): 545—551.

[142] 龚兆仁, 程乾生. 空间 Lp1(μ1)和 Lp2(μ2)中范数的关系[J]. 黑龙江大学自然科学学报, 1993, 10(2): 49—53.

[143] Cheng Q S. Equivalence of Non-Gaussian Linear Processes[J]. Advances in Mathematics(China), 1993, 22(3): 285.

[144] Cheng Q S. On the unique representation of non-Gaussian linear processes[J]. The Annals of Statistics, 1992, 20(2): 1143—1145.

[145] Cheng Q S. Rank of class of autocorrelation matrices spectral estimation[J]. Chinese Science Bulletin, 1991, 36(1): 72—74.

[146] Cheng Q S. Parameter estimation in exponential models[J]. Journal of Time Series Analysis, 1991, 12(1): 27—40.

[147] 程乾生. 多谱估计的参数方法[J]. 电子学报, 1991, 19(1): 98—104.

[148] Cheng Q S. The Uniqueness Problem of Non-Gaussian Linear Processes[J]. Advances in Mathematics (China), 1991, 20(4): 499—500.

[149] Cheng Q S. Parameter estimation of exponential models for spectra:

Proceedings of the International Conference on Acoustics, Speech, and Signal Processing (ICASSP),1990[C]. 1990,5:2611—2614.

[150] Cheng Q S. Minimum entropy deconvolution of one-and multi-dimensional non-gaussian linear random processes[J]. Science in China: Ser. A, 1990,33(10):1153—1162.

[151] Cheng Q S. Maximum standardized cumulant deconvolution of non-Gaussian linear processes[J]. The Annals of Statistics,1990,18(4):1774—1783.

[152] 程乾生. 在谱估计中一类相关矩阵的秩[J]. 科学通报,1990,35(7):552—554.

[153] 程乾生. 一维和多维非高斯线性随机过程的最小熵反褶积[J]. 中国科学:A 辑,1990,20(6):577—584.

1985—1989 年

[154] 徐雷,程乾生,孙靖,胡震中. 应用模式识别和专家系统方法进行地震层序分析和地震相识别:勘探地球物理国际讨论会,北京,中国,1989[C]. 1989:491—496.

[155] 程乾生. 最大熵指数模型的参数估计[J]. 信号处理,1989,5(2):65—67.

[156] Cheng Q S. Convolution decomposition of 1-D and 2-D linear stationary signals: Proceedings of the International Conference on Acoustics, Speech,and Signal Processing,1988[C]. 1988:898—899.

[157] 程乾生. S 阶偶项信号和最小平方偶项卷积反演方法[J]. 电子学报,1987,15(4):34—40.

[158] 程乾生. 信噪比与分辨率及线性滤波与反滤波[J]. 石油地球物理勘探,1987,22(6):636—643.

[159] 程乾生. 确定性有理信号参数的估计:最小预测误差均方比(MPEMR)方法[J]. 信号处理,1986(4):220—227.

[160] 程乾生. 最小交叉熵谱的简单推导[J]. 信号处理,1986,2(2):120—120.

[161] Cheng Q S. Z-transform models and data extrapolation formulas in the maximum entropy methods of power spectral analysis[J]. Science Bulletin, 1985,30(4):436—440.

[162] 程乾生,朱迎善. 平稳 ARMA 模型的判别条件[J]. 黑龙江大学自然

科学学报,1985(1):40—46.

[163] 程乾生.有附加噪声时信号的最大互信息谱估计和最大联合熵谱估计[J].信号处理,1985,1(3):182—190.

1980—1984 年

[164] 龚兆仁,程乾生.随机信号的 ARMA 模型[J].黑龙江大学自然科学学报,1984(4):65—67.

[165] 程乾生.对三角级数的广义 Padé 有理逼近[J].计算数学,1984,6(2):182—193.

[166] 许文源,程乾生.最小平方解的有界性问题[J].计算数学,1984,6(4):351—359.

[167] 程乾生.在功率谱分析最大熵方法中的 Z 变换模型和数据开拓[J].科学通报,1984,29(19):1210—1213.

[168] Cheng Q S. A property of multi-dimen-sional function in disks[J]. A Monthly Journal of Science, 1983(8):1139—1140.

[169] Cheng Q S. Modified-generalized pad rational approximation[J]. A Monthly Journal of Science, 1983(6):856—857.

[170] Cheng Q S. Multidimensional all-pass filters and minimum-phase filters[J]. A Monthly Journal of Science,1983,28(5):588—591.

[171] Cheng Q S. Uniform convergence and mean square convergence of filter factors[J]. Acta Mathematicae Applicatae Sinica,1983,6(3):267—275.

[172] 程乾生.修正的广义 Padé 有理逼近程乾生[J].科学通报,1983,28(1):63—63.

[173] 程乾生.修正的广义 Padé 有理逼近[J].科学通报,1983(1):63.

[174] 程乾生,许承德.由矩阵测度 F 形成的 $L\sim 2$(F)空间的强结构性质和多维平稳过程的强分解性质[J].数学学报,1983,26(4):424—432.

[175] 程乾生.滤波因子的均匀收敛和均方收敛性质[J].应用数学学报,1983(3):267—275.

[176] Chen D,Wu Z,Yang X,Xie Z,Cheng Q S,Jiang S. New Ring Signals beyond the 6 Uranian Ring[J]. Current Topics in Chinese Science:Section E:Astronomy,1982(1):149.

[177] Cheng Q S. The convolution-type matrix and the property of the complex space l_2[J]. Science in China:Ser. A,1982(2):125—137.

[178] 程乾生,袁燕姝.符号位地震记录振幅恢复理论的两个基本定理[J].

地球物理学报,1982(2):147-152.

[179] 程乾生.多维函数在多重圆内的一个性质[J].科学通报,1982,27(22):1407-1407.

[180] 程乾生.多维全通滤波器和最小相位滤波器[J].科学通报,1982,27(23):1414-1414.

[181] 程乾生.新的地震记录褶积模型[J].石油地球物理勘探,1982(3):16-24.

[182] 程乾生.符号位地震信号检测[J].石油物探,1982(1):37-46.

[183] 龚兆仁,童光荣,程乾生.多维平稳序列对线性系统的滤波问题[J].武汉大学学报(理学版),1982(2):19-30.

[184] 龚兆仁,朱迎善,程乾生.随机信号在线性系统下的最佳滤波问题[J].哈尔滨科学技术大学学报,1981(1):10-22.

[185] 陈道汉,武志贤,杨修义,谢衷洁,程乾生,蒋世仰.天王星ε环之外的环信息[J].科学通报,1981(5):292.

[186] 程乾生,邵幼英.符号位地震记录的真振幅恢复[J].石油地球物理勘探,1981(2):24-36.

[187] 程乾生.对"石油物探"两篇关于符号位文章的商榷[J].石油物探,1981(3):104-105.

[188] 程乾生.信号的褶积分解及其在地震勘探中的应用[J].石油物探,1981,20(4):118-124.

[189] 程乾生.最小熵反褶积的理论问题[J].石油物探,1981,20(1):54-64.

[190] 程乾生,许文源.l_2空间Wold分解的连续性问题和一类投影算子的性质[J].数学学报,1981,24(6):844-850.

[191] 谢衷洁,程乾生.具有平稳干扰的极大信噪比滤波问题[J].应用数学学报,1981,4(4):362-380.

[192] 程乾生.褶积型矩阵和复l_2空间的性质[J].中国科学,1981(7):795-806.

[193] 程乾生.对"地震记录道的振幅包络、瞬时相位、瞬时频率"一文的注记[J].石油物探,1980(2):118-119.

[194] 程乾生,何国华.最大方差模反褶积[J].石油物探,1980,19(2):11-18.

[195] 程乾生.关于多维平稳序列奇异性和WOLD分解的谱表示[J].数学学报,1980,23(5):684-694.

1979年以前

[196] 程乾生.希尔伯特变换与信号的包络,瞬时相位和瞬时频率[J].石油

地球物理勘探,1979,14(3):1—14.

[197] 谢衷洁,程乾生.在一类非平稳干扰下的极大信噪比线性滤波问题[J].数学学报,1979,22(6):693—712.

[198] 程乾生,谢衷洁.谱估计中的最佳高分辨时窗函数[J].应用数学学报,1979,2(2):119—131.

[199] 陈道汉,杨修义,武志贤,吴月珍,蒋世仰,黄永伟,叶基棠,瞿迪生,谢衷洁,程乾生.天王星环掩星的光电观测与天王星环信息的数理统计检测[J].中国科学:A辑,1978(3):325—331.

[200] 宏油兵(闵嗣鹤),舒立华(程乾生).独立自主发展地震数字处理(续完)[J].数学学报,1976,19(1):63—72.

[201] 宏油兵(闵嗣鹤),舒立华(程乾生).独立自主发展石油地震数字处理[J].数学学报,1975,18(4):231—246.

[202] Shu L H(Cheng Q S). Energy transmission properties of a pure phase series(chinese)[J]. Acta Mathematicae Applicatae Sinica,1974,17(1):20—27.

[203] 程乾生."关于极限定理(Ⅰ)"一文的一点注记[J].吉林大学自然科学学报,1964(4):83.

编 后 记

2010年12月3日,程乾生教授在广西崇左讲学期间因猝死不幸逝世。根据家属的意愿丧事从简,程乾生教授的遗体于2010年12月5日在南宁火化。为了悼念程乾生教授,由北京大学数学科学学院信息科学系和北京大学民盟第七支部联合发起,于12月15日在北京大学数学科学学院举行了程乾生教授追思会。当日会后,北京大学数学科学学院信息科学系的教师和程乾生教授的研究生及生前好友都积极提议筹备出版程乾生教授的学术论文集。通过论文集的编辑和出版,我们能够对程乾生教授的教育和科研历程做全面的了解和总结,领略其研究的精髓,发扬其思想与精神,并为同行提供珍贵的学术资料,为后来者提供重要的启发和参考。随后,北京大学数学科学学院信息科学系、程乾生教授的研究生代表和北京大学出版社共同成立了论文集筹备组,并积极进行了论文集的编辑和出版工作。

程乾生教授自1963年北京大学数学力学系毕业后留校任教以来,积极投身到教学和科研中,特别是改革开放以来,培养了大量的博士和硕士研究生,并在信号与信息处理、系统工程和管理、决策与分析等领域做出了大量的研究成果,发表了大量的学术论文。论文集筹备组对程乾生教授所发表的论文进行了全面的收集和整理,得到了所有发表论文的清单及论文复印件,其中论文清单已列于书末,而预印件论文未列入其中。然后,根据所研究的问题,将论文分为四个专题,并从中挑选出24篇最具代表性的论文作为论文集的入选论文。大部分论文是程乾生教授独立完成的,但也包括一部分由程乾生教授指导研究生共同完成的论文。随后,根据版权要求对这24篇论文所涉及的版权问题进行了版权引进。最后,北京大学出版社对这些论文进行了重新的编辑和校验工作。在论文的编辑中,我们完全保持了原论文的文字和内容,图和表格基本保持不变,反应出论文的原貌。

首先感谢北京交通大学袁保宗教授对论文集的支持和关心,并为本书作序。袁保宗教授是程乾生教授的老朋友,一起创建了中国电子学会信号处理分会,德高望重,学术造诣深厚,并多次担任信号处理分会的主任委员。

感谢北京大学出版社王明舟社长和孙晔副社长对《程乾生学术论文集》出版的关心、帮助和支持。孙晔副社长多次亲临现场参加编选工作。

感谢程乾生教授的夫人李明西女士授权出版本书,感谢程乾生教授的妹夫庄大蔚教授对本书出版的关心和帮助,并为论文集提供了许多重要的资料。希望本书的出版能为程乾生教授的亲属带来一点慰藉。

参加本书选编工作的有马尽文、于剑、裘宗燕、李宏伟、肖灵、闫峻、周洲。感谢各位的辛勤工作,才保证了该书的顺利出版。

感谢北京大学出版社鼎力支持《程乾生学术论文集》的出版。

2015 年 6 月